WELCOME TO THE UNIVERSE

PRINCETON
UNIVERSITY
PRESS

PRINCETON
& OXFORD

WELCOME
TO THE
UNIVERSE

AN ASTROPHYSICAL TOUR

NEIL deGRASSE TYSON

MICHAEL A. STRAUSS

J. RICHARD GOTT

Published by Princeton University Press
41 William Street, Princeton, New Jersey 08540

In the United Kingdom: Princeton University Press
6 Oxford Street, Woodstock, Oxfordshire OX20 1TR

press.princeton.edu

Jacket and book design by Chris Ferrante

Library of Congress Cataloging-in-Publication Data

Names: Tyson, Neil deGrasse. | Strauss, Michael Abram. | Gott, J. Richard.
Title: Welcome to the universe : an astrophysical tour / Neil deGrasse Tyson, Michael A. Strauss, and
    J. Richard Gott.
Description: Princeton : Princeton University Press, 2016. | Includes bibliographical references and index.
Identifiers: LCCN 2016013487 | ISBN 9780691157245 (hardback : alk. paper)
Subjects: LCSH: Cosmology—Popular works. | Stars—Popular works. | Relativity (Physics)—Popular works.
Classification: LCC QB982 .T974 2016 | DDC 523.1—dc23 LC record available at
    https://lccn.loc.gov/2016013487

British Library Cataloging-in-Publication Data is available

This book has been composed in Adobe Text Pro and Trade Gothic LT Std

Printed on acid-free paper. ∞

Printed in the United States of America

10 9 8 7 6 5 4 3 2 1

*To the memories of Lyman Spitzer, Jr., Martin Schwarzschild,
Bohdan Paczyński, and John Bahcall, indelible influences on
the three of us in astrophysics research and education.*

CONTENTS

PREFACE

When my granddaughter Allison was born, one of the first things I said to her was "Welcome to the universe!" It's something my coauthor Neil Tyson has said many times on radio and TV. Indeed it is one of Neil's signature sayings. When you are born, you become a citizen of the universe. It behooves you to look around and get curious about your surroundings.

Neil felt a call from the universe on a first visit to the Hayden Planetarium in New York City when he was 9 years old. As a city kid, he saw the glories of the nighttime sky for the first time displayed on the planetarium dome and decided at that moment to become an astronomer. Today he is the director of that institution.

In fact, we are all touched by the universe. The hydrogen in your body was forged in the birth of the universe itself, while the other elements in your body were made in distant, long-dead stars. When you call a friend on your mobile phone, you should thank astronomers. Mobile phone technology depends on Maxwell's equations, whose verification depended on the fact that astronomers had already measured the speed of light. The GPS that tells your phone where you are and helps you navigate relies on Einstein's theory of general relativity, which was verified by astronomers measuring the deflection of starlight passing near the Sun. Did you know there is an ultimate limit to how much information can ever be stored in a 6-inch-diameter hard drive and that it depends on black hole physics? At a more mundane level, the seasons you experience every year depend directly on the tilt of Earth's axis relative to the plane of its orbit around the Sun.

This book aims to better acquaint you with the universe in which you live. The idea for this book started when the three of us taught a new undergraduate course on the universe for nonscience majors at Princeton University—for students who perhaps had never taken a science course before. For this purpose, Neta Bahcall, our colleague and director of undergraduate studies, selected Neil deGrasse Tyson, Michael Strauss, and me. Neil's genius at explaining science to nonscientists was apparent, Michael had just discovered the most distant quasar yet found in the universe, and I had just received the university's President's Award for Distinguished Teaching. The course was launched with great fanfare and attracted so many students that we couldn't hold it in our own building and had to move it to the biggest lecture hall in the Physics Department. Neil talked about "Stars and Planets," Michael talked about "Galaxies and Quasars," and I talked about "Einstein, Relativity, and Cosmology." The course was mentioned in *Time* magazine, when *Time* honored Neil as one of the 100 most influential people in the world in 2007. Among other features of this book, you will get to know Neil as a professor, telling you things he tells his students.

FIGURE 0.1. The three authors, left to right: Strauss, Gott, and Tyson.
Photo credit: Princeton, Denise Applewhite

After we had taught the course for a number of years, we decided to put its ideas down in the form of a book for readers who hungered for a deeper understanding of the universe.

We give you a tour of the universe from an astrophysical point of view, from the point of view of trying to understand what is going on. We tell you how Newton and Einstein got their greatest ideas. You know Stephen Hawking is famous. But we tell you what made him famous. The great movie of his life story, *The Theory of Everything*, won Eddie Redmayne a best actor Oscar for his compelling portrayal of Hawking. It shows Hawking having his greatest idea by simply staring into the fireplace and having it suddenly come to him. We tell you what the movie left out: how Hawking didn't believe the work of Jacob Bekenstein, but he ended up reaffirming it and taking it to an entirely new conclusion.

And that's the same Jacob Bekenstein who found the ultimate limit for how much information could be stored on your 6-inch-diameter hard drive. It's all connected. In this book, of all the topics in the universe, we focus particularly on those we are most passionate about, and we hope our excitement will be contagious.

Much has been added to astronomical knowledge since we began, and this book reflects that. Neil's views on the status of Pluto have been ratified by the International Astronomical Union, in a historic vote in 2006. Thousands of new planets have been discovered circling other stars. We discuss them. The standard cosmological model, including normal atomic nuclei, dark matter, and dark energy, is now known with exquisite accuracy, thanks to results from the Hubble Space Telescope, the Sloan Digital Sky Survey, and the Wilkinson Microwave Anisotropy Probe (WMAP) and Planck satellites. Physicists have discovered the Higgs Boson at the Large Hadron Collider in Europe, bringing us one step closer to the hoped for theory of everything. The Laser Interferometer Gravitational-Wave Observatory (LIGO) experiment has made a direct detection of gravitational waves from two inspiraling black holes.

We explain how astronomers have determined how much dark matter there is, and how we know that it is not made of ordinary matter (with atomic nuclei containing protons and neutrons). We explain how we know the density of dark energy, and how we know that it has a negative pressure. We cover current speculations on the origin of the universe and on its future evolution. These questions bring us to the frontiers of physics knowledge today. We have included spectacular images from the Hubble Space Telescope, the WMAP satellite, and the New Horizons spacecraft—showing Pluto and its moon Charon.

The universe is awesome. Neil shows you that in the very first chapter. This leaves many people thrilled, but feeling tiny and insignificant at the same time. But our aim is to empower you to understand the universe. That should make you feel strong. We have learned how gravity works, how stars evolve, and just how old the universe is. These are triumphs of human thought and observation— things that should make you proud to be a member of the human race.

The universe beckons. Let's begin.

J. RICHARD GOTT
Princeton, New Jersey

PART I

STARS, PLANETS, AND LIFE

1

THE SIZE AND SCALE
OF THE UNIVERSE

NEIL deGRASSE TYSON

We begin with the stars, then ascend up and away out to the galaxy, the universe, and beyond. What did Buzz Lightyear say in *Toy Story*? "To Infinity and Beyond!"

It's a big universe. I want to introduce you to the size and scale of the cosmos, which is bigger than you think. It's hotter than you think. It is denser than you think. It's more rarified than you think. Everything you think about the universe is less exotic than it actually is. Let's get some machinery together before we begin. I want to take you on a tour of numbers small and large, just so we can loosen up our vocabulary, loosen up our sense of the sizes of things in the universe. Let me just start out with the number 1. You've seen this number before. There are no zeros in it. If we wrote this in exponential notation, it is ten to the zero power, 10^0. The number 1 has no zeros to the right of that 1, as indicated by the zero exponent. Of course, 10 can be written as 10 to the first power, 10^1. Let's go to a thousand—10^3. What's the metric prefix for a thousand? *Kilo-* kilogram—a thousand grams; kilometer—a thousand meters. Let's go up another 3 zeros, to a million, 10^6, whose prefix is *mega-*. Maybe this is the highest they had learned how to count at the time they invented the megaphone; perhaps if they had known about a billion, by appending three more zeroes, giving 10^9, they would have called them "gigaphones." If you study file sizes on your computer, then you're familiar with these two words, "megabytes" and "gigabytes." A gigabyte is a billion bytes.[1] I'm not convinced you know how big a billion actually is. Let's look around the world and ask what kinds of things come in billions.

First, there are 7 billion people in the world.

Bill Gates? What's he up to? Last I checked, he's up to about 80 billion dollars. He's the patron saint of geeks; for the first time, geeks actually control the world. For most of human history that was not the case. Times have changed. Where have you seen 100 billion? Well, not quite 100 billion. McDonald's. "Over 99 Billion Served." That's the biggest number you ever see in the street. I remember when they started counting. My childhood McDonald's proudly displayed "Over 8 Billion Served." The McDonald's sign never displayed 100 billion, because they allocated only two numerical slots for their burger count, and so, they just stopped at 99 billion. Then they pulled a Carl Sagan on us all and now say, "billions and billions served."

Take 100 billion hamburgers, and lay them end to end. Start at New York City, and go west. Will you get to Chicago? Of course. Will you get to California? Yes, of course. Find some way to float them. This calculation works for the diameter of the bun (4 inches), because the burger itself is somewhat smaller than the bun. So for this calculation, it's all about the bun. Now float them across the ocean, along a great circle route, and you will cross the Pacific, pass Australia, Africa, and come back across the Atlantic Ocean, finally arriving back in New York City with your 100 billion hamburgers. That's a lot of hamburgers. But in fact you have some left over after you have circled the circumference of Earth. Do you know what you do with what you have left over? You make the trip all over again, 215 more times! Now you still have some left over. You're bored going around Earth, so what do you do? You stack them. So after you've gone around Earth 216 times, then you stack them. How high do you go? You'll go to the Moon, and back, with stacked hamburgers (each 2 inches tall) after you've already been around the world 216 times, and only then will you have used your 100 billion hamburgers. That's why cows are scared of McDonald's. By comparison, the Milky Way galaxy has about 300 billion stars. So McDonald's is gearing up for the cosmos.

When you are 31 years, 7 months, 9 hours, 4 minutes, and 20 seconds old, you've lived your billionth second. I celebrated with a bottle of champagne when I reached that age. It was a tiny bottle. You don't encounter a billion very often.

Let's keep going. What's the next one up? A trillion: 10^{12}. We have a metric prefix for that: *tera-*. You can't count to a trillion. Of course you could try. But if you counted one number every second, it would take you a thousand times

31 years—31,000 years, which is why I don't recommend doing this, even at home. A trillion seconds ago, cave dwellers—troglodytes—were drawing pictures on their living room walls.

At New York City's Rose Center of Earth and Space, we display a timeline spiral of the Universe that begins at the Big Bang and unfolds 13.8 billion years. Uncurled, it's the length of a football field. Every step you take spans 50 million years. You get to the end of the ramp, and you ask, where are we? Where is the history of our human species? The entire period of time, from a trillion seconds ago to today, from graffiti-prone cave dwellers until now, occupies only the thickness of a single strand of human hair, which we have mounted at the end of that timeline. You think we live long lives, you think civilizations last a long time, but not from the view of the cosmos itself.

What's next? 10^{15}. That's a quadrillion, with the metric prefix *peta-*. It's one of my favorite numbers. Between 1 and 10 quadrillion ants live on (and in) Earth, according to ant expert E. O. Wilson.

What's next? 10^{18}, a quintillion, with metric prefix *exa-*. That's the estimated number of grains of sand on 10 large beaches. The most famous beach in the world is Copacabana Beach in Rio de Janeiro. It is 4.2 kilometers long, and was 55 meters wide before they widened it to 140 meters by dumping 3.5 million cubic meters of sand on it. The median size of grains of sand on Copacabana Beach at sea level is 1/3 of a millimeter. That's 27 grains of sand per cubic millimeter, so 3.5 million cubic meters of that kind of sand is about 10^{17} grains of sand. That's most of the sand there today. So about 10 Copacabana beaches should have about 10^{18} grains of sand on them.

Up another factor of a thousand and we arrive at 10^{21}, a sextillion. We have ascended from kilometers to megaphones to McDonald's hamburgers to Cro-Magnon artists to ants to grains of sand on beaches until finally arriving here: 10 sextillion—

the number of stars in the observable universe.

There are people, who walk around every day, asserting that we are alone in this cosmos. They simply have no concept of large numbers, no concept of the size of the cosmos. Later, we'll learn more about what we mean by the *observable universe*, the part of the universe we can see.

While we're at it, let me jump beyond this. Let's take a number much larger than 1 sextillion—how about 10^{81}? As far as I know, this number has no name.

It's the number of atoms in the observable universe. Why then would you ever need a number bigger than that? What "on Earth" could you be counting? How about 10^{100}, a nice round-looking number. This is called a *googol*. Not to be confused with Google, the internet company that misspelled "googol" on purpose.

There are no objects to count in the observable universe to apply a googol to. It is just a fun number. We can write it as 10^{100}, or if you don't have superscripts, this works too: 10^100. But you can still use such big numbers for some situations: don't count *things*, but instead count the ways things can happen. For example, how many possible chess games can be played? A game can be declared a draw by either player after a triple repetition of a position, or when each has made 50 moves in a row without a pawn move or a capture, or when there are not enough pieces left to produce a checkmate. If we say that one of the two players must take advantage of this rule in every game where it comes up, then we can calculate the number of possible chess games. Rich Gott did this and found the answer was a number less than 10^(10^4.4). That's a lot bigger than a googol, which is 10^(10^2). You're not counting things, but you are counting possible ways to do something. In that way, numbers can get very large.

I have a number still bigger than this. If a googol is 1 followed by 100 zeros, then how about 10 to the googol power? That has a name too: a *googolplex*. It is 1, with a googol of zeroes after it. Can you even write out this number? Nope. Because it has a googol of zeroes, and a googol is larger than the number of atoms in the universe. So you're stuck writing it this way: 10^{googol}, or $10^{10^{100}}$ or 10^(10^100). If you were so motivated, I suppose you could attempt to write 10^{19} zeros, on every atom in the universe. But you surely have better things to do.

I'm not doing this just to waste your time. I've got a number that's bigger than a googolplex. Jacob Bekenstein invented a formula allowing us to estimate the maximum number of different quantum states that could have a mass and size comparable to our observable universe. Given the quantum fuzziness we observe, that would be the maximum number of distinct observable universes like ours. It's 10^(10^124), a number that has 10^{24} times as many zeros as a googolplex. These 10^(10^124) universes range from ones that are scary, filled with mostly black holes, to ones that are exactly like ours but where your nostril is missing one oxygen molecule and some space alien's nostril has one more.

So, in fact, we do have some uses for some very large numbers. I know of no utility for numbers larger than this one, but mathematicians surely do.

A theorem once contained the badass number $10^{\wedge}(10^{\wedge}(10^{\wedge}34))$. It's called *Skewe's number*. Mathematicians derive pleasure from thinking far beyond physical realities.

Let me give you a sense of other extremes in the universe.

How about density? You intuitively know what density is, but let's think about density in the cosmos. First, explore the air around us. You're breathing 2.5×10^{19} molecules per cubic centimeter—78% nitrogen and 21% oxygen.

A density of 2.5×10^{19} molecules per cubic centimeter is likely higher than you thought. But let's look at our best laboratory vacuums. We do pretty well today, bringing the density down to about 100 molecules per cubic centimeter. How about interplanetary space? The solar wind at Earth's distance from the Sun has about 10 protons per cubic centimeter. When I talk about density here, I'm referencing the number of molecules, atoms, or free particles that compose the gas. How about interstellar space, between the stars? Its density fluctuates, depending on where you're hanging out, but regions in which the density falls to 1 atom per cubic centimeter are not uncommon. In intergalactic space, that number is going to be much less: 1 per cubic meter.

We can't get vacuums that empty in our best laboratories. There is an old saying, "Nature abhors a vacuum." The people who said that never left Earth's surface. In fact, Nature just *loves* a vacuum, because that's what most of the universe is. When they said "Nature," they were just referring to where we are now, at the base of this blanket of air we call our atmosphere, which does indeed rush in to fill empty spaces whenever it can.

Suppose I smash a piece of chalk against a blackboard and pick up a fragment. I've smashed that chalk into smithereens. Let's say a smithereen is about 1 millimeter across. Imagine that's a proton. Do you know what the simplest atom is? Hydrogen, as you might have suspected. Its nucleus contains one proton, and normal hydrogen has an electron occupying an orbital that surrounds it. How big would that hydrogen atom be? If the chalk smithereen is the proton, would the atom be as big as a beach ball? No, much bigger. It would be 100 meters across—about the size of a 30-story building. So what's going on here? Atoms are pretty empty. There are no particles between the nucleus and that lone electron, flying around in its first orbital, which, we learn from quantum mechanics, is spherically shaped around the nucleus. Let's go smaller and smaller and smaller, to get to another limit of the cosmos, represented by the measurement of things that are so tiny that we can't even measure them. We do not yet know what the diameter of the electron is. It is smaller than we are able to measure. However, superstring theory suggests that it may be a tiny vibrating string as small as 1.6×10^{-35} meters in length.

Atoms are about 10^{-10} (one ten-billionth) of a meter. But how about 10^{-12} or 10^{-13} meters? Known objects that size include uranium with only one electron, and an exotic form of hydrogen having one proton with a heavy cousin of the electron called a *muon* in orbit around it. About 1/200 the size of a common hydrogen atom, it has a half-life of only about 2.2 microseconds due to the spontaneous decay of its muon. Only when you get down to 10^{-14} or 10^{-15} meters are you measuring the size of the atomic nucleus.

Now let's go the other way, ascending to higher and higher densities. How about the Sun? Is it very dense or not that dense? The Sun is quite dense (and crazy hot) in the center, but much less dense at its edge. The average density of the Sun is about 1.4 times that of water. And we know the density of water—1 gram per cubic centimeter. In its center, the Sun's density is 160 grams per cubic centimeter. But the Sun is quite ordinary in these matters. Stars can (mis)behave in amazing ways. Some expand to get big and bulbous with very low density, while others collapse to become small and dense. In fact, consider my proton smithereen and the lonely, empty space that surrounds it. There are processes in the universe that collapse matter down, crushing it until it reaches the density of an atomic nucleus. Within such stars, each nucleus rubs cheek to cheek with the neighboring nuclei. The objects out there with these extraordinary properties happen to be made mostly of neutrons—a super-high-density realm of the universe.

In our profession, we tend to name things exactly as we see them. Big red stars we call *red giants*. Small white stars we call *white dwarfs*. When stars are made of neutrons, we call them *neutron stars*. Stars that pulse, we call them *pulsars*. In biology they come up with big Latin words for things. MDs write prescriptions in a cuneiform that patients can't understand, hand them to the pharmacist, who understands the cuneiform. It's some long fancy chemical thing, which we ingest. In biochemistry, the most popular molecule has ten syllables—deoxyribonucleic acid! Yet the beginning of all space, time, matter, and energy in the cosmos, we can describe in two simple words, *Big Bang*. We are a monosyllabic science, because the universe is hard enough. There is no point in making big words to confuse you further.

Want more? In the universe, there are places where the gravity is so strong that light doesn't come out. You fall in, and you don't come out either: *black hole*. Once again, with single syllables, we get the whole job done. Sorry, but I had to get all that off my chest.

How dense is a neutron star? Let's take a thimbleful of neutron star material. Long ago, people would sew everything by hand. A thimble protects your fingertip from getting impaled by the needle. To get the density of a neutron star, assemble a herd of 100 million elephants, and cram them into this thimble. In other words, if you put 100 million elephants on one side of a seesaw, and one thimble of neutron star material on the other side, they would balance. That's some dense stuff. A neutron star's gravity is also very high. How high? Let's go to its surface and find out.

One way to measure how much gravity you have is to ask, how much energy does it take to lift something? If the gravity is strong, you'll need more energy to do it. I exert a certain amount of energy climbing up a flight of stairs, which sits well within the bounds of my energetic reserves. But imagine a cliff face 20,000 kilometers tall on a hypothetical giant planet with Earthlike gravity. Measure the amount of energy you exert climbing from the bottom to the top fighting against the gravitational acceleration we experience on Earth for the whole climb. That's a lot of energy. That's more energy than you've stored within you, at the bottom of that cliff. You will need to eat energy bars or some other high-calorie, easily digested food on the way up. Okay. Climbing at a rapid rate of 100 meters per hour, you would spend more than 22 years climbing 24 hours a day to get to the top. That's how much energy you would need to step onto a single sheet of paper laid on the surface of a neutron star. Neutron stars probably don't have life on them.

We have gone from 1 proton per cubic meter to 100 million elephants per thimble. What have I left out? How about temperature? Let's talk hot. Start with the surface of the Sun. About 6,000 kelvins—6,000 K. That will vaporize anything you give it. That's why the Sun is gas, because that temperature vaporizes everything. (By comparison, the average temperature of Earth's surface is a mere 287 K.)

How about the temperature at the Sun's center? As you might guess, the Sun's center is hotter than its surface—there are cogent reasons for this, as we'll see later in the book. The Sun's center is about 15 million K. Amazing things happen at 15 million K. The protons are moving fast. Really fast, in fact.

Two protons normally repel each other, because they have the same (positive) charge. But if you move fast enough, you can overcome that repulsion. You can get close enough so that a brand-new force kicks in—not the repulsive electrostatic force, but an attractive force that manifests over a very short range. If you get two protons close enough, within that short range they will stick together. This force has a name. We call it the *strong force*. Yes, that's the official name for it. This strong nuclear force can bind protons together and make new elements out of them, such as the next element after hydrogen on the periodic table, helium. Stars are in the business of making elements heavier than those they are born with. And this process happens deep in the core. We'll learn more about that in chapter 7.

Let's go cool. What is the temperature of the whole universe? It does indeed have a temperature—left over from the Big Bang. Back then, 13.8 billion years ago, all the space, time, matter, and energy you can see, out to 13.8 billion light-years, was crushed together. The nascent universe was a hot, seething cauldron of matter and energy. Cosmic expansion since then has cooled the universe down to about 2.7 K.

Today we continue to expand and cool. As unsettling as it may be, the data show that we're on a one-way trip. We were birthed by the Big Bang, and we're going to expand forever. The temperature is going to continue to drop, eventually becoming 2 K, then 1 K, then half a kelvin, asymptotically approaching absolute zero. Ultimately, its temperature may bottom out at about 7×10^{-31} K because of an effect discovered by Stephen Hawking that Rich will discuss in chapter 24. But that fact brings no comfort. The stars will finish fusing all their thermonuclear fuel, and one by one they will die out, disappearing from the sky. Interstellar gas clouds do make new stars, but of course this depletes their gas supply. You start with gas, you make stars, the stars evolve during their lives, and leave behind a corpse—the dead end-products of stellar evolution: black holes, neutron stars, and white dwarfs. This keeps going until all the lights of the galaxy turn off, one by one. The galaxy goes dark. The universe goes dark. Black holes are left, emitting only a feeble glow of light—again predicted by Stephen Hawking.

And the cosmos ends. Not with a bang, but with a whimper.

Way before that happens, the Sun, to talk about size, will grow. You don't want to be around when that happens, I promise you. When the Sun dies, complicated thermal physics happens inside, forcing the outer surface of the

Sun to expand. It will get bigger and bigger and bigger and bigger, as the Sun in the sky slowly occupies more and more and more of your field of view. The Sun eventually engulfs the orbit of Mercury, and then the orbit of Venus. In 5 billion years, the Earth will be a charred ember, orbiting just outside the Sun's surface. The oceans will have already come to a rolling boil, evaporating into the atmosphere. The atmosphere will have been heated to the point that all the atmospheric molecules escape into space. Life as we know it will cease to exist, while other forces, after about 7.6 billion years, cause the charred Earth to spiral into the Sun, vaporizing there.

Have a nice day!

What I've tried to give you is a sense of the magnitude and grandeur of what this book is about. And everything that I've just referenced here appears in much more depth and detail in subsequent chapters. Welcome to the universe.

2

FROM THE DAY AND NIGHT SKY TO PLANETARY ORBITS

NEIL DeGRASSE TYSON

In this chapter, we will cover 3,000 years of astronomy. Everything that happened from antiquity, the time of the Babylonians, up until the 1600s. This is not going to be a history lesson, because I'm not going to cover all the details of who thought and who discovered what first. I just want to give you a sense of what was learned during all that time. It begins with people's attempts to understand the night sky.

Here's the Sun (figure 2.1). Let's draw Earth next to it; it's not drawn to scale in either size or distance, but is simply meant to illustrate certain features of the Sun–Earth system. Way out, of course, are the stars in the sky. I'm going to pretend that the sky is just stars, dots of light on the inside of a big sphere, which will make some other things easier to describe.

Earth, as you probably know, spins on an axis, and that axis is tilted relative to our orbit around the Sun. That angle of tilt is 23.5°. How long does it take for us to spin once? A day. How long to go around the Sun once? A year. Thirty percent of the general public in America who were asked that second question got the wrong answer.

A spinning object in space is actually quite stable, so that, as it orbits, its orientation in space remains constant. If we move the Earth around the Sun, from June 21 to December 21, as it comes around the other side of the Sun (to the right in figure 2.1), Earth will still preserve that spin orientation in space— its axis points in that same direction for the entire journey around the Sun.

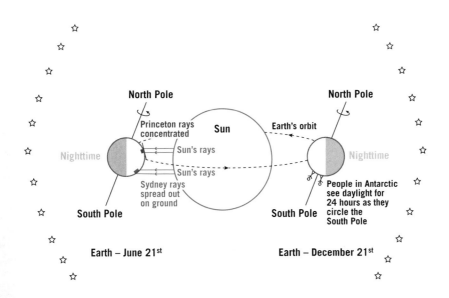

FIGURE 2.1. Earth circles the Sun, providing different nighttime views as the seasons change. Because of the tilt of Earth's axis relative to its orbit, on June 21, the Northern Hemisphere receives the Sun's rays more directly, while Australia and the entire Southern Hemisphere receive them obliquely. On December 21, people south of the Antarctic Circle see daylight for 24 hours as they circle around the South Pole as Earth rotates. *Credit:* J. Richard Gott

This makes for some interesting features. For example, on June 21, a vertical line perpendicular to the plane of the Earth's orbit divides the Earth into day and night. What can you say about the part of the Earth to the left of that line, away from the Sun? That's nighttime. But on December 21, when Earth is on the opposite side of its orbit, nighttime is now on the opposite side—to the right in the illustration. All the people on Earth who look up in the nighttime can see only that part of the sky opposite the Sun. The nighttime sky on June 21 is different—the stars on the far left—from the nighttime sky you see on December 21—the stars on the far right. During the summer nights, we see the "summer" constellations, such as the Northern Cross and Lyra, whereas during the winter nights, we see the "winter" constellations, such as Orion and Taurus.

Let's look at something else. On December 21, if it's nighttime to the right of the vertical line and Earth turns on its axis, what about the upside-down people standing in Antarctica, south of the Antarctic circle? They go around the South Pole. Does a person there see darkness? Nope. On December 21, a person there sees 24 hours without dark—24 hours of sunlight—as Earth rotates. There is no nighttime for anybody within the entire South Polar cap of Earth on that day. That's true for anyone between the Antarctic Circle and the South Pole. Following this argument, if I come up to the North Pole and I watch people North of the Arctic circle revolve around the North Pole—Santa Claus and his friends—they never rotate into the daytime side of Earth. For them, on December 21, there are 24 hours of darkness. As you might suspect, on June 21, the reverse happens: it's the people south of the Antarctic Circle who have no day at this time of the year and the people in the Arctic who have no night.

Let's observe from Princeton, New Jersey—it's close to New York City, but with no skyscrapers or bright city lights to interfere with the view. The town's latitude on Earth is about 40° North. At dawn on June 21, the Northern Hemisphere rotates New Jersey into daytime, receiving quite direct sunlight, whereas the sunlight hitting the Southern Hemisphere is rather oblique to Earth's surface.

Noon is when the Sun reaches its highest point in the sky. Did you know that nowhere in continental United States is the Sun ever directly overhead at any time of day, on any day of the year? Odd because if you grab people in the street and ask, "Where is the Sun at 12 noon?" most will answer, "It's directly overhead." In this case and in many others, people simply repeat the stuff they think is true, revealing that they've never looked. They've never noticed. They've never conducted the experiment. The world is full of stuff like that. For example, what do we say happens to the length of daylight in winter? "The days get shorter in the winter, and longer in the summer." Let's think about that. What's the shortest day of the year? December 21, which is the solstice and also the first day of winter in the Northern Hemisphere. If the first day of winter is the shortest day of the year, what must be true for every other day of winter? They must get longer. So days get longer in the winter, not shorter. You don't need a PhD or a grant from the National Science Foundation to figure that out. Hours of daylight get longer during the winter and shorter during the summer.

What's the brightest star in the nighttime sky? People say the North Star. Have you ever looked? Most haven't. The North Star (also known as Polaris) is not in the top 10. It's not in the top 20. It's not in the top 30. It's not even in the top 40. Australia sits too far to the south for anybody there to see the North Star. They don't even have a South Pole star to look at. And while we're talking celestial hemispheres, don't ever be jealous of the constellations in the southern sky. Take the Southern Cross; you may have heard about it. People write songs about it. But did you know that the Southern Cross is the smallest constellation out of all 88 of them? A fist at arm's length covers the entire constellation completely. Meanwhile, the four brightest stars of the Southern

Cross make a crooked box. There is no star in the middle to indicate the center of the cross. It's more accurately thought of as the Southern *Rhombus*. In contrast, the Northern Cross covers about 10 times the area in the sky and has six prominent stars—it looks like a cross, with one star in the middle. In the North we've got some great constellations.

The North Star is actually the 45th brightest star in the nighttime sky. So do me a favor and grab people in the street, ask them that question, and then set them straight. If you must know, the brightest star in the nighttime sky is Sirius, the Dog Star.

Now let's compare what happens to the sunlight at two locations on Earth. Look at the ground at noon in Princeton on June 21—sunlight hits it from a very high angle (see figure 2.1). The two parallel rays traveling from the Sun hit Princeton only a short distance apart on the ground. The ground at Sydney, Australia, at noon takes in a similar pair of rays, except that the rays are coming in at a much lower angle and are spread much farther apart on the ground. What's going on here? Which place is getting its ground heated more efficiently? Princeton, of course. The energy impinging on Princeton's ground is more concentrated, because of how directly the rays intersect Earth's surface, making Princeton's ground hotter. It's summertime in Princeton on June 21. At this same time of year, it's winter in Sydney, Australia. The reverse will apply 6 months later on December 21.

The Sun heats the ground; the ground heats the air. The Sun does not appreciably heat the air itself, which is transparent to most of the energy that comes from the Sun. The Sun's energy peaks in the visible part of the spectrum, and you already know that you can see the Sun through the atmosphere. From this we conclude the obvious fact that the Sun's visible light is not being absorbed by the air, otherwise, you wouldn't see the Sun at all. If you are indoors in a room with no windows, you can't see the Sun, because the roof of your building is absorbing the visible light from the Sun. You must either look out a transparent window or go outside to see the Sun. So, in sequence, light from the Sun passes through the transparent air and hits the ground. The ground absorbs the light from the Sun, and then reradiates that energy as invisible infrared light, which the atmosphere can and does absorb—we'll talk more about these other parts of the spectrum in chapter 4.

The ground absorbs visible light from the Sun, gets hotter, and then the ground heats the air through the infrared energy it emits. This doesn't happen instantaneously. It takes time. But how much time? What's the hottest time of day? It's not 12 noon—the time of peak ground heating. The hottest time of day is never 12 noon. It's always a few hours later because of this effect: 2 pm, 3 pm. Even as late as 4 pm in some places.

So that's summertime in the Northern Hemisphere. In summer the North Pole of Earth's axis tilts toward the Sun, and of course this is winter for those in the Southern Hemisphere. For the same reason that the hottest time of day is after 12 noon, the hottest time of year in the Northern Hemisphere is after June 21. That's why the season of summer *starts* on June 21, and it gets hotter after that. Similarly, December 21 is the start of winter in the Northern Hemisphere, and it gets colder after that.

Three months later, on March 21, spring starts. Every part of Earth rotates both into sunlight and out of sunlight on the first day of Northern Hemisphere spring (March 21) and on the first day of Northern Hemisphere fall (September 21). So everybody on Earth gets equal amounts of darkness and lightness on those two days—the equinoxes.

Earth's North Pole points toward Polaris, the North Star. A cosmic coincidence? Not really, because we don't point exactly there. You can fit 1.3 full-moon widths between the actual spot in the sky where our axis points (the North Celestial Pole) and the position of Polaris.

Let's go back to Princeton, as shown in figure 2.2. Standing there at night, you'll see any star on one side of the sky at that instant. In the figure, these stars are marked "Stars visible above Princeton's horizon." Princeton's horizon is drawn—this line is tangent to the surface of Earth where you're standing. When you look up, you'll see stars making circles around Polaris, as Earth turns (shown on the right in figure 2.2). (Polaris is so close to the North Celestial Pole that it barely moves.) So there's a cap in the sky where these stars make circles around Polaris but never actually set below your horizon. These are called *circumpolar stars*.

Suppose you look at a star farther away from Polaris. That star sets, then comes around and rises again. That's what the sky looks like, the view from Earth. One of the more familiar *asterisms* (star patterns) of the night sky is the Big Dipper, well-known because its stars are bright and it goes around Polaris (see figure 2.2). It dips down, just skimming the horizon (as seen from Princeton), and then comes back up again. Anything much farther from Polaris than the Big Dipper actually sets. How high is Polaris, in angle, as seen from

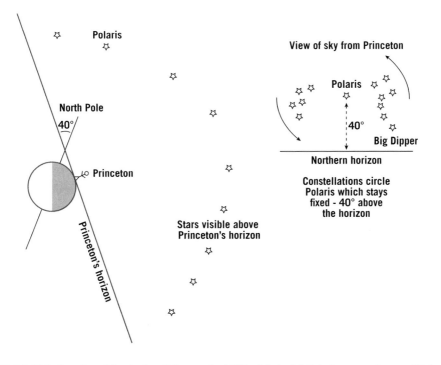

FIGURE 2.2. Nighttime view of the sky from Princeton (at 40° North latitude). Polaris stays stationary, 40° above the northern horizon. The Big Dipper revolves counterclockwise around it. *Credit:* J. Richard Gott

Princeton? We can figure this out. First, let's say we have gone to visit Santa Claus at the North Pole. Where is Polaris in the sky? If you're visiting Santa Claus, Polaris will be (almost directly) straight overhead. It's always straight overhead there. A star halfway up in the sky as seen from the North Pole circles Polaris as Earth turns, always staying above the horizon. A star right on the horizon circles along the horizon, so every star you can see stays above the horizon. No star rises, no star sets; they all circle Polaris overhead, and you see the entire Northern Hemisphere of the sky. That's Santa's view.

What's the latitude of the North Pole? Ninety degrees. What's the altitude of Polaris from the horizon as seen from the North Pole? It's 90°—the same number. That's not a coincidence. Polaris is 90° up, and you're at 90° latitude. Let's go down to the equator. What's the latitude of the equator? Zero degrees. Polaris is now on the horizon, 0° up. What's my latitude in Princeton? Forty degrees. So, from Princeton, the altitude of Polaris is 40° above the horizon.

People who navigate by the stars know that the altitude of Polaris you observe is equal to your latitude on Earth. Christopher Columbus set sail at a fixed latitude that he maintained for his entire journey across the Atlantic Ocean.

Go back and look at his maps. That's how they navigated; they kept at that latitude by keeping Polaris at the same altitude above the horizon during the trip.

Did you ever play with a top when you were a kid, and watch the top wobble? Earth wobbles. We are a spinning top, under the influence of the gravitational tug from the Sun and Moon. We wobble. The time it takes to make one complete wobble is 26,000 years. We rotate once in a day and wobble once in 26,000 years. That has an interesting consequence. First, consider the sphere of stars that I drew around the solar system. As Earth moves around the Sun, the Sun occupies a different place against the background of stars. On June 21, in our earlier figure 2.1, the Sun sits between us and the stars on the far right, which means that the Sun passes in front of those stars as seen by us on June 21. But on December 21, the Sun is between us and the stars to the far left. In between times, the Sun occupies a place in front of different sets of stars throughout the year, as it circles the sky. Long ago, when most of the world was illiterate, when there was no evening television, no books or internet, people put their culture onto the sky. Things that mattered in their lives. The human mind is very good at making patterns where none really exist. You can easily pick patterns out of random assortments of dots. Your brain says, "I see a pattern." You can try this experiment: if you're good at programming a computer, take dots and place them randomly on a page. Take about a thousand dots, look at them, and you may think, "Hey, I see . . . Abraham Lincoln!" You'll see stuff. In a similar way, these ancient people put their culture on the sky when they had no other idea what was going on. They didn't know what the planets were doing; they didn't understand laws of physics. They said, "Hmm! The sky is bigger than I am—it must influence my behavior." So they supposed, "There's a crustacean-looking constellation of stars over here, and it's got some personality traits; the Sun was in that part of the sky when you were born. That must have something to do with why you're so weird. And then over here we've got some fishes, and over there we've got some twins. Because we don't have HBO, let's weave our own storylines and pass these stories on from person to person." In so doing, ancient peoples laid out the zodiac, the constellations in front of which the Sun appears to move throughout the year.

There were twelve of these zodiac constellations; you know them all— Libra, Scorpio, Aries, and so on. And you know them because they're in nearly every daily news feed. Some person you've never met makes money telling you about your love life. Let's try to understand that. First of all, it's not really twelve constellations that the Sun moves through, it's thirteen. They don't tell you that, because they couldn't make money off of you if they did. Do you know what the thirteenth constellation of the zodiac is? Ophiuchus. It sounds like a disease, as in: "Do you have Ophiuchus today?" I know you know what your sign is, so don't lie and say, "I never read my horoscope." Most Scorpios are actually Ophiuchans, but we don't find Ophiuchus in the astrology charts.

Well, let's keep this up for a minute. When did they lay out the zodiac? It was encoded 2,000 years ago. Claudius Ptolemy published maps of it. Two thousand years is 1/13 of 26,000 years. Almost 1/12. Do you realize that because of Earth's wobble (we call it *precession*, the official term), the month of the year in which the Sun is seen against a particular constellation in the zodiac shifts? Every single zodiacal constellation that has been assigned to the dates identified in the newspapers is off by an entire month. So, Scorpios and Ophiuchans are currently Librans.

Therein lies the greatest value of education. You gain an independent knowledge of how the universe works. If you don't know enough to evaluate whether or not others know what they're talking about, there goes your money. Social anthropologists say that state lotteries are a tax on the poor. Not really. It's a tax on all those people who didn't learn about mathematics, because if they did, they would understand that the probabilities are against them, and they wouldn't spend a dime of their hard-earned money buying lottery tickets.

Education is what this book is all about. Plus a dose of cosmic enlightenment.

Let's discuss the Moon, and then get straight to Johannes Kepler and then to my man, Isaac Newton, whose home I visited when filming *Cosmos: A Space-time Odyssey*.

But first, we've got Earth going around the Sun, and of course we have the Moon going around Earth, so let's show that in figure 2.3. We put the Sun way off in the distance to the right and Earth in the center of the diagram, and we show the Moon at different positions as it circles Earth. We are looking down on the north pole of the Moon's orbit, as sunlight comes in from the right.

Both Earth and the Moon are always—at all times—half illuminated by the Sun. If you're standing on Earth, looking at the Moon when it is opposite the Sun, what do you see? What phase is the Moon? Full. The big pictures in figure 2.3 show the appearance of the Moon as seen from Earth at each point in its orbit.

FIGURE 2.3. The Moon's phases as it circles Earth. The Sun, at right, always illuminates half of Earth and half of the Moon. The diagram shows the sequence (counterclockwise) of positions the Moon occupies as it orbits Earth. We are looking down on the orbit from the north. The Moon always keeps the same face toward Earth. Notice that at new moon, its back side, never seen from Earth, is illuminated. The large photographs show the appearance of the Moon at each position as seen from Earth. *Photo credit:* Robert J. Vanderbei

Why don't we have a lunar eclipse every month, when Earth is between the Sun and Moon like this? It is because the Moon's orbit is tipped at about 5° relative to Earth's orbit around the Sun. So, in most months, the Moon passes north or south of Earth's shadow in space, preserving our normal view of the *full Moon*. Once in a while, when the Moon is full as it crosses the plane of Earth's orbit, it will pass into Earth's shadow, and we have lunar eclipse.

Now let the Moon continue 90° counterclockwise in its orbit. The Moon is now in *third-quarter* phase. Colloquially known as half-Moon—you see the moon half illuminated. Bring the Moon 90° further along, counterclockwise in its orbit, and the Moon passes between Earth and the Sun. Only the side of the Moon facing the Sun is lit up and you can't see that, so when standing on

Earth, you can't see the Moon at all. We call it *new Moon*. The Moon usually passes north or south of the Sun, during this phase. Occasionally, when it passes directly in front of the Sun, we get a solar eclipse.

So far, we have full Moon, third-quarter Moon, and new Moon. Come around another 90°, and we get *first-quarter Moon*, when it is half illuminated again. We also have in-between phases. Crossing from new Moon to first-quarter Moon, what do you see? Only a little smidgen. A crescent. It's called a *waxing* crescent Moon, because it grows thicker every day. And just before new Moon, we get a *waning* crescent. These crescents face opposite directions as the Moon shrinks and then grows again.

Between first-quarter and full Moon we have something called *waxing gibbous*. It's a pretty awkward looking phase, and is almost never drawn by artists, even though half the time we ever see the Moon it's in a gibbous phase—not quite full, not quite a quarter Moon. If artists were drawing the sky randomly throughout the year, we might expect to see a gibbous Moon half the time in their work, yet they typically choose to draw either a crescent Moon or a full Moon. They are not capturing the full reality that lay in front of them.

Of course, this entire cycle takes a month, formerly known as a "moonth." If the full Moon is opposite the Sun, what time of day does it rise? If it is opposite the Sun and the Sun is setting, then we conclude the full Moon is rising, at sunset. And if the Sun is rising, the full Moon is setting.

The situation is different at other times of the month. When the third-quarter Moon is high in the sky, the Sun is rising. Notice in the diagram, where Earth is rotating counterclockwise, you are getting rotated into sunlight when the third-quarter Moon is high in the sky. Imagine taking your brain and your eyes into that picture, looking around, and then stepping back in the real world, to check your result.

I have an app on my computer, such that every time I bring up the desktop, the Moon is there, showing its phase, day by day. That's my lunar clock. It connects me to the universe even when I'm staring at my computer screen.

Let's get back to the solar system—mid-to-late 1500s. In Denmark, there lived a wealthy astronomer named Tycho Brahe. The crater *Tycho* on the Moon is named after him.

I spent an hour once with someone who was native to Denmark, learning how to pronounce this astronomer's name correctly: tī'kō brä. I worked hard on that. But of course in America, we pronounce it however it looks to us.

Tycho Brahe cared a lot about the planets, enough to keep track of them. He built the best naked-eye instrument of the day, maintaining the most accurate measurements of planetary positions ever. Telescopes were not invented until 1608, so Tycho used sighting instruments, while writing down the positions of stars on the sky and of planets as a function of time. Tycho had an enormous database, and a brilliant assistant, the German mathematician Johannes Kepler.

Kepler took the data, and he figured stuff out. Kepler said to himself, "I understand what the planets are doing. In fact I can create laws that describe exactly what the planets are doing." Before Kepler, the organization of the universe was plain and obvious: "Look, the stars revolve around us. The Sun rises and sets. The Moon rises and sets. We must be at the center of the universe." This not only felt good to believe, it also *looked* that way. It stoked the human ego, and the evidence supported it, so no one doubted it—until the Polish astronomer Nicolaus Copernicus came along. If Earth were in the middle, what are the planets doing? You look up, and from day to day you watch Mars move against the background stars. Hmm. Right now it's slowing down now. Oh wait, it stopped. Now it's going backward (that's called *retrograde* motion), then it's going forward again. Why should it do that?

Copernicus wondered—if the Sun were in the middle, and Earth went around the Sun, what then? Well, these forward and backward motions get explained in a snap. The Sun is in the middle, Earth goes around the Sun in an orbit, like a racecar going around a racetrack. Mars, the next planet out from the Sun, orbits more slowly, like a slower racecar in an outer lane. When Earth passes Mars on the inside track, Mars seems to be going backward in the sky for a while. If you are in the fast lane on the highway and pass a slower car in the next lane, that car appears to drift backward relative to you. If you put the Sun in the middle, and made Earth and Mars go around the Sun in simple circular orbits, it explained the retrograde motion; it explained what was going on in the nighttime sky. Planets farther from the Sun orbited more slowly. Copernicus published all this in a tome called *De Revolutionibus orbium coelestium*. If you try to buy the first edition of that book at auction, it will cost you over two million dollars, as it is one of the most important books ever written in human history.

It was published in 1543, and it got people thinking. Copernicus had been afraid to publish the book at first, and had been showing his manuscript to colleagues privately. You couldn't just start saying to everyone that Earth was no longer in the center of universe. The powerful Catholic Church had other ideas about things, asserting that Earth was in the center.

Aristotle had said so. In ancient Greece, Aristarchus had correctly deduced that Earth orbited the Sun—but Aristotle's view won out, and the church still subscribed to it, since it was consistent with Scripture. So, when did Copernicus publish his book? When he lay on his deathbed. You can't be persecuted when you're dead. He reintroduced the Sun-centered universe, called the *heliocentric model*.

"Helio-" means Sun. Before then, we had *geocentric* models. That came from Aristotle, Ptolemy, and later, by decree, the church.

And then came Kepler. Kepler, who agreed with Copernicus, up to a point. Copernicus invoked orbits that were perfect circles. But because these didn't quite match the observed motions of the planets, Copernicus had adjusted them by adding smaller epicyclic circles (as Ptolemy had done as well). Still, they didn't exactly match the positions of the planets in the sky. Kepler figured that the Copernican model needed fixing. And from the data—planetary position measurements over time—left to him by Tycho Brahe, he deduced three laws of planetary motion. We call them *Kepler's laws*.

The first one says: *Planets orbit in ellipses, not circles* (see figure 2.4). What's an ellipse? Mathematically, a circle has one center, and an ellipse sort of has two centers: we call them *foci*. In a circle, all points are equidistant from the center, whereas in an ellipse, all points have the same sum of distances to the two foci. In fact, a circle is the limiting case of an ellipse, in which the two foci occupy the same spot. An elongated ellipse has foci that are far apart. As I bring the foci together, I get something that more closely resembles a perfect circle.

According to Kepler, planets orbit in ellipses with the Sun at one focus. This is already revolutionary. The Greeks said, if the universe is divine, it must be perfect, and they had a philosophical sense of what being perfect meant. A circle is a perfect shape: every point on a circle is the same distance from its center; that's perfection. Any movement in the divine universe must trace perfect circles. Stars move in circles, they thought. This philosophy had endured for thousands of years. Then here comes Kepler saying, no, people, they are not circles. I've got the data, left to me by Tycho, to show they're ellipses.

He further showed that as planets orbit, the speed of a planet varies with its distance from the Sun. Imagine an orbit that is a perfect circle. There's no reason for the speed to be any different on one part of the circle than another;

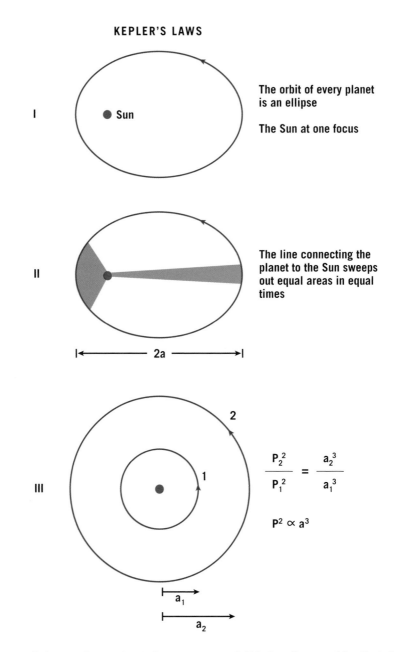

KEPLER'S LAWS

I   The orbit of every planet is an ellipse

The Sun at one focus

II   The line connecting the planet to the Sun sweeps out equal areas in equal times

III

$$\frac{P_2^{\,2}}{P_1^{\,2}} = \frac{a_2^{\,3}}{a_1^{\,3}}$$

$$P^2 \propto a^3$$

FIGURE 2.4. Kepler's Laws. The quantity *a* is the *semi-major axis*, half the long diameter of the elliptical orbit. For a circular orbit, with zero eccentricity, the semi-major axis is the same as the radius. *Credit:* J. Richard Gott

the planet should just keep the same speed. But not so with the ellipse. Where would the planet have the most speed? As you might suspect, when the planet is closest to the Sun. Kepler found that a planet travels fast when it is close to the Sun and more slowly when it is farther away.

Thinking about the problem geometrically, Kepler said, "let's measure how far the planet goes, for example, in a month." When it is close to the Sun and moving fast, in a month it will sweep out a certain area of its orbit, in a stubby, fat fan (see figure 2.4). Call this area A1. Let's do that same experiment in another part of the orbit, when it is farther away from the Sun. Kepler observed that it is moving more slowly when it is farther away, and therefore it's not going to travel as far in the same amount of time. As it travels a shorter distance, it will trace out a long, thin, fan-shaped area, A2, in the same 1-month period. Kepler was clever enough to notice that the area it swept out in a month was the same whether it was close or far from the Sun: A1 = A2. He therefore made a second law: *Planets sweep out equal areas in equal times.*

This has a fundamental derivation, which comes about from the conservation of *angular momentum*. If you've never seen that term before, you can understand it intuitively.

Ice skaters use it. Notice how spinning figure skaters start with their arms out. What do they do? They pull their arms in, shortening the distance between their arms and their axis of rotation, and their rotation speeds up in response. As the planet on an elliptical orbit moves closer to the Sun, shortening its distance to the Sun, it speeds up.

We call it conservation of angular momentum. Kepler didn't have this vocabulary available to him at the time. But that is, in fact, what he had found.

Kepler's third law was brilliant, just brilliant (see figure 2.4 again). It took him a long time. The first two laws he just banged right out, practically overnight. The third law took him 10 years, and he struggled with it. He was trying to figure out a correspondence between the distance of a planet from the Sun and the time it takes to go around the Sun, its orbital period. The outer planets take longer to make a complete orbit than the inner planets do.

How many planets were then known? Mercury; Venus; Earth; Mars; Jupiter; and everybody's favorite planet, Saturn.

Third graders used to name Pluto as their favorite planet—which put me on their bad list when, at the Rose Center for Earth and Space, we downgraded Pluto's planetary status to that of an ice ball in the outer solar system.

The Greek word *planetos* meant "wanderer." To the ancient Greeks, Earth was not considered a planet, because we were at the center of the universe. And the Greeks recognized two other planets I haven't listed; what would they be? They were also moving against the background stars: the Sun and the

Moon. By the definition from ancient Greece, these were the seven planets. And the seven days of the week owe their names to the seven planets or the gods related to them. Some are obvious, like *Sunday* and *Monday*. Saturday is *Saturn*-day. You have to go to other languages to get the rest; we have Frigga for Friday, for example. Frigga (or sometimes Freyja) was the Norse goddess associated with Venus.

At last, Kepler figured out an equation. It's the first equation of the cosmos. Kepler started by measuring all distances in Earth–Sun units.

We call these Astronomical Units, or AU. The distance of a planet from the Sun varies with time. An ellipse is a flattened circle; it has a long axis and a short axis, which are called the *major* and *minor axes,* respectively. Kepler (brilliantly) figured out that he should take half the major axis of its orbit as his measure of a planet's distance from the Sun. We call this the semi-major axis; it's the average of the planet's maximum and minimum distances from the Sun.

And if we measure time in Earth-years, we have an equation that was the dawn of our power to understand the cosmos. If we use the symbols P, a planet's orbital period in Earth-years, and a, the average of the planet's minimum and maximum distances from the Sun in AU, we get:

$$P^2 = a^3,$$

Kepler's third law. Let's see if that works for Earth. Let's try out the equation. Earth has period 1. And its average min-max distance is 1. So the equation gives $1^2 = 1^3$. Or $1 = 1$. It works. That's good.

If this is a Solar-System-wide law, it should work for any planet (or other object orbiting the Sun) that was known then, or would later be discovered. How about Pluto? Kepler didn't know about Pluto. Let's do Pluto. Pluto's average min-max distance from the Sun is 39.264 AU. So the law says $P^2 = 39.264^3$. What's 39.264 cubed? It's 60,531.8. You can check that on a calculator. The orbital period P has to equal the square root of 60,531.8, which is 246.0, rounded to four digits. What is the actual period for Pluto in its orbit? 246.0 years.

Kepler was badass.

When Isaac Newton invented the universal law of gravitation, he invoked $P^2 = a^3$ to figure out how gravitational attraction fell off with distance. It fell off like one over the square of the distance. To arrive at his answer, he used calculus— which, conveniently, he had just invented. Newton generalized Kepler's law to find a law that no longer applied to just the Sun and planets. It applied to any two bodies in the universe, based on a newly revealed gravitational force attracting them toward each other given by:

$$F = Gm_a m_b / r^2$$

G is a constant, m_a and m_b are the masses of the two bodies, and r is the distance between their two centers.

From that equation, you can derive Kepler's third law, $P^2 = a^3$, as a special case. You can also derive Kepler's first and second laws: that the general orbit of a planet around the Sun is an ellipse with the Sun at one focus, and that a planet sweeps out equal areas in equal times! That's how powerful Newton's law of gravitation is, and it's even bigger than this. It is the entire description of the force of gravity between any two objects any place in the universe, no matter what kinds of orbit they have. Newton expanded our understanding of the cosmos and came out with a description of the planets that went far beyond anything Kepler imagined. Newton derived this formula before he turned 26. Newton discovered the laws of optics, labeled the colors of the spectrum, and he determined that amazingly, the colors of the rainbow, when combined, gave you white light. He invented the reflecting telescope. He invented calculus. He did all this.

The next chapter is all about him.

3

NEWTON'S LAWS

MICHAEL A. STRAUSS

Copernicus made the great breakthrough of explaining planetary motions in terms of the *heliocentric* universe, by placing the Sun at the center of what we now call the *solar system*. The various planets, including Earth, are all moving in orbits around the Sun. We are sitting on a moving platform. To figure out how fast Earth is going, we need to determine how far it goes in a specific interval of time; its speed is then that distance divided by that time.

As we saw in chapter 2, Kepler showed that Earth's orbit is an ellipse. In fact, the orbits of most of the planets in our solar system are close to circles, so we will take the approximation, for the time being, that Earth is moving in a circle, around which it travels in a year. The radius of that circle, the distance from the Sun to Earth, is a quantity that we find ourselves using constantly in astronomy. As described in the last chapter, it is officially named the Astronomical Unit, or AU, for short. One AU is approximately 150 million kilometers, or 1.5×10^8 km.

We thus go around the circumference of a circle 150 million km in radius in one year. The circumference of a circle is 2π times its radius. Everyone knows that π is approximately 3. That's the kind of approximation astronomers like to use when making rough estimates. We need to divide the circumference by the time, which is 1 year.

We would like to express that year in seconds, which will be useful for present purposes. The number of seconds in a year is 60 (seconds in a minute), times 60 (minutes in an hour), times 24 (hours in a day), times 365 (days in a year). You could multiply that out on a calculator, but recall that, in chapter 1, Neil said that he drank champagne on his billionth second when he was about

31 years old. Thus, a year is about 1/30 of a billion, which is about 30 million seconds. We will write this as approximately 3.0×10^7 seconds in a year.

Putting this all together, we find that the speed at which Earth is orbiting the Sun is $2\pi r/(1 \text{ year}) = 2 \times 3 \times (1.5 \times 10^8 \text{ km})/(3 \times 10^7 \text{ sec}) = 30 \text{ km/sec}$. That's how fast we are going around the Sun right now. We are trucking! We think of ourselves as sitting still, which may explain why it was so natural for the ancients to imagine that they were at the center of the universe. It seemed so obvious. But in fact there is a great deal of motion going on. Earth is rotating on its axis once a day. It is going around the Sun once a year, traveling at 30 km/sec. We'll see in Part II of this book that the Sun is moving as well (carrying Earth and the other planets with it) in a variety of additional motions.

Copernicus told us that the various planets are orbiting the Sun. Kepler used Tycho Brahe's data to determine the orbits of the various planets and learn about their properties. As mentioned in chapter 2, he abstracted three laws from these orbits. Isaac Newton, one of the greatest heroes of our story, was able to deduce from Kepler's third law that gravity was a radial force between pairs of objects, proportional to one over the square of the distance between them.

Newton was perhaps the greatest physicist, maybe the greatest scientist of any type, who ever lived. He made an amazing number of fundamental discoveries. He wanted to understand how everything moved: not just the planets orbiting the Sun, but a ball tossed in the air or a rock rolling down a hill.

In science, one takes a large number of observations and tries to abstract from them a small number of laws that encompass and explain these observations. Newton came up with his own three laws of motion. The first is the law of *inertia*. What does inertia mean? In everyday usage, if you say "I have a lot of inertia today," it means you really don't want to get going; you are sitting still, and you want to continue to be a couch potato and not budge. It takes something else to get you going. An object at rest (like a couch potato) will remain at rest unless acted on by a force.

Let's talk about what the force is. Newton's law of inertia comes in two parts. The first part states that *an object that is at rest will remain at rest, unless acted on by an external force.* That makes sense. Consider an apple sitting on the table. It has no net force acting on it, and it remains at rest.

The second part of Newton's law of inertia is less intuitive: *an object with uniform velocity will remain at that uniform velocity, unless acted on by an external force.* Uniform velocity means that it goes at a certain speed and in a certain direction, neither of which change. If I roll a ball along the floor, it doesn't continue at a constant speed and in a constant direction forever, but rather slows down and stops, because a force is acting on it: friction between the ball and the floor. Friction is ubiquitous in everyday circumstances. Consider throwing a piece of paper through the air: it slows and then flutters to the floor. Actually, two forces are acting on it: (1) gravity, about which we will have a great deal to say in a moment, and (2) the force due to the resistance of the air itself. The paper has a large surface area for the air to strike, making air resistance important.

The idea that an object in motion will continue to move with constant velocity unless acted on by a force is not intuitive, because friction is all around us. It's hard to find everyday situations in which there's no friction, and therefore no force. A figure skater has little friction between the ice and her skates, and thus she can effortlessly glide for a long time across the ice. In the limit of no friction at all, an object given a push would retain a constant velocity. Galileo figured this out. Outer space offers the most dramatic examples of being away from all frictional forces. In space, you really can send something off with uniform velocity and know that it will keep on going, because there is nothing in its path to stop it. Newton formulated this all into a basic law.

Newton's *second law of motion* tells us about what happens when an object is being acted on by a force. An object can be acted on by a variety of forces, but whatever those forces are, it is the sum of all the forces that causes a deviation from this uniform velocity. We use the term *acceleration* to quantify this deviation: acceleration is the change in velocity per unit of time. The second law, then, relates the acceleration of an object to a force acting on it. When you push an object with some force, the object will accelerate. If the object has a small mass, the acceleration will be large, whereas if it is very massive, the acceleration will be smaller for the same amount of force. This relationship gives us Newton's most famous equation, $F = ma$; force equals mass times acceleration.

Newton's *third law of motion* can be phrased colloquially as, "I push you, you push me." That is, if one body exerts a force on another, that second body pushes back on the first with an equal and opposite force. If you push down on a tabletop with your hand, you feel a pressure back on your hand; the table is pushing back on you. Every force is paired with an equal and opposite force.

Consider an apple sitting in your hand. It is clearly sitting still. Does it have any forces acting on it? Yes, gravity from Earth. It should be accelerating downward, but clearly it's not. The reason is that your hand is holding the apple, pushing upward on it (using your arm muscles). In response, by Newton's third law, the apple is pushing down on your hand; that's what we refer to as the apple's *weight*. The gravitational force from Earth pulling downward on the apple and the force of your hand pushing back up on the apple balance each other out; the sum of these two forces is zero. Zero force means zero acceleration by Newton's second law, so the apple, which starts at rest, is not going anywhere.

Actually, the story is a bit more interesting than that. Earlier we calculated that Earth is going around the Sun in a circle, at 30 km/sec, and thus the apple is moving at that same speed. To think about this, we need to take a detour to talk about the nature of circular motion.

Motion at constant speed of 30 km/sec in a circle is *not* uniform velocity, because the direction of Earth's motion is constantly changing as it circles the Sun. If it didn't change direction, Earth would just go off on a straight line, not a circle. The acceleration that arises from going around in a circle is familiar from everyday life. Various rides in amusement parks send you around in a circle, and you can feel the acceleration viscerally.

Newton used the tools of differential calculus, which he had just invented, to determine the acceleration of an object moving in a circle of radius r at a constant speed v. That acceleration is v^2/r, directed toward the center of the circle. The apple in your hand, which we considered to be standing still, is in fact moving at 30 km/sec in an enormous circle; it's being accelerated. From Newton's second law, we know that there must be a force acting on it. That force is the gravitational attraction of the Sun. The Sun is pulling Earth around in an orbit, and it's pulling our apple around as well. The apple is subject to the force of the Sun's gravity just as you and I are.

We're moving at 30 km/sec around the Sun. Given that enormous speed, you might expect the resulting acceleration to be large, but the acceleration is actually quite small, because the radius of the circle is so enormous. Let's calculate just how small. The velocity of Earth is 30 km/sec, or 30,000 meters/sec, and the radius of Earth's orbit is 150,000,000,000 meters. Using our formula v^2/r, the acceleration a equals $(30{,}000 \text{ meters/sec})^2/150{,}000{,}000{,}000 \text{ meters}$ = 0.006 meters/sec², or 0.006 meters per second per second. That means that every second, the velocity changes by 6 millimeters per second. That is tiny. Galileo found that the acceleration of an object falling to the ground under the influence of Earth's gravity is about 9.8 meters per second per second, a much larger value. Therefore, although we're moving around the Sun at very high speed, Earth is being accelerated by only a small amount. On an amusement ride, in contrast, we're not going anywhere near 30 km/sec, but the radius r of the circle we're moving around is tiny; when we divide by that small value of r in the formula v^2/r, the resulting acceleration gets quite large, and we are immediately aware of the pull of this acceleration. (For example, a ride moving you at 10 meters per second with a radius of 10 meters, would give an acceleration of 10 meters per second per second.)

When we try to observe the acceleration due to the Sun, our situation is more subtle. The Sun is gravitationally accelerating *everything* on Earth—you, the book you're holding, the apple in your hand—all at the same rate. We are all in a free-fall orbit around the Sun. We don't detect any motion relative to the objects around us. It seems to us that we are stationary; we don't notice that we are moving, nor do we notice that we're being accelerated.

But the fact remains, Earth is being accelerated toward the Sun by an amount v^2/r. Newton then used Kepler's third law to figure out how the acceleration produced by the Sun varies with radius. The orbital period P of the planet is

$$P = (2\pi r/v);$$

that is, the orbital period, P, is the distance the planet travels in completing one orbit ($2\pi r$) divided by its velocity (v). Thus,

- P is proportional to r/v, and
- P^2 is proportional to r^2/v^2.

Kepler told us that P^2 is proportional to a^3, where a is the semi-major axis of the planet's orbit. In this case, Earth's orbit is nearly circular, so we can say approximately that $r = a$, and therefore, substituting r for a, we find:

- P^2 is proportional to r^3.

Since P^2 is also proportional to r^2/v^2,

- r^2/v^2 is proportional to r^3.

Dividing by r, we get:

- r/v^2 is proportional to r^2.

Inverting, we find,

- v^2/r (the acceleration) is proportional to $1/r^2$.

With these few steps of reasoning, Kepler's third law, and a little algebra, we've shown that the gravitational acceleration, and thus the force, exerted by the Sun on a body at distance r away is proportional to one over the square of that distance: Newton's "inverse-square" law of gravity. We have it in Newton's own words:

> 'I was in the prime of my age for invention & minded Mathematicks & Philosophy more than any time since' [my] ensuing deduction 'from Kepler's rule of the periodical times of the Planets being in a sesquialterate proportion of their distances from the centers of their Orbs' that 'the forces which keep the Planets in their Orbs must [be] reciprocally as the squares of their distances from the centers about which they revolve.'[1]

Newton applied this understanding of gravity to Earth and the Moon. Consider the famous falling apple that inspired Newton. It lies one Earth radius from the center of Earth and falls toward Earth with an acceleration of 9.8 meters per second per second. The Moon lies at a distance of 60 Earth radii. If the gravitational attraction of Earth falls off like $1/r^2$ (as is true for the Sun), then at the orbit of the Moon, Earth's gravitational attraction should cause an acceleration $(60)^2$ times smaller than the 9.8 meters per second per second at Earth's surface, or about 0.00272 meters per second per second.

Just as we did for the motion of Earth around the Sun, we can calculate the acceleration of the Moon as it undergoes circular motion around Earth, using its period (27.3 days) and the radius of its orbit (384,000 kilometers). Plugging in the numbers to v^2/r gives an acceleration of 0.00272 meters per second per second. Eureka! It agrees beautifully with the prediction from the apple. As Newton himself said, he found the two results to "answer pretty nearly." The same force that pulls the apple toward Earth also pulls the Moon toward Earth, curving its path away from a straight-line trajectory to keep the Moon in an approximately circular orbit around Earth. The gravity exerted by Earth that causes the apple to fall to Earth extends to the orbit of the Moon. Newton discovered this while staying at his grandmother's house when Cambridge

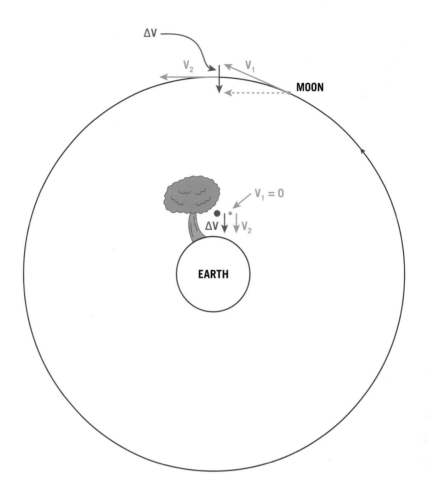

FIGURE 3.1. Acceleration of the Moon and Newton's apple, falling from its tree. Note that in each case, the acceleration (change of velocity) is directed toward the center of Earth. *Credit:* J. Richard Gott

University was closed during the plague years. But he didn't publish his results. Perhaps he was upset that the agreement between the prediction and the observation was not perfect, a slight discrepancy caused by the fact that Newton did not have a really accurate measurement of the radius of Earth to work with. In any case, it was only many years later that he would be prodded by Edmund Halley (of comet fame) into publication.

Newton worked out what is sometimes rather grandly called the *universal law of gravitation*, introduced in chapter 2. Consider two objects, say, Earth and the Sun. The distance between them (1 AU, or 1.5×10^8 km) is about 100 times the diameter of the Sun itself (1.4×10^6 km). They have masses M_{Earth} and M_{Sun}, respectively.

Newton found that the force of gravity between the two bodies is proportional to each of their masses, and to the inverse square of the distance r between them (using the reasoning from Kepler's third law, as just described). "Proportional" here means that the force will involve a constant of proportionality, which we call G, or Newton's constant, in Sir Isaac's honor. Here's Newton's formula for the force between the Sun and Earth:

$$F = GM_{Sun}M_{Earth}/r^2 .$$

The force is attractive: the two bodies attract each other, and thus the force is directed from each object toward the other.

By Newton's third law of motion, this formula covers both the gravitational force of the Sun on Earth and the force of Earth on the Sun. But the Sun's mass is much, much larger than Earth's mass. Newton's second law tells us that the acceleration is the force divided by the mass. As a consequence, the acceleration of Earth is much, much larger than that of the Sun, and therefore, the motion of the Sun due to this force is tiny compared with that of Earth. (They both orbit their mutual center of mass, but this is inside the surface of the Sun. The Sun executes a tiny circular motion about this center of mass, while Earth makes a grand circuit around the Sun.)

Here is another fascinating consequence of Newton's formula. By Newton's second law, the force of gravity, which we have just written down, is equal to the mass of Earth (M_{Earth}) times its acceleration, and for circular motion, the acceleration is equal to v^2/r. So in this case, $F = ma$ can be rewritten as:

$$GM_{Sun}M_{Earth}/r^2 = M_{Earth}v^2/r .$$

Note that the mass of Earth appears on both sides of this equation, and thus we can divide it out, leaving:

$$GM_{Sun}/r^2 = v^2/r .$$

What this means is that the acceleration of Earth ($GM_{Sun}/r^2 = v^2/r$) does not depend on Earth's mass. That's a remarkable fact. The acceleration of gravity does not depend on the mass of the object being accelerated, either for orbits around the Sun or for objects falling in Earth's gravitational field, because the mass of the object appears on both sides of the $F = ma$ equation and thus factors out. If I drop both a book and a piece of paper, they will feel the same acceleration, and should fall at the same rate, even though the book is much more massive. That's what Galileo said would occur in the vacuum. Does it work in practice? No, a book and a piece of paper fall at different rates, because of air resistance. Air resistance exerts a force on both the book and the paper, but since the book is much more massive than the piece of paper, the acceleration of the book due to the air resistance is small—essentially negligible. However, if I put the piece of paper on top of a big book, so that the book blocks the air resistance to the paper, and drop them again, the paper will stay sitting on the book as they fall together at the same rate. Try the experiment yourself!

When the Apollo 15 astronauts went to the Moon, they brought along a hammer and a feather to do an experiment to test this principle. The Moon has effectively no atmosphere: a very good vacuum exists above its surface, and hence there is no appreciable air resistance. When the astronauts dropped the feather and the hammer simultaneously, they fell at exactly the same rate, just as Newton (and Galileo) had predicted. You can see the video record of their lunar experiment online.

You may know that Aristotle got this wrong. Aristotle said that more massive objects would be subject to a greater acceleration and fall faster. He said that because it seemed logical to him, but in fact he never did an experiment to see whether his idea was correct. He could have taken big rocks and little rocks (neither of which is much affected by air resistance) and dropped them to discover that they would fall at the same rate. The bottom line here is that in science, it is crucial to check your intuitions with experiment!

Let's do a related problem. Consider the gravitational force exerted by Earth on an apple held in your outstretched hand. Newton's formula includes the distance r from the apple to Earth. We might naively think we should use the distance from the apple to the floor, about 2 meters. But that turns out not to be right. Newton realized that you must take into account the gravitational attraction from each and every gram of Earth: not just the piece at our feet, but also those parts on the other side of the globe. It took him about 20 years to figure out how to do this calculation. He needed to add up the forces from every separate chunk of Earth, each at their own distance and direction from this apple. To add up all these forces, he needed to invent a new branch of mathematics, now called *integral calculus*. The net result of the calculation is that gravity for a spherical object (like Earth) acts as if all its mass were concentrated at its center, a very nonintuitive concept. To do the calculation of the gravitational force on the apple, you need to imagine that the full mass of

Earth lies at a point 6,371 km beneath your feet, the distance from the surface to the center of Earth. We've already invoked this process when we discussed Newton's comparison of the falling apple to the orbiting Moon.

But an apple falling (straight down) surely doesn't seem to be the same as the orbital motion of the Moon. Why does the Moon go in circles, whereas the apple simply hits the ground? To put the apple into orbit, I'd have to throw it hard horizontally, so hard that it could go all the way around Earth. Consider the case of the Hubble Space Telescope, just a few hundred kilometers above Earth's surface. It travels completely around Earth, a circumference of 25,000 miles, in about 90 minutes. If we convert this to a speed, it turns out to be about 5 miles per second. So, to get an apple into orbit, I'd have to throw it horizontally at about 5 miles per second.

Imagine standing on top of a high mountain (above the frictional effects of the atmosphere) and throwing objects horizontally at ever greater speeds. Throw that apple as hard as you can; it quickly falls to the ground. Get a major league pitcher to toss it; it will go somewhat farther, but it will still fall down. Now let's get Superman to throw it. As he throws harder and harder, the apple will go farther and farther before its downward-curving trajectory hits the surface of Earth. But Earth's surface is not flat; it also curves downward in the distance. Superman can indeed throw an object at 5 miles per second. The object will also fall under the influence of gravity, but its curved trajectory now matches the curvature of Earth such that it never hits the surface and will end up in a circular orbit. The object in orbit is falling the entire time, albeit with plenty of sideways motion. When you drop an apple, it falls down due to the acceleration of Earth's gravity. That same gravity is causing both the Hubble Space Telescope to orbit Earth and the Moon to go around Earth (in a much higher orbit, therefore moving more slowly). In low-Earth orbit, you are falling at the same rate that Earth curves around, and you never hit the ground. Newton understood this, and proposed the idea of an artificial satellite in orbit around Earth—270 years ahead of its actually being done!

If you've ever been in an elevator that suddenly jerks down quickly, for a very brief period you're falling, and everything around you is falling with you. When you drop an apple, you yourself don't fall with it, because the force of the ground on the soles of your feet keeps you up. You are standing at rest relative to your surroundings, but the apple feels the acceleration and it falls. If you were knocked off your feet and fell with the apple, I would see the apple falling with you (at least until you and the apple both hit the floor).

You have probably seen images of astronauts in the International Space Station in orbit around Earth. Earth's gravity is acting on the astronauts and the International Space Station alike. But everything in the space station is falling at the same rate—recall our calculation that the acceleration of gravity does not depend on the mass of the object in orbit. With everything falling at the same rate, the astronauts feel weightless. "Weight" means what a bathroom scale registers when you stand on it (or equivalently, how much the bathroom scale pushes back on you, by Newton's third law). But if the scale is falling just as you are, you are not pushing down on the scale, and it registers your weight as zero. You are weightless.

This doesn't mean that your *mass* is zero, however. Mass and weight are not the same thing! Mass, according to Newton, is the quantity that goes into his second law of motion (relating forces, masses, and acceleration); it's also the quantity that gives rise to gravity. When people talk about losing "weight" what they really want to do is lose mass. Fat has mass, and they wish to get rid of some of that. Then, with the same amount of force, they can accelerate faster, and get around more easily.

Let's now take stock of what Newton accomplished. From observations of the motions of the planets known at the time, Kepler had abstracted three laws to describe their orbits. Then Newton came along and thought about this in a whole different way; with his three laws of motion, he attempted to understand how *everything* moves, not just the six planets known at the time to orbit the Sun. In addition, he developed a physical understanding of the force of gravity, the most important force in astronomy. Using Kepler's third law, he showed that the force of gravity must fall off like $1/r^2$. He found that the gravitational force between two bodies was attractive: the gravitational force of the Sun on a planet was $F = GM_{Sun}M_{Planet}/r^2$. Putting these together, we saw that we could understand Kepler's third law in terms of Newton's laws of motion and law of gravity. Newton came up with a much broader understanding of the physics behind Kepler's third law than Kepler had done.

In a final triumph, Newton showed that his law of gravitation predicted that a planet would trace out a perfect elliptical orbit with the Sun at one focus, and that a line connecting the planet to the Sun would sweep out equal areas in equal times. All three of Kepler's laws can now be seen as a direct consequence of Newton's one law of gravitational attraction, together with his three laws of motion.

Newton's laws of gravity were the first laws of physics we understood. Importantly, they could be used to make predictions that could be tested. Halley used Newton's laws to discover that several comet appearances over the centuries (including one in 1066 recorded in the Bayeux Tapestry) were actually all the same comet on a highly elliptical orbit. It returned approximately once every 76 years. It was perturbed by Jupiter and Saturn as it crossed their orbits, and its somewhat variable time of arrival could be predicted with Newton's laws—whereas with Kepler's laws, it would have been exactly periodic. Halley predicted the comet would return again in 1758. Halley died in 1742 and didn't live to see the event, but when it did reappear in 1758 as he had predicted, they named it after him the next year: Halley's comet. Its closest approach to the Sun was predicted by Alexis Clairaut, Jérôme LaLande, and Nicole-Reine Lepaute, using Newton's laws, with an accuracy of 1 month. This was a remarkable confirmation of Newton's laws of gravity.

Newton's laws had another great success. The planet Uranus was not following Newton's laws exactly; its orbit seemed to be perturbed. Urbain Le Verrier found that this could be explained if Uranus was being pulled by the gravity of another unseen planet farther out from the Sun. He predicted where this planet could be found, and in 1846, Johann Gottfried Galle and Heinrich Louis d'Arrest, using Le Verrier's calculations, found it only 1° in the sky away from where Le Verrier had predicted it would be. Newton's laws had been used to discover a new planet: Neptune. Newton's reputation soared.

We'll find ourselves using these basic notions of forces and gravity again and again throughout this book for understanding the universe.

HOW STARS RADIATE
ENERGY (I)

NEIL DeGRASSE TYSON

We now attempt to understand the distances to the stars. We've already seen that the distance from the Sun to Earth, 150 million km (or 1 AU), is about 100 times the diameter of the Sun itself. Imagine that we scale the Earth–Sun distance down to 1 meter; the Sun itself is then 1 centimeter across. The nearest stars are about 200,000 AU away, so to scale, that is 200 km. The space between stars is enormous compared to their size. We will find it convenient to refer to these distances not in kilometers or centimeters, but in terms of the time it takes light to traverse them.

The speed of light, which we refer to with the letter c, is 3×10^8 meters/sec, another number worth keeping in mind. In chapter 17, we will see in great detail why this speed represents the cosmic speed limit. It's as fast as anything can go. Since we observe stars by their light, it provides the most natural distance units. One light-second is the distance that light travels in 1 second: 3×10^8 meters, or 300,000 km—about seven times Earth's circumference. The Moon is 384,000 km away, and light travels that distance in 1.3 seconds. We say the Moon is about 1.3 light-seconds away. The distance from Earth to the Sun (1 AU) is about 8 light-minutes; taking light about 8 minutes to travel that distance. The nearest stars are about 4 light-years away. A light-year is thus a measure of distance, not of time—the distance light travels in 1 year. One light-year is about 10 trillion km. The light we see today from the nearest stars left them 4 years ago. In the universe, we're always looking back in time. We are seeing these nearby stars, not as they are at the present moment, but as they were 4 years ago.

This is true in everyday life as well. The speed of light, expressed in other units, is about 1 foot per nanosecond, so two people sitting at a table are seeing each other with a delay of a few nanoseconds. Of course, this is much too small for us to notice, but all our visual contact has a time delay built into it.

How can we measure the distances to the nearest stars? Four light-years is enormous. We can't simply stretch a tape measure between here and a star. In that effort, we need to introduce the concept of *parallax*. Earth orbits around the Sun (see figure 4.1). Earth is on one side of the Sun in January and 6 months later, in July, Earth is on the other side of the Sun. Toward Earth's right in the figure there's a nearby star, and then way out on the right is a field of more distant stars. They're so far away that I am going to stick all of them way off to the right. Then imagine that I take a picture of the nearby star in January. I am going to see all kinds of stars on that photograph, and one of them will be the star in question (filled in). See the view from Earth in January in figure 4.1. Alone, this picture tells us nothing, of course. Remember, I don't know which stars are close and which ones are far away—I don't know anything about this yet. But we wait 6 months, and we take that picture again from the opposite side of Earth's orbit in July, when Earth has moved to a new position. Now we see all that identical background, but our (filled-in) star has appeared to move

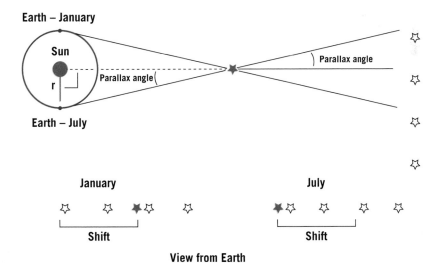

FIGURE 4.1. Parallax. As Earth circles the Sun, a nearby star shifts position in the sky relative to distant stars.
Credit: J. Richard Gott

from where it once was, to its new location as viewed from Earth in July. It has shifted. Everything else basically stays in the same place. What will happen in another 6 months? It shifts back, from whence it came. That shifting just repeats itself, back and forth, depending on when in the year we observe the star.

Flash the two pictures back and forth, one after the other. If you are flashing them and the two photographs are identical except for one star that moves, then *that star* is the one that is closer than all the others. If this star were even closer, then this shift on the picture would be bigger. Closer stars "shift" more. I put "shift" in quotes because the star is just sitting there—we are the ones moving back and forth around the Sun; the shift is really due to a change of our perspective.

INSTRUCTIONS FOR SEEING STEREOSCOPIC ART IN THREE DIMENSIONS

Given that we sense depth in the real world when our two eyes see things from slightly different perspectives, we can trick ourselves into seeing a three-dimensional scene even in the flat pages of a book—all we need is to present two images side by side, one as seen by the point of view of the right eye and one as seen by the point of view of the left eye. In this stereo pair (see figure 4.2), the image for the right eye is on the left and the image for the left eye is on the right, so you will be looking at this cross-eyed. It is easier than you think. Hold the book with one hand about 15 inches in front of your eyes. Hold the index finger of your other hand straight up about half-way between your eyes and the page. Look at the page. You will also see two blurry transparent images of your finger (one as seen by the right eye and one seen by the left eye). Move your finger back and forth until these two transparent images of your finger are perfectly centered at the bottom of each image on the page. You might have to tilt your head left or right to get the two images of your finger level with each other. Now focus your attention on your finger. You should see one image of your finger and three blurry versions of the pictures on the page. Carefully shift your attention to the middle picture, without uncrossing your eyes. It should come into focus as a beautiful 3D image, with the bright foreground star Vega jumping right off the page in front of the other stars! You can see that different stars are at different distances. Your brain is automatically measuring the shifts and doing the parallax calculation. This, of course, is how we generate 3D vision. Our brain is constantly comparing the views from our two eyes and doing parallax calculations to determine the distances to the objects we see. Alternatively, start by just looking at your finger—your eyes will naturally cross to look at it. Behind it will appear the three blurry images, Shift your gaze to the center one, and it will appear in 3D. Keep trying—it takes a bit of practice. Not everyone can see it, but if you can, it is a spectacular effect, and worth the effort to master. We will use this technique later in the book once again in *figure 18.1*.

You can demonstrate this for yourself. Close your left eye and hold your thumb out at arm's length. Line your thumb up with an object in the distance using your right eye only. Now wink to the other eye. What happens? Your thumb appears to move. Now take your thumb and position it only half an arm's length from your eyes and repeat that exercise. Your thumb shifts even more. People discovered this effect and realized that it works for stars: the nearby star is your thumb, and the diameter of Earth's orbit is the separation of your two eyes. Obviously, if you use your own eyes to try to measure the distance to a star, it will not be effective, because the couple of inches between your eyeballs is not enough to get nicely different angles on the star. But the diameter of Earth's orbit is 300 million kilometers. That's a nice broad distance for winking at the universe and deriving a measure of how close a star is to you.

In figure 4.2 we have a simulation of this showing the constellation Lyra. The stars in the two pictures have been shifted proportional to their observed parallax as if representing two photos taken at two times 6 months apart in Earth's orbit. We have just exaggerated the amount of the shift so that you can see it easily.

The brightest star in the picture, Vega, is only 25 light-years away. It is much closer than its fellow stars in the constellation of Lyra in the center. If you

FIGURE 4.2. Parallax of Vega. Two simulated pictures of the constellation Lyra as if taken 6 months apart from Earth as it circles the Sun. Each of the stars in the picture has a parallax shift inversely proportional to its distance. (The parallax shifts have been exaggerated by a large factor to make them visible.) Vega (the brightest star in Lyra), a foreground star only 25 light-years away, shifts the most. You can see Vega's parallax shift by comparing its position in the two images. You can also see this as a 3D image that jumps off the page by following instructions in the text to view the two pictures as a cross-eyed stereo pair. *Photo credit:* Robert J. Vanderbei and J. Richard Gott

compare the two pictures carefully, looking for differences, you will see that Vega has shifted more than the other stars.

The farther away a star is, the smaller the shift becomes. But for many relatively nearby stars, we can measure their distances using this technique. To do this, we need to apply a few basic facts of geometry. In figure 4.1, we saw the nearby star in front of one set of stars in January and then we saw the star shift in front of other stars in July. By convention, half of this shift is called the *parallax angle*, corresponding to the shift you would see if you moved only 1 AU instead of 2 AU. We know the radius of Earth's orbit (1 AU) in kilometers. We can measure the parallax angle. Consider the triangle formed by Earth the Sun, and the star. It is a right triangle with its 90° angle at the Sun. The shift in angle you observe during the year when looking at the nearby star is exactly the same shift that an observer sitting on that nearby star would see, looking back at you along the same two lines of sight. That means that the parallax angle (half the total shift) you observe will be equal to the angle between the Sun and Earth (in July) as seen by an observer on the star (see figure 4.1 again). Thus, the Earth–Sun–Star triangle has a 90° angle (at the Sun), an angle equal to the parallax angle (at the star), and angle (at Earth) of 90° minus the parallax angle; this is true because, according to Euclidean geometry, the sum of angles in a triangle must equal 180°.

You know one leg of the triangle (the Earth–Sun distance), and if you know the angles in the triangle, you can determine the length of the leg of the triangle connecting the Sun and the star. That gives you a direct measure of the star's distance. Let's invent a new unit of distance. Let's designate a *distance* such that a star at that distance would have a parallax angle of 1 arc second. One arc second is of course 1/60 of a minute of arc, which is itself 1/60 of a degree. An arc second is 1/3,600 of a degree. There exists a distance a star can have where the parallax angle is 1 arc second. That distance is called 1 *parsec*. Is that name cool or what? A parallax angle of 1 second of arc is $1/(360 \times 60 \times 60)$ of the circumference of a circle. If the star is at a distance d, the circumference of that circle is $C = 2\pi d$. The Earth–Sun distance $r = 1$ AU subtends $1/(360 \times 60 \times 60)$ of that circumference, making $1 \text{ AU}/2\pi d = 1/(360 \times 60 \times 60)$. Therefore, for a parallax of 1 second of arc, $d = 206{,}265$ AU = 1 parsec. It's all just Euclidean geometry.

If you watch *Star Trek*, you hear them use this unit of distance. What distance is that in light-years? It's 3.26 light-years. The unit of parsec is cute and fun to say, but in this book we mostly stick to light-years. In case you ever encounter this term *parsec*, now you know where it comes from. Astronomers coined the word by combining those two other terms, *par*allax and arc *sec*ond. A star that has a parallax of ½ arc second is 2 parsecs away. A star that has a parallax of 1/10 of an arc second is 10 parsecs away. Easy. We have several made-up terms in astronomy that get a lot of mileage—*quasar*, for instance. It comes from *quas*i-stell*ar* radio source. *Pulsar* is from *puls*ating st*ar*—we made that one up, and people love it. There is a Pulsar watch.

What is the star nearest to Earth? The Sun. If you said Alpha Centauri, I tricked you. The star system nearest to the Sun is Alpha Centauri. *Alpha* says it's the brightest star of its constellation, the southern constellation Centaurus, but it's actually a three-star system, one of whose stars is closest to our solar system. A triple star system—very cool. There is Alpha Centauri A, a solar-type star, 123% the diameter of the Sun; Alpha Centauri B, 86.5% the diameter of the Sun; and Proxima Centauri, a dim red star, only 14% the diameter of the Sun. Of these 3 stars, the one closest to our Sun is Proxima Centauri. That's why we call it *Proxima*. Its distance from us is about 4.1 light-years, giving a parallax of 0.8 arc seconds.

One arc second is really, really small. For most images you will ever see of the night sky taken from professional telescopes on Earth, the apparent size of a star on the image is typically one arc second. That's typical for ground-based telescopes. The Hubble Space Telescope does ten times better than that. The atmosphere wreaks havoc and blurs the images when we use telescopes here on Earth. Starlight comes in as a sharp point of light, minding its own business, and then it hits the atmosphere, gets bounced and jiggled and smeared, and finally ends up being this blob. On Earth we say, "Oh, isn't that pretty? The star is twinkling!" But twinkling is nasty to an astronomer trying to observe the star, and one arc second is a typical width of that twinkling image.

Notice that one parsec is less than the distance to the nearest star. That's why it took thousands of years for the parallax to be measured. Not until 1838 did the German mathematician Friedrich Bessel measure our first stellar parallax. (If the atmosphere smears an image to a width of 1 second of arc, an observer looking through a telescope had to take many measurements to achieve an accuracy *below* 1 arc second.) In fact, arguments put forth by Aristarchus more than 2,000 years ago to say that Earth was in orbit around the Sun were squashed by the lack of observed parallax at that time. The Greeks were smart folks. "Okay," they said, "you don't like our geocentric universe with the Sun going around Earth? You want Earth to go around the Sun?" They knew that, if Earth indeed went around the Sun, you would have a different angle for viewing close stars when Earth was on one side of the Sun compared with the other. They said we ought to be able to see this parallax effect. Telescopes weren't invented yet, so they just looked very carefully, and kept on looking. No matter how hard they squinted, they couldn't find any difference. In fact, because this effect couldn't be measured without telescopes, they used it as potent evidence against the Sun-centered, heliocentric universe. But absence of evidence isn't always the same as evidence of absence.

Even after watching all those stars in the nighttime sky, and noticing that fuzzy nebulous objects lurked among them, we had no real sense of the universe until the early decades of the twentieth century. That's when we obtained data by passing starlight through a prism and looked at the resulting features. From there we learned that some stars can be used as "standard candles." Think about it. If every star in the night sky were exactly the same—if they were cut out with some cookie cutter, and flung into the universe—the dim ones would always be farther away than the bright ones. Then it would be simple. All the bright stars would be close. Dim stars would be far. But it's just not the case. Among this zoo of stars, no matter where we find them, we search for and find stars of the same kind. So, if a star has some peculiar feature in its spectrum, and if a star of that same variety is close enough to observe its parallax, it's a happy day. We can now calibrate the star's luminosity and use it to find out whether other stars like it are one-fourth as bright or one-ninth as bright, and then we can calculate how far away they are. But we need that standard candle, that yardstick. And we didn't have such yardsticks until the 1920s. Until then we were pretty ignorant about how far away things are in the universe. In fact, books from that period describe the universe as simply the extent of the stars, with no knowledge or account of a larger universe beyond.

When trying to understand stars, you need some additional mathematical tools for your utility belt. One tool is going to be *distribution functions*. They are powerful and useful mathematical ideas. I want to ease into them, so let's introduce a simple version of a distribution function, something that *USA Today* might refer to as a bar chart, since they're big on charts and graphs. For example, we could plot the number of people in a typical college classroom as a function of age (figure 4.3).

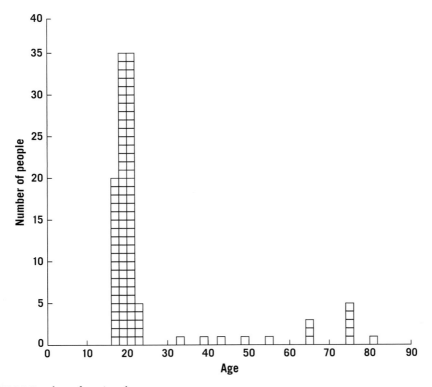

FIGURE 4.3. Bar chart of ages in a class.
Credit: J. Richard Gott

To make such a chart, one would start by asking the people in the class if anyone was 16 years old or younger. If no one answered, the chart would get a value of zero for those ages. Next, one would ask how many are 17–18 years old. Let's say 20 people. Make the bar for 17–18 years exactly 20 units tall. And 19–20 years old? Thirty-five people. Keep going until all the people have been tallied.

Let's take a step back and look at figure 4.3. There are things it can tell us about the distribution of people in this typical class. For example, most people cluster around age 20, which tells anyone looking at this chart that it's probably a college class. Then there is a gap and a few stragglers and another bump in the mid-70s—we have two bumps, two modes. We would call this a *bimodal distribution*. Most individuals in this older group are not actually undergraduate students; they are probably auditors in the class, and because people who audit daytime college classes are not those who have to work 9 to 5 to pay the rent, they're probably retired. You can gain insight into a population just by looking at this distribution. If we did this for the students in an entire college campus, then we would probably fill in some of the empty spots, but I bet it would take pretty much this same shape: mostly undergraduates, some older people, and occasionally you will find that precocious 14-year-old—maybe one in a thousand—because every freshman class seems to have one. This bar chart has bins 2 years wide. If I could increase the sample size enough, to include all college students in the country, I could make each bin only 1 day wide. I could collect so much data that I could fill this chart in and it would not be so jagged. With that much data, my bins would be so narrow that I could step back and put a smooth curve on it. If you go from a bar chart to a smooth curve and you can represent this with some mathematical form, your bar chart has become a distribution function.

What is the total number of people in the class? That's easy—just step along the scale and add up the numbers. In this case, we get 109. If you have smooth functions, you can use integral calculus to add up the area under the curve and give you the total number of things represented in it. Isaac Newton invented differential and integral calculus by the time he was 26—and was in my opinion the smartest person ever to walk the face of Earth!

How does this apply to stars? Let's look at the Sun. I'm going to say, "Sun, tell me something. I want to know how many particles of light you are emitting." Isaac Newton also came up with the idea of corpuscles of light—particles—long before Einstein, I might add. I have a word for these particles, *photons*—not protons, but photons. *Pho-* as in *photograph*, as in "photon torpedoes." Trekkies know that term.

Photons come in all flavors. Isaac Newton took white light and passed it through a prism. He listed the colors of the rainbow he saw: red, orange, yellow, green, blue, indigo (a big dye color back in Newton's day, so he included it in the spectrum), and violet. Today we typically mention only six colors in the rainbow. But as an *hommage* to Isaac, I usually include indigo, plus you get to spell out "Roy G Biv"—a good way to remember the rainbow colors.

The English astronomer William Herschel discovered that another whole other branch of the spectrum—something today we call *infrared*, which our eyes are not sensitive to. On the scale of energy it falls "below" red. Herschel passed sunlight through a prism and noticed that a thermometer placed off the red end of the visible spectrum got hot. Off the other side of the visible spectrum, you can also go beyond violet to get *ultra*violet, or UV. You've heard of these bands of light before, because they show up in everyday life. UV radiation gives you a suntan or a sunburn; infrared heaters in a restaurant keep your French fries warm until you buy them.

The spectrum is therefore much richer than what shows up in the visible part. Further beyond ultraviolet, we have X-rays. There are X-ray photons. Beyond X-rays, we have gamma rays. You've heard of all of these. Let's go the other way, toward the infrared. Below infrared? Microwaves. Below them? Radio waves. Microwaves used to be considered a subset of radio waves, but now they're treated as a separate part of the spectrum in their own right. These are all the parts of the spectrum for which we have words. There is nothing beyond gamma rays—we just continue to call them gamma rays—and nothing beyond radio waves.

A photon is a particle. We can also think of it as being a wave, a wave–particle duality. Well, you say, which is it? Is it a wave or a particle? That question has no meaning. We should instead be asking ourselves why our brains can't wrap themselves around something that has a dual reality inherent in it. That's the problem. We could make up a word such as "wavicle." This term was introduced some time ago, but it never caught on because people still want to know which it is. The answer depends on how you measure it. We can think of it as a wave, and waves have wavelengths. Except we don't use L to denote the length of the wave; we use the Greek letter that has the same sound as the L, lambda. We use lowercased lambda, which looks like this: λ, the preferred symbol for wavelength.

How big are radio waves? Think about them this way: in the old days, if you wanted to change the channel on your TV, you had to get off the couch, walk up to it, and turn a knob. This was so long ago. That same TV had a "rabbit-ear" antenna on it—two extendable wires that went up like a V—if the reception didn't come in right, you moved the two wires of the antenna. These antennas had a certain length to them, about a meter. In fact, TV waves are about a meter in wavelength. The antennas received TV waves from the air. That's why when you go to a TV studio, there is a sign that says "On the Air," because it's broadcasting through the air to your house. Of course, much of it now comes via cable, but the sign today doesn't say "On the Cable." And in any case, light (including radio waves) passes through the vacuum of space with no trouble. So the air is irrelevant, which always left me wanting to change the "On the Air" signs to "On the Space."

How about mobile phones? How big are their antennas? Quite small. They use microwaves, which are only a centimeter in length. Nowadays, the antenna is built into the phone itself, but in the old days, you would extend a short, stubby antenna every time you used your mobile phone.

How big are the holes on the screen of your microwave oven? They have holes so you can see the food cooking inside. Maybe you didn't notice, but these holes are only a couple of millimeters across. That's smaller than the actual wavelength of the microwaves that heat your food. So the 1-centimeter microwave trying to get out of the oven sees a hole only a couple of millimeters wide, and it can't get out. It can't find any exit from the microwave oven. Do you know who else uses microwaves? Police do when they point a radar beam at drivers to measure their speed. Microwaves reflect off the metal of your car. Here's one way to thwart that: you know those black canvas bug protectors that some people, usually guys with sports cars, put over the front end of their cars? They absorb microwaves very well, so if you beam microwaves at it, the signal that returns to the police radar gun is so weak that usually you can't get a reading back. Of course a car's windshield is transparent to microwaves. How do you know microwaves pass through glass? Where do people put their radar detector? Typically, inside the car on the dashboard. So obviously, microwaves pass through the glass. In the same way, you can cook food in a glass container in a microwave oven, because microwaves pass through unobstructed. Police use something called the *Doppler shift* to get your speed, which we'll be discussing a bit later. For now, you just need to know that in this case, it's a measure of the change in the wavelength of a signal reflected off a moving body. You get the most accurate reading if you take the measurement in the exact path of the object in motion. In practice, radar gun detectors do not measure the correct speed for your car, because to do so, the police officer would have to stand in the middle of the traffic lane, which they don't tend to do. Instead, they stand to the side, which means the speed they get is always less (unfortunately) than your actual speed. So if they catch you speeding you have no argument. Just pay your ticket and move on.

The police radar gun sends a signal that reflects off your car. Imagine you are looking at your own reflection in a mirror that is 10 feet away, and the mirror is moving toward you at 1 foot per second. Your reflected image starts out 20 feet away from you (the light goes out 10 feet and back 10). But one second later, the mirror is only 9 feet from you and you see your reflected image only 18 feet away. You see your reflected image rushing toward you at 2 feet per second. Likewise, the policeman is observing the reflection of his own radar gun rushing toward him at twice *your* speed. Try explaining that to the judge! Of course, radar guns are calibrated to report half the Doppler shift they measure—to properly report the velocity of the mirror—your car. By the way, radar is an acronym of "radio detection and ranging," from back when microwaves were considered part of the radio wave family.

Since we're talking about microwaves, as it happens, the water molecule, H_2O, is very responsive to microwaves; the microwaves in your microwave oven flip the molecule back and forth at the frequency of the wave itself. If you have a bunch of water molecules, they will all do it. Billions of trillions of them. Before long, the water gets hot because of the friction between the molecules as they undergo these flips. Anything you put in a microwave oven that has water in it will get hot. Everything you eat other than salt has water in it. That's why microwave ovens are so effective at cooking your food, and why it doesn't heat your glass plate if you don't have food on it.

The human body reacts to infrared radiation. Your skin absorbs it, creating heat, and you feel warm. Visible light we know well. Depending on what shade of skin color you have, you will be more or less sensitive to ultraviolet light. It can damage the lower layers of your skin and give you skin cancer. Ozone in the atmosphere protects us from most of the Sun's ultraviolet rays. The oxygen in the air is in molecular form: O_2 plus some ozone O_3 (molecules composed of two and three oxygen atoms, respectively). Ozone lives in the upper atmosphere and is just waiting to break apart. In comes an ultraviolet photon, which gets absorbed, breaking the ozone apart. The ultraviolet light is gone—it just got eaten by the ozone. If you take away the ozone, there will be nothing to consume the ultraviolet, and it will come straight down, sending skin cancer incidence up. Mars has no ozone, so the surface of Mars is constantly bathed by ultraviolet light from the Sun. That's why we suspect, and I think correctly, that Mars has no life on its surface today, even if there may be life below the surface. Anything biological exposed to that much ultraviolet radiation would have decomposed.

Almost everyone has been X-rayed. Can you remember what the X-ray technician does before he or she turns the switch to expose you? The technician lays you out and says, "okay, hold still," and then goes outside, behind some lead shielding, closes the door, and then turns the switch. Your technician doesn't want X-ray exposure. You should take a hint that what's about to happen is not good for you. But usually, *not* taking the X-rays is worse than taking them if you need the X-rays for a diagnosis—if your arm is broken the X-ray image can tell you. X-rays penetrate much deeper than your skin; they can trigger cancerous growths in your internal organs. But if the X-ray dose you receive is low, the risk is small.

Gamma rays are worse. They go right to your DNA and can mess you up. Even comic books know that gamma rays are bad for you. Remember the Incredible Hulk? How did he become the Hulk? What happened to him? Wasn't he doing some experiment that exposed him to a high dose of gamma rays? And now when he gets angry, he gets big, ugly, and green. So watch out for gamma rays—we don't want that happening to you. As you move along the spectrum to shorter wavelengths, from UV to X-rays to gamma rays, the energy contained in each photon goes up, and its capacity to do damage increases.

In modern times, radio waves are all around us. All the time. And there's a simple experiment you can do to prove it. Turn on a radio, and tune into a station. Any station, at any time. They are all around you, constantly broadcasting. How do you know you are constantly bathed in microwaves—all the time? Your mobile phone can ring anytime while you are just sitting there. Presuming you never crawl into the high-intensity field of a microwave oven, microwaves are harmless compared to what is going on at the high-energy part of the spectrum.

All these photons travel at one speed through empty space. The speed of light. It's not just a good idea—it's the law. Visible light, as we have defined it, sits in the middle part of the electromagnetic spectrum, but it is all light, traveling at 300,000 km/sec (299,792,458 meters/sec, to be exact). It's one of the most important constants of nature that we know.

The photons of all bands of light move at the same speed, yet they have different wavelengths. As I stand watching them pass, their *frequency* is defined as the number of wave crests that go by per second. If the waves are of shorter wavelength, many more crests will go by in a second. So high frequency corresponds to short wavelengths, and conversely, low frequencies correspond to longer wavelengths. It's a perfect situation for an equation: the speed of light (c) equals frequency times wavelength (λ). For frequency, we use the Greek letter *nu*: ν. Our equation becomes

$$c = \nu\lambda.$$

Suppose we had radio waves with a wavelength of 1 meter. The speed of light is approximately 300,000,000 meters/sec, which is equal to ν times 1 meter, making the frequency 300,000,000 crests or (cycles) per second (or 300 megacycles).

In fact, frequency and the energy in a photon are bound in an equation, too. The energy E of a photon is equal to $h\nu$:

$$E = h\nu.$$

Einstein discovered this equation. The equation uses Planck's constant h, named for the German physicist Max Planck. It serves as a *proportionality constant* in the equation, telling us how the frequency and energy of a photon are related. The higher its frequency is, the higher the energy of an individual photon will be. While X-ray photons pack a large punch, radio wave photons each carry only a tiny amount of energy.

Time to query the Sun. How many photons of each of these wavelengths are you giving me? How many green photons are coming from your surface, how many red ones, how many infrared, microwave, radio wave, and gamma ray photons? I want to know. So many photons emerge from the Sun that I can do much better than a simple bar chart, because I am flooded with data. I can make a smooth curve, and when I do, I will plot intensity versus wavelength. In this case, *intensity*, plotted vertically, represents the number of photons per second coming out of the Sun per square meter of the Sun's surface, per unit wavelength interval, at the wavelength of interest, *times* the energy each photon carries. We could have just counted photons, but in the end, we're typically interested in the energy they carry. This vertical axis gives us the *power* (energy per unit time) emerging from the Sun's surface per unit area per unit wavelength. Horizontally, I have wavelength increasing to the right. So let's put in X-rays, UV, visible (the rainbow-colored band), infrared (IR), and microwaves (labeled µwave). Figure 4.4 shows the distribution function of intensity from the Sun.

The hot Sun emits radiation at a temperature of about 5,800 K. The distribution was figured out by Max Planck. It peaks in the visible part of the spectrum, and that's no accident—our eyes have evolved to detect the maximum amount of sunlight out there. To compare with other stars, let's pick an average

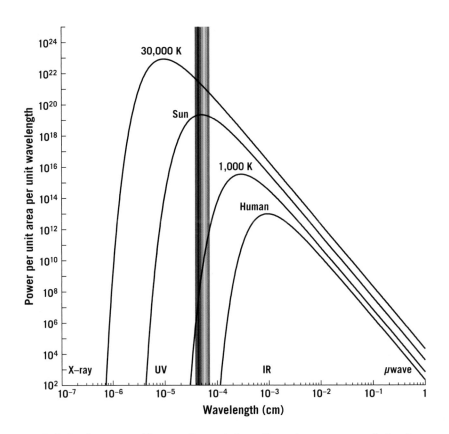

FIGURE 4.4. Radiation from stars and humans. The vertical coordinate plots energy per unit time (i.e., power) emitted by various objects per unit wavelength per unit surface area. The horizontal coordinate is wavelength. We show a 30,000 K star, the Sun (5,800 K), a 1,000 K brown dwarf star, and a human (310 K). Wavelengths corresponding to X-rays, UV, visible light (rainbow-colored bar), infrared, and microwaves (μwaves) are shown. *Photo credit:* Michael A. Strauss

square meter to use as an example. The actual size of the patch doesn't matter, as long as we use the same size patch from one example to the next. Sometimes people say we have a yellow Sun, but it's not yellow. If you want to call it yellow because it peaks near yellow, you could justifiably argue that it peaks at green, but no one says we have a green star. Besides the yellow, you must add in as much violet, indigo, blue, green, and red light as the curve shows the Sun is emitting. Add them all together, and you have about equal amounts of every one of these colors. Think back to Isaac Newton. What is this? White light. If you pass equal amounts of the colors of the visible spectrum *back* through a prism, what will emerge is white light. Newton actually did this experiment. Therefore, the Sun, radiating roughly equal amounts of all these colors, gives us white light. No matter how the Sun is drawn in a textbook, no matter what people in the street tell you, we have a white star—it's just that simple. By the way, if the Sun were truly yellow, then white surfaces would look yellow in full sunshine, and snow would look yellow (whether or not you were near a fire hydrant).

The Sun's surface temperature is about 5,800 K. Temperature on the Kelvin scale (K) is Celsius (C) plus 273. Ice freezes at 0° C (or 273 K). Water boils at 100° C (or 373 K). Celsius and K values are separated by only 273, and as we get to higher and higher temperatures, tracking that difference becomes less and less meaningful. In any case, 5,800 K is very hot. It will vaporize you. And to round things out, 0 K (you may have heard it called *absolute zero*) is the coldest possible temperature. Molecular motion stops at 0 K.

Let's find another star. Here is a "cool" one that checks in at a mere 1,000 K (see figure 4.4). Where does the 1,000 K star peak? In the infrared. Can your eyes see infrared? No. Is this star invisible? No. A small part of the star's radiation emerges in the visible spectrum. The intensity is falling sharply in the visible part of the spectrum as one goes from red to blue—it is emitting much more red light than blue light. This star will look red to our eyes. Now let's look at a star with a temperature of 30,000 K. As a reminder, I am asking the same kind of questions about its light distribution that we asked about the age distribution for students in a college class. Where does that star peak? In the ultraviolet. It gives off more UV than any other kind of light. We can't see ultraviolet, but can you see this star? Of course you can. It's got a lot of energy coming out in the visible part of the spectrum, too, with more energy emerging in the visible part of the spectrum per square meter of its surface than the Sun emits. Unlike the Sun, however, its mixture of colors isn't equal, but rather it's tipped toward the blue. If I add its colors together, I will get blue. Blue hot is in fact the hottest of hots. All astrophysicists know that the coolest glowing temperature is red, and the hottest glowing temperature is blue. If romance novels were astrophysically accurate they would describe "Blue-Hot Lovers," not "*Red*-Hot Lovers."

Our 30,000 K star peaks in the UV. If I picked an even hotter star, its color would also be blue. A blue color just means that the blue receptors in your eye are getting more radiation than your green or red receptors. A 30,000 K star is blue, a 5,800 K star is white, and a 1,000 K star is red.

How about the human body? What temperature are you? Unless you have a fever, you are 98.6° F, or about 310 K. The spectrum of your emission peaks in the infrared. How much visible light do you give off normally? You can see other people with your eyes, only because they reflect visible light. But if you turn off all the lights in a room, everything goes black. You can't see the people. You'll notice that if the lights are off, the curve for 310 K tells us that humans give off virtually no radiation in the visible. But they, being at a temperature of 310 K, are still emitting infrared light. Bring out an infrared camera, or infrared night goggles, and you can see the people, radiating strongly in the infrared. We put the whole universe on such a chart in the next chapter.

5

HOW STARS RADIATE ENERGY (II)

NEIL deGRASSE TYSON

I'd like to plug you into the rest of the universe. In chapter 4, we looked at curves showing the thermal emission of radiation from stars. Figure 5.1 is similar, except that we have added something. The vertical coordinate is intensity (power per unit surface area per unit wavelength), and the horizontal coordinate is wavelength—increasing to the right. The interval of wavelengths that we call "visible light" is identified with a rainbow-colored bar as before.

This figure shows thermal emission curves for the Sun at 5,800 K, a hot star at 15,000 K, a cooler one at 3,000 K, and a human at 310 K. The human emission curve peaks at about 0.001 centimeters. Way below this curve and off to the right is something new, an emission curve whose temperature is 2.7 K, which is the temperature of the whole universe! That's the famous background radiation coming to us from all parts of the sky. Because it peaks in the microwave part of the spectrum, it is called the *cosmic microwave background* (CMB). It was discovered in New Jersey, at Bell Laboratories, in the mid-1960s. Arno Penzias and Robert Wilson used a radio telescope—they called it the "microwave horn antenna." When they aimed it up at the sky, no matter which direction they pointed, they detected this microwave signal, from everywhere in the sky, which corresponds to something radiating at a temperature of about 3 K (the modern, more accurate, value is 2.725 K). And it's the thermal radiation left over from the Big Bang. We will have much more to say about this in chapter 15.

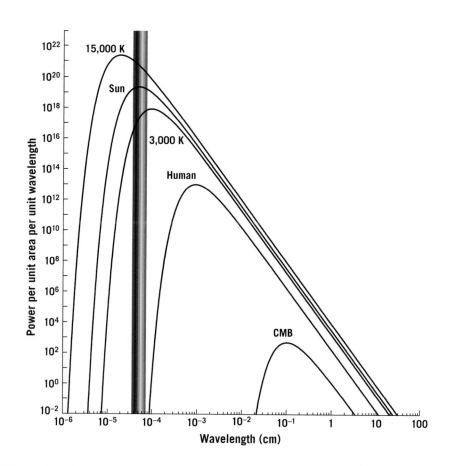

FIGURE 5.1. Thermal emission in the universe. The spectra of blackbodies of different temperatures, as a function of wavelength. The vertical coordinate plots energy per unit time (i.e., power), per unit wavelength, emitted per unit surface area of the object at the quoted temperature; the units are arbitrary. The curves correspond to stars of surface temperature 15,000 K (which will appear blue-white), 5,800 K (the Sun, which appears white), and 3,000 K (which will appear red). The visible part of the spectrum is shown as a colored bar; also shown is a human (310 K) and the cosmic microwave background (CMB, 2.7 K), about which we will learn much more in chapter 15. *Photo credit:* Michael A. Strauss

As before, we can query these graphs in different ways. Where does each curve peak? They peak in different places. How much total energy is emitted per second? We need a way to add up the area under each curve to determine how much total energy is being emitted per second. First, we need to define some terms.

A *blackbody* is an object that absorbs all incident radiation. A blackbody that is at a certain temperature will emit what we call *blackbody radiation*, which follows the curves we've been showing. The term "blackbody" looks like a misnomer, but it is not. We agree that these stars aren't black: one star glows blue, one star glows white, and one star glows red. Yet all qualify as blackbodies, as I've drawn them in the figure. A blackbody is quite simple; it eats any and all energy hitting it. I don't care what you feed it—that doesn't matter—it will eat it. You can feed it gamma rays or radio waves. Black things absorb all energy that falls on them. That's why black clothing is not a common fashion option in the summer. Blackbodies then reradiate these curves—it is that simple. The curve's shape and position depend only on the temperature of the blackbody.

You can heat something, increase its temperature, and all you need to do is ask, what is your new temperature? Then, return to your curves, and see where this new temperature fits in. I have a wonderful equation that describes these curves. They are distribution functions, called *Planck functions*, after Max Planck, whom we've met previously, and who was the first person to write down the equation for these curves. To the right of the equal sign, we have energy per unit time per unit area coming out per unit wavelength interval at a particular wavelength λ; we call this quantity intensity (I_λ), which depends only on the temperature T of the blackbody:

$$I_\lambda(T) = (2hc^2/\lambda^5)/(e^{hc/\lambda kT} - 1)$$

Let's understand the parts that make up this landmark equation. First of all, λ (lambda) is wavelength, no secrets there. The constant e is the base of the natural logarithms, and it has its own button on every scientific calculator, which is usually shown as e^x ("e to the x"). The value of e is 2.71828 . . . ; it's a number like π whose digits go on forever. It's just a number. The letter c is the speed of light, which we've seen before. The letter k is the Boltzmann constant. The letter T is simply temperature, and h (introduced in chapter 4) is Planck's constant. If you assign a temperature T to an object, the only unknown in this equation is λ, the wavelength. So, as you run λ from very small values to very large values, you get a value for intensity I_λ as a function of wavelength that will precisely track these curves. Max Planck introduced this equation in 1900, and it revolutionized physics.

With his new constant, Planck gave birth to the *quantum*, which makes Max Planck the first parent of quantum mechanics. Look at just the first term in parenthesis, which is $2hc^2/\lambda^5$. As wavelengths get longer, what happens to the energy being emitted? It drops. The $1/\lambda^5$ term goes to zero as λ becomes large. For large λ, the term $hc/\lambda kT$ becomes small. Mathematicians will tell you that e^x becomes approximately $1 + x$ as x becomes small. So for large λ, the term

$hc/\lambda kT$ becomes small, and $e^{hc/\lambda kT}$ is approximately $1 + hc/\lambda kT$, and if we subtract 1 from that it makes the term $(e^{hc/\lambda kT} - 1)$ equal to $hc/\lambda kT$. Thus, in the limit as λ becomes large, the whole expression becomes $I_\lambda(T) = (2hc^2/\lambda^5)/(hc/\lambda kT)$ $= 2ckT/\lambda^4$. People were familiar with this relation before Planck. It is called the Raleigh-Jeans Law after its inventors Lord Raleigh and Sir James Jeans. As λ gets larger and larger the intensity I_λ starts dropping off, like $1/\lambda^4$ in a very well-defined way. What happens when you move toward smaller and smaller wavelengths? As λ^4 gets smaller and smaller, $1/\lambda^4$ gets bigger and bigger, making the equation blow up (and disagree with experiments). This was once called "the ultraviolet catastrophe." Something was wrong. Wilhelm Wien figured out a law that had an exponential cutoff at small wavelengths that fit the data at small wavelengths but didn't fit the data at large wavelengths. We had no real understanding of these blackbody curves until 1900, when Max Planck found a formula that fit at both the small and large wavelength limits and everywhere in between. The formula includes a constant h that *quantizes energy*, so that you only get energy in discrete packets. If you get it in discrete packets, then as you get to smaller and smaller wavelengths, the exponential in Planck's formula kicks in and squashes the $1/\lambda^5$ term. When λ gets small, $hc/\lambda kT$ gets big, and e raised to that power $(e^{hc/\lambda kT})$ gets really big, really fast. It dominates the -1, so that you can forget about the -1 term, and with the $e^{hc/\lambda kT}$ in the denominator, the answer gets small. It's a contest between these two parts of the equation: the $1/\lambda^5$ term and the $1/e^{hc/\lambda kT}$ term. As λ goes to zero, the $1/e^{hc/\lambda kT}$ goes to zero much faster than the $1/\lambda^5$ term is blowing up, making the whole curve go to zero. Without the exponential term, the formula would blow up to infinity as the wavelength went to zero, and we knew from experiments that this was not the way matter behaved. The quantum was needed to understand thermal radiation, and this equation captures how these curves work.

The formula's got it all. It can tell you where the curve peaks. Isaac Newton invented math that allowed you to figure out where a function peaks: it's where the slope of the curve goes to zero at the curve's maximum. You can use Newton's calculus to take the derivative of the function and determine this location. When we do that, we get a very simple answer: $\lambda_{peak} = C/T$, where C is a new constant, which we can find from the constants in the initial equation: $C = 2.898$ millimeters when T is expressed in kelvins. Where is the peak? If the temperature is $T = 2.7$ K, as in the CMB, then λ_{peak} is a little over 1 millimeter or 0.1 centimeter. We can confirm this by checking the CMB curve in figure 5.1.

The human is about a hundred times hotter than that; the human emission peaks at about 0.001 centimeter (also shown in figure 5.1), in the infrared.

It's beautiful. As temperature gets higher, the wavelength at which the curve peaks gets smaller and smaller. That is borne out just by looking at how this equation $\lambda_{peak} = C/T$ behaves. With T in the denominator, it says that something twice as hot will peak at one-half the wavelength. (Wilhelm Wien figured this out—we call it "Wien's Law.")

How do I get the total energy per unit time per unit area coming out from under one of these curves? I want to add up the contributions from all the different wavelengths, the total area under a particular curve. I can use calculus again and integrate to find the area—once more, thank you, Isaac Newton. If we integrate the Planck function over all wavelengths, we get another beautiful equation:

Total energy radiated per second, per unit area = σT^4, where $\sigma = 2\pi^5 k^4/(15c^2 h^3) = 5.67 \times 10^{-8}$ watts per square meter, with the temperature T given on the Kelvin scale. This law is called the *Stefan–Boltzmann law*. Josef Stefan and Ludwig Boltzmann were two towering figures in nineteenth-century physics. Sadly, Boltzmann committed suicide at age 62. But we have this law. If we integrate the Planck function, we get the value of the constant σ (Greek sigma). That's profound. How did Stefan and Boltzmann figure out this law, when Planck had not yet derived his formula? Stefan found it experimentally, while Boltzmann derived it from a thermodynamic argument.

With total energy radiated per second per unit area = σT^4, if I double the temperature, the energy flux being radiated increases by a factor of $2^4 = 16$. Triple the temperature and what do you get? $3^4 = 81$. Quadruple the temperature: $4^4 = 256$. And that trend is borne out in figure 5.1, which shows how much bigger these curves become as the temperature increases.

Here's one way to remember why this formula works: Imagine taking some thermal radiation and putting it in a box. Now slowly squeeze the box until it has shrunk by a factor of 2. The number of photons in the box stays the same, but the volume of the box shrinks by a factor of 8, making the number of photons per cubic centimeter in the box go up by a factor of 8. But squeezing the box shrinks the wavelength of each photon by a factor of 2 as well. This makes the thermal radiation in the box hotter by a factor of 2, because its peak wavelength has shrunk by a factor of 2. It also doubles the energy of each photon, doubling the energy in the box. The increase in energy for each photon comes from the energy you invest in squeezing the box, pushing against the radiation pressure inside. That means that the energy density in the box is $8 \times 2 = 16$ times what it was before, and 16 equals 2^4. Therefore, the energy density of thermal radiation is proportional to the fourth power of the temperature, or T^4.

Let's define some additional terms. *Luminosity* is the total energy emitted per unit time by a star. Luminosity is measured in watts, as in a light bulb. A 100-watt light bulb has a luminosity of 100 watts. The Sun has a luminosity of 3.8×10^{26} watts. It's a potent light bulb.

I propose a puzzle. Suppose the Sun has the same luminosity as another star that has a surface temperature of 2,000 K. How hot is the Sun? For this example, let's just round the temperature to 6,000 K. The other star is only 2,000 K, so I know if it is that much cooler, it cannot be emitting nearly as much energy per unit area per unit time as the Sun, but then I declare that the Sun has the same luminosity as this star—how is that possible? I take that other star and I get a 1-square-inch patch of it, a 2,000 K patch, and I get a 1-square-inch patch from the Sun, at 6,000 K—three times hotter. How much more energy per unit time is being emitted by a 1-square-inch patch on the Sun than by a 1-square-inch patch on the other star? Eighty-one times more energy. How can this other star be emitting the same total energy per second as the Sun? Something else must be different about these two stars besides their temperatures for them to be equal at the end of the day. The other star, the cool star, must have much more *surface area* from which to radiate than does the Sun. In fact, it must have 81 times the surface area of the Sun. It must be a red giant star, with 81 times the surface area to make up for its deficiency in each little square-inch tile on its surface. Now let's use our equations. What is the surface area of a sphere? It's $4\pi r^2$, where r is the radius of the sphere. You may have learned that equation in middle school. What comes next is so beautiful. If luminosity is energy emitted per unit time, and the energy emitted per unit time *per unit area* is equal to σT^4, then I have an equation for the luminosity of the Sun:

$$L_{Sun} = \sigma T_{Sun}{}^4 \times (4\pi r_{Sun}{}^2).$$

I have a similar equation for the other star. Let's denote the other star's luminosity by an asterisk, L_*. The equation for its luminosity is $L_* = \sigma T_*{}^4 \times (4\pi r_*{}^2)$. Now I have an equation for each of them. Furthermore, I have declared that L_{Sun} is equal to L_*. I have declared that, in posing the example, I don't actually need to know the surface area of the Sun, because this problem is talking about the ratios of things. We can get tremendous insight into the universe simply by thinking about the ratios of things.

Let's divide the two equations: $L_{Sun} / L_* = \sigma T_{Sun}^4 \times 4\pi r_{Sun}^2 / (\sigma T_*^4 \times 4\pi r_*^2)$. What do I do next? I cancel identical terms in the numerator and denominator of the fraction on the right side of the equation. First, I'll cancel out the constant σ. I don't even care what its particular value is, because when I'm comparing two objects and the constant shows up for both stars, I can cancel the constants out. The number 4 cancels, and π cancels too. Continuing, on the left of the equation, what is L_{Sun}/L_*? It is 1, because I stated that the two stars have equal luminosities; their ratio is 1. So, I am left with a simpler equation: $1 = T_{Sun}^4 r_{Sun}^2 / T_*^4 r_*^2$. The temperature of the Sun is 6,000 K, and the temperature of the other star is 2,000 K. Of course, $6,000^4$ divided by 2000^4 is the same thing as 3^4, which is 81. Now I have $1 = 81 r_{Sun}^2 / r_*^2$. Let's multiply both sides of the equation by r_*^2. Thus, $r_*^2 = 81 r_{Sun}^2$. Take the square root of both sides of the equation: $r_* = 9 r_{Sun}$. The radius of the cooler star with the same luminosity as the Sun is 9 times that of the Sun! That's our answer. If we are thinking in terms of area, this star has a surface area 81 times as large as that of the Sun, because the square of the radius is proportional to the area. These are immensely fertile equations.

I could have given a different example. I could have started with a star of the same temperature as the Sun, but 81 times as luminous. Both stars have the same amount of energy coming out per second per square inch of surface, so the other star must have 81 times the surface area of the Sun and 9 times the radius of the Sun. The equation has the same terms, but we're putting different variables into different parts of the equation. That's all we're doing here.

Recall (from chapter 2) that the hottest part of the day on Earth is not at noon, but sometime after noon, because the ground absorbs visible light. That visible light slowly raises the temperature of the ground, and the ground then radiates infrared to the air. The ground is behaving as a blackbody—absorbing energy from the Sun, and then reradiating it according to the recipe given by the Planck function. The ground has a temperature of roughly 300 K. (That's 273 K plus the ground temperature in Celsius—if it's 27° C, that makes the ground an even 300 K.)

You can ask the question, what is your own body's luminosity? Plug in your Kelvin body temperature, which is 310 K, take it to the fourth power, multiply by sigma—and you will get how much energy you emit per unit time, per unit area. If you multiply that by your total skin area (about 1.75 square meters for the average adult), you will get your luminosity—your wattage. It is not coming out in visible light. It is coming out mostly in infrared, but you for sure have a wattage. Let's get the answer. The Stefan–Boltzmann constant σ is 5.67×10^{-8} watts per square meter if the temperature is measured in K. Multiply by $(310)^4$. The value of 310^4 is 9.24×10^9. Multiply that by 5.67×10^{-8}, giving 523 watts per square meter. Multiply by your area of 1.75 square meters, and you get 916 watts. That's a lot. Remember, though, that if you are sitting in a room that is 300 K (80° F), your skin is absorbing about 803 watts of energy, by the same formula. Your body has to come up with about 100 watts of energy to keep yourself warm. You get that by eating and metabolizing food. Warmblooded animals that keep their body temperature higher than their surroundings need to eat more than coldblooded animals do. When you put air conditioning in a room, there are two major questions to ask: How big is the room? What other sorts of energy will be released in the room? This will include asking, for example, how many light bulbs will be on in the room and how many people will be in it, because every person is equivalent to a certain wattage light bulb that the air conditioning must fight against to maintain the temperature. To determine what air conditioning flow is required to keep the proper temperature, you have to account for how many people (with their watts) will gather in the room.

Let me toss in one more notion, called *brightness*. The brightness of a star you observe is the energy received per unit time per unit area hitting your telescope. Brightness tells you how bright the star *looks* to you. This depends on the star's luminosity, as well as on its distance from you. Let's think intuitively about brightness. How bright does an object appear to you? It should make sense to you that if you see an object shining with a particular brightness and then I move it farther away, its brightness will decrease. The luminosity, however, is energy emitted per unit of time by the object; it has nothing to do with its distance from you—it is simply what is emitted. It has nothing to do with your measuring it. A 100-watt light bulb has a luminosity of 100 watts, no matter where in the universe you put it. However, brightness will depend on an object's distance from an observer.

Brightness is simple, and I love it. Are you ready? Let me draw a contraption that I never built, but you can patent it if you like. It's a butter gun: you load it with a stick of butter and it has a nozzle at the front where the butter sprays out (see figure 5.2).

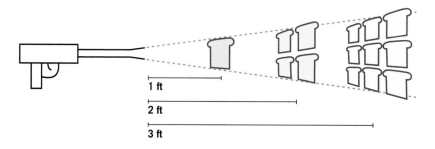

FIGURE 5.2. Butter gun. It can spray one slice of bread 1 foot away, four slices of bread 2 feet away, or nine slices of bread 3 feet away. *Credit:* J. Richard Gott

Put a slice of bread 1 foot away from the butter gun. I have calibrated this butter gun such that, at a distance of 1 foot, I butter the entire slice of bread, exactly covering it. If you're one of those people who like to butter up to the edge, this invention is for you. Now let's say I want to save money, as any good businessperson wants to do: I'd like to take the same amount of butter and butter more slices of bread. But I still want to spread it evenly. The first slice of bread was 1 foot away—now let's go 2 feet away. The spray of butter is spreading out. At twice the distance, the butter gun covers an area that is two slices of bread wide and two slices of bread tall. The spray covers a 2 × 2 array of slices, buttering 4 slices of bread. Just by doubling the distance, you can now spray 4 slices of bread. If I go three times the distance, you can bet that the spray will cover 3 × 3 = 9 slices of bread. One slice, four slices, nine slices. How much butter is one slice of bread 3 feet away getting compared to the single slice only 1 foot away? Only one ninth. It is still getting butter, but only a ninth as much. This is bad for the customer but good for my bottom line. I assert that there is a deep law of nature expressed in this butter gun. If, instead of this being butter, it were light, its intensity would drop off at exactly the same rate that this butter drops off in quantity. After all, light rays travel in straight lines just like the butter, and spread out in just the same way. At 2 feet away, the light from a light bulb would be 1/4 as intense as it was at 1 foot away. At 3 feet away, it would be 1/9 as intense; at 4 feet away, 1/16 as intense; and at 5 feet away, 1/25 as intense, and so on. It goes as *one over the square of the distance*—an inverse square. In fact, we have obtained an important law of physics, telling us how light falls off in intensity with distance, the *inverse square law*. Gravity behaves this way too. Do you remember Newton's equation, Gm_am_b/r^2? That r squared in the denominator shows it's an inverse-square relation, because it is behaving like our butter gun. Gravity and butter are acting alike.

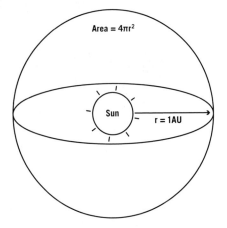

FIGURE 5.3. Sun in a sphere. The Sun's rays spread out over an area of $4\pi r^2$ as it passes through a sphere of radius r. *Credit:* J. Richard Gott

Imagine a light source like the Sun emitting light in every direction (figure 5.3). Let's further imagine I surround the Sun with a big sphere having a radius r equal to the radius of Earth's orbit (1 AU).

The Sun is emitting light in every direction, and I am intersecting some of the Sun's light. I'm only getting a little piece of all the light that's penetrating a Sun-centered sphere with a radius equal to the distance where I find myself. What is the area of that big sphere? It's $4\pi r^2$, where r is the radius of the sphere. Of all the light the Sun emits, the fraction hitting my detector is equal to the area of my detector divided by the area of that huge sphere ($4\pi r^2$). If I move twice as far away, my detector stays the same size, but the radius of my sphere will be twice as big (2 AU), and the area the Sun's rays are passing through will be four times as great. I will detect one quarter as many photons in my detector as I did when I was 1 AU away. Brightness is given in watts per square meter falling on my detector. To calculate the brightness that I observe at a radius r from the Sun, I start with the Sun's luminosity (in watts) and divide by this spherical area—$4\pi r^2$. This gives the watts per square meter from the Sun falling on me. I multiply by the area of my detector (say, my telescope), and I get the energy per second falling on it. If L is the luminosity of the Sun, the brightness (B) of the Sun as seen by me is $B = L/4\pi r^2$, where r is my distance from the Sun. As my distance increases, the denominator ($4\pi r^2$) gets larger, reducing the brightness. On Neptune, which is 30 times as far from the Sun as Earth is, the Sun appears only 1/900 as bright as it does here.

Suppose two stars have the same brightness in the sky, but I know that one is 10,000 times more luminous than the other. What must be true about these stars? The more luminous star must be farther away. How many times farther away? 100 times. How did I get 100? Yes, 100 squared is equal to 10,000.

You have just learned some of the most profound astrophysics of the late nineteenth and early twentieth centuries. Boltzmann and Planck, in particular, became scientific heroes for coming to the understanding that you have just gotten in this chapter and the previous one.

6

STELLAR SPECTRA

NEIL DeGRASSE TYSON

What's actually happening inside a star? A star is not just a flashlight that you switch on and light emerges from its surface. Thermonuclear processes are going on deep in its core, making energy, and that energy slowly makes its way out to the surface of the star, where it is then liberated and moves at the speed of light to reach us here on Earth or anywhere else in the universe. It's time to analyze what goes on when this bath of photons moves through matter, which doesn't happen without a fight.

We first must learn what the photons are fighting on their way out of the Sun. Our star, and most stars, are made mostly of hydrogen, which is the number one element in the universe: 90% of all atomic nuclei are hydrogen, about 8% are helium, and the remaining 2% comprise all the other elements in the periodic table. All the hydrogen and most of the helium are traceable to the Big Bang, along with a smidgen of lithium. The rest of the elements were later forged in stars. If you are a big fan of the argument that somehow life on Earth is special, then you must contend with an important fact: if I rank the top five elements in the universe—hydrogen, helium, oxygen, carbon, and nitrogen—they look a lot like the ingredients of the human body. What is the number one molecule in your body? It's water—80% of you is H_2O. Break apart the H_2O, and you get hydrogen as the number one element in the human body. There's no helium in you, except for when you inhale helium from balloons, and temporarily sound like Mickey Mouse. But helium is chemically inert. It's in the right-hand column of the periodic table: with an outer electron shell that's closed—all filled up, with no open parking spaces to share electrons with other atoms—and therefore, helium doesn't bond with anything. Even if helium were available to you, there's nothing you could do with it.

Next in the human body, we have oxygen, prevalent once again from the water molecule H_2O. After oxygen comes carbon—the entire foundation of our chemistry. Next we have nitrogen. Leaving out helium, which does not bond with anything, we are a one-to-one map of the most abundant cosmic elements into human life on Earth. If we were made of some rare element, such as an isotope of bismuth, you would have an argument that something special happened here. But, given the cosmically common elements in our bodies, it's humbling to see that we are not chemically special, but at the same time, it's quite enlightening, even empowering, to realize that we are truly stardust. As we'll discuss in the next few chapters, oxygen, carbon, and nitrogen are all forged in stars, over the billions of years that followed the Big Bang. We are born of this universe, we live in this universe, and the universe is in us.

Consider a gas cloud—something with the cosmic mixture of hydrogen, helium, and the rest—and let's watch what happens. Atoms have a nucleus in the center composed of protons and neutrons, with electrons orbiting them. It's constructive, if pictorially misleading, to imagine a simple, classical-quantum atom like Neils Bohr proposed about a hundred years ago. It has a ground state, the tightest orbit an electron could have: let's call this ground state *energy level 1*. The next possible orbit out would be an excited state, and this would be *energy level 2*. Let's draw a two-level atom just to keeps things simple (see figure 6.1). An atom has a nucleus and a cloud of electrons, which we say are "in orbit" around the nucleus, but these are not the classical orbits that we know from gravity and planets and Newton; in fact, rather than use the word "orbits," we introduce a new word derived from it: *orbitals*. We call them *orbitals*, because they are like orbits, but they can take a variety of different shapes. Actually they are "probability clouds" where we are likely to find the electrons. Electron clouds. Some are spherically shaped, some are elongated. There are families of them, and some have higher energies than others. We are going to abstract that and simply talk about energy levels, when we are actually representing orbitals, places occupied by electrons surrounding nuclei of atoms.

The nucleus is the dot in the center. Energy level $n = 1$ corresponds to an electron in a spherical orbital closest to the nucleus. Energy level $n = 2$ is a spherical orbital farther away from the nucleus. Energy level $n = 2$ corresponds to an electron that is less tightly bound to the nucleus. Electrons and protons

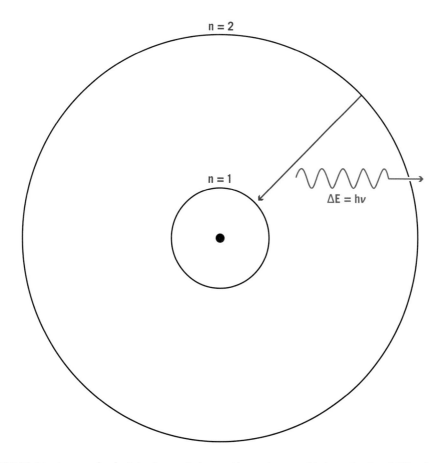

FIGURE 6.1. Atomic energy levels. A simple atom is shown with two electron orbitals, $n = 1$ and $n = 2$. If the electron starts further out in energy level 2, and drops down to the lower energy level 1, it emits a photon with an energy $\Delta E = h\nu$, where $\Delta E = E_2 - E_1$ is the difference in energy between level 2 and level 1. After the electron is in energy level 1, it can absorb a photon with energy $\Delta E = h\nu$ and jump back up to energy level 2. *Credit:* Michael A. Strauss

attract: it takes energy to move the electron away from the nucleus to a more distant orbital. Energy level 2 is higher in energy than energy level 1.

Suppose there is an electron sitting in the ground state, energy level 1. This electron cannot hang out anywhere between energy level 1 and 2. There is no place for it to sit. This is the world of the *quantum*. Things do not change continuously. For the electron to jump up to the next level, you have to give it energy. It must absorb energy somehow, and for the moment, a nice source of energy is a photon. A photon comes in, but it is not going to be just any photon. It would only be a photon having energy equal to the difference in energy between the two levels. If the electron sees it, it eats the photon and jumps up to energy level 2. If the photon has slightly more energy or slightly less energy, it goes right on by, unconsumed. Now, unlike humans, atoms have no interest in staying excited—this electron in energy level 2, if given enough time, will spontaneously drop down to the lower energy level 1 (as shown by the blue arrow in figure 6.1).

In some cases, enough time means a hundred millionth of a second. Electrons don't spend much time at all staying excited in an atom. So, when they drop back down, what must happen? They must spit out a photon—a new photon, of exactly the same energy that entered the first time. Jumping up involves absorption of a photon. Falling back down involves emission of a photon, as shown in red in figure 6.1. The energy E of this photon, by Einstein's famous equation, equals $h\nu$, where h is Planck's constant and ν is the frequency of the photon. The energy of the emitted photon exactly equals the energy difference between the two energy levels, ΔE. (A capital Greek letter delta, Δ, commonly symbolizes a difference or change in a quantity.) This gives the equation $\Delta E = h\nu$, which allows us to calculate the frequency of the photon emitted when the electron drops from level 2 to level 1.

Ever play with one of those glow-in-the-dark Frisbees? To make it glow in the dark, you must expose it to some light first. You stick it in front of a light bulb. What's going on? The electrons embedded in the toy's atoms and molecules are rising to higher energy levels (these larger atoms have many energy levels) as they absorb photons of light. The designers chose material that takes some time for these electrons to cascade back down, and as they do, they emit visible light, but not forever. It stops glowing after the electrons have descended back to their original states. Glow-in-the-dark Frisbees, and those glow-in-the-dark skeleton costumes you had as a kid, all operate on the same principle.

The energy an electron absorbs can come from a photon, but there are other possible sources of that energy. The electron could be kicked by another atom by bumping into it, and when it's kicked, the electron can be sent up to a higher energy level. In this case, kinetic energy is doing the job. How does that work in a cloud of hydrogen gas? First, we have to ask, what is the temperature of this hydrogen cloud? The temperature in kelvins is proportional to the average kinetic energy of the molecules or atoms in the cloud. The bulk motion of the cloud is not contributing to this measurement. Kinetic energy, of course, is energy of motion, so the higher the temperature, the faster these particles are jostling back and forth. If I'm an electron in the ground state, and I'm getting kicked in the pants, I can ask what the energy is of that kick. If that kick, that energy, would only get me part way up to energy level 2, I stay put. But, if that kick is exactly the right amount of energy to reach the second level, I'll take that energy, absorb it, and jump up to level 2.

Depending on the temperature, you can sustain an entire population of atoms with some fraction of their electrons in a higher state. You can keep it in equilibrium by arranging conditions so that every time an electron drops, you kick it back up. It's like the juggler who is keeping all the balls in air. And it's all a function of the temperature. At low temperatures, the great majority of electrons hang out in energy level $n = 1$ with only very few electrons in energy level $n = 2$. As the temperature gets hotter, more electrons get kicked up into energy level $n = 2$.

Let's put all this together. Consider an interstellar gas cloud being illuminated by the light from a 10,000 K star. Most atoms will have multiple energy levels with great complexity; that's the natural order of things—by comparison, hydrogen's energy levels are simple. It is the mixture of all these that wreaks havoc on the pure thermal spectrum that is emitted by the 10,000 K star. So let's take a look at what havoc is wrought.

First, I give you the full-blown hydrogen atom. It has an infinite number of energy levels corresponding to concentric orbitals that go farther and farther out: $n = 1$ (the ground level—innermost orbital), $n = 2$ (the first excited level), $n = 3, n = 4, n = 5, n = 6, \ldots, n = \infty$. The energy level diagram looks like a ladder, so we call it a *ladder diagram*. The lower energy levels, which are more tightly bound to the nucleus, are lower down in the diagram (figure 6.2).

For hydrogen, the first excited state $n = 2$ is three-quarters of the way up, followed by $n = 3$, then $n = 4$, then $n = 5$, and so on. The energy tops out at zero. An electron with a high n occupies a very large orbital that is only weakly bound to the proton. In atoms, we measure energy in *electron volts*, eV. That's the energy it takes to move an electron across a voltage difference of 1 volt. Let's say you have a flashlight that operates using a 9 volt battery. Each electron generates 9 eV of energy in the form of light and heat as it passes through the wires of the flashlight. This flashlight might be passing 6.24×10^{18} electrons per second through its wires, generating $9 \times (6.24 \times 10^{18})$ eV per second (or 9 watts) of light and heat energy. One electron volt is thus a very small amount of energy; it is just a convenient unit for talking about the small amounts of energy associated with electronic transitions. For example, -13.6 eV in the figure is the energy of level $n = 1$. This is shown as a negative energy. You have

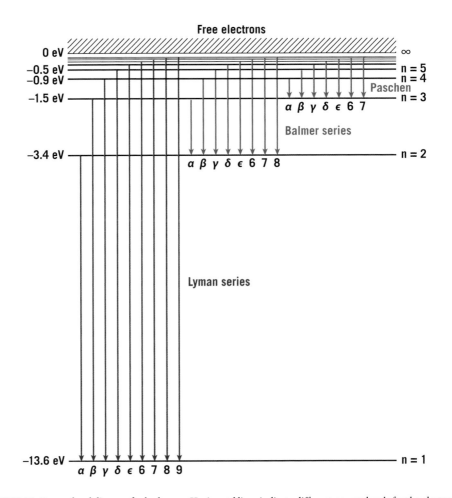

FIGURE 6.2. Energy level diagram for hydrogen. Horizontal lines indicate different energy levels for the electron in a hydrogen atom, in electron volts (eV). Arrows indicate transitions an electron can make from one energy level to another, emitting a photon of energy equal to the difference in energy. Transitions are shown to the first energy level (the Lyman series, which gives photons in the ultraviolet part of the spectrum), to the second energy level (the Balmer series, which gives visible light photons), and to the third energy level (the Paschen series, in the near infrared). The diagram shows electrons dropping down and emitting photons. If an electron is in level $n = 3$ and drops to level $n = 2$, as indicated by the red arrow, it will emit an Hα (Balmer series) photon with an energy of 1.9 eV. *Credit:* Michael A. Strauss

to add 13.6 eV to that electron in energy level $n = 1$ to remove it from the atom. We say 13.6 eV is the *binding energy* of the ground state $n = 1$. What happens if an electron in the ground state sees a photon with energy greater than 13.6 eV? Can it absorb that photon? Here comes a photon with that much energy—what will the electron do with it? If the electron absorbs this photon, it will have enough energy to jump up above $n = \infty$. What is above $n = \infty$? That would be freedom. If an electron pops up there to an energy above zero, the electron escapes from the atom and leaves the proton by itself. We say we have *ionized* the atom—stripped it of an electron. (The atom now has a net charge, making it an *ion*.) The departed electron has an energy above zero; that "excess" energy above zero becomes its kinetic energy of motion as it escapes the atom. As you might suspect, an atom can also get ionized by being slammed by another atom.

With this knowledge of energy levels, we can now understand how the light comes out of a 10,000 K star. At a temperature of 10,000 K, it is hot enough that a small but significant fraction of the hydrogen atoms have electrons in the first excited state $n = 2$. That's why I chose this star, because a temperature of 10,000 K maximizes the situation we are about to describe. Deep inside the star there is this thermal radiation spectrum, a beautiful Planck curve. It is trying to emerge from the outer layer of the star; that smooth, 10,000 K thermal spectrum hits those hydrogen atoms in the outer layer with some electrons in the first excited state, and those electrons are hungry. I can ask the question, in that thermal spectrum, how much energy do the individual photons have? The energies of many of those photons are found in the visible part of the spectrum, it just so happens, and the 10,000 K hydrogen gas has some hungry hydrogen atoms with electrons in the $n = 2$ level, and they are absorbing appropriate photons like mad and getting kicked up to higher energy levels for having done so.

But not all photons are being absorbed—only those whose wavelengths would kick an electron exactly up to a particular higher energy level. For example, an electron in level $n = 2$ (at an energy of –3.4 eV) can absorb a photon with just enough energy to make it jump up to level $n = 3$ (at an energy of –1.5 eV; see figure 6.2). The energy difference between those two levels is 1.9 eV. That's how much energy the electron must get to jump up. Such an electron will absorb a photon with energy 1.9 eV. This photon is called an *H-alpha*, or Hα, photon. It has a wavelength of 6,563 Ångstroms, or 656.3 nanometers—its color is Burgundy red. That photon gets taken out, kicking an electron up from level 2 to level 3; that photon is now gone from the spectrum. Many electrons doing this causes a dip in the Planck spectrum at a wavelength of 6,563 Ångstroms, called an *H-alpha* (Hα) *absorption line*. Photons with a wavelength of 4,861 Ångstroms can boost an electron from level 2 to level 4; that causes another dip in the spectrum called the *H-beta* (Hβ) absorption line. There are more lines: *H-gamma* (Hγ) at 4,340 Ångstroms, *H-delta* (Hδ) at 4,102 Ångstroms, and so on, where photons are being taken out, sending electrons from $n = 2$ to levels $n = 5$, $n = 6$, ... In comes a continuous spectrum, out goes what we call an *absorption spectrum*, with narrow lines knocked out where those photons are getting eaten, making deep narrow valleys in the spectrum that are called *absorption lines*. The entire group of them is called the *Balmer series*: Hα, Hβ, Hγ, Hδ, Hε, followed by H6, H7, H8, and so on (nobody is expected to remember that many Greek letters). The spacing of these lines relates to those differences in energy on the ladder diagram. Figure 6.3 shows the spectrum of an actual 10,000 K star. The inset shows a close-up of the shorter-wavelength portion.

If we look at a star whose surface is a bit hotter, say, 15,000 K, the story changes dramatically: the electrons have so much energy from getting kicked in the pants that they leave the hydrogen atoms altogether: the electrons and the protons are running around separately—the atoms have become *ionized*. Ionized hydrogen no longer possesses discrete energy levels, and will no longer absorb Balmer photons. This is why the Balmer series is seen strongly in 10,000 K stars but not in hotter stars.

So far, we have only been considering what happens to hydrogen. Throw in calcium, and carbon, and oxygen, and everybody is getting a piece of the action. I will give you my favorite analogy—a tree. You can think of the outermost layer

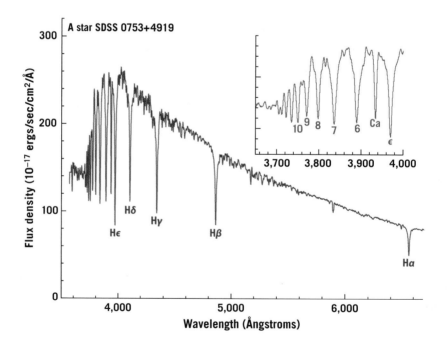

FIGURE 6.3. Stellar spectrum showing Balmer absorption lines. Spectrum of an A star from the Sloan Digital Sky Survey, showing the Balmer series of absorption lines of hydrogen; they are called Hα, Hβ, Hγ, and so forth. The lines pile up at the shortest wavelengths; the inset shows an expanded view, labeling the lines up to H10 (by convention, numbers are used rather than Greek letters beyond Hε). There is also one line due to singly ionized calcium, marked "Ca." *Credit:* Sloan Digital Sky Survey and Michael A. Strauss

of the star as a tree. Do you know what is coming toward the tree (from inside the star)? Mixed nuts. We have a mixed-nut cannon (the interior of the star) firing mixed nuts (photons of different frequencies) into the tree, and in the tree we have squirrels. My squirrels like acorns ($H\alpha$ photons)—these are acorn squirrels. They see all these mixed nuts coming through, but they are only grabbing the nuts they like, the acorns; on the other side (outside the star), out come mixed nuts minus the acorns (the thermal radiation minus the $H\alpha$ photons). Now let's bring in another species: let's get some macadamia chipmunks. What comes out the other side? Mixed nuts minus acorns and minus macadamia nuts. For every species of rodent we stick in that tree, each preferring to eat a different kind of nut, you can infer who is hanging out in the tree based on what is missing on the other side, if you know what they eat.

This is precisely the problem we face in astrophysics. Because we can't go in the star (you wouldn't want to go in there anyway; it's too hot), we analyze it from afar, observing the light to see what gets taken out of the continuous thermal spectrum. We look at its spectrum and ask, does it match the lines of hydrogen? Mostly, but it has other elements too. Go to the lab, check out calcium, and the other elements, to see what frequencies they absorb in a laboratory setting. Then check each element to see whether it matches this star's pattern, because each element leaves a unique fingerprint. These energy levels, these ladder diagrams, are unique to each element and molecule. (For example, figure 6.3 shows an absorption line due to calcium, labeled Ca, in addition to the lines from hydrogen.)

For the general case, let's not even think of a star, but a cloud of gas in interstellar space, a hydrogen cloud with a continuous spectrum of energy coming in from a bright nearby star. Light from the star enters the cloud and comes out on the other side, so there is an absorption spectrum with lines missing. We must now account for the energy somehow; light at those wavelengths got absorbed, with electrons rising to higher energy levels. These electrons will fall back down, emitting photons as they fall. Thus, it's a temporary affair between the electron and the photon. As the electron returns to its original energy level, a photon just like the one the electron absorbed gets sent out in a random direction. It's as if the squirrels and chipmunks had indigestion and spit out the nuts they had just eaten in random directions. If the gas cloud is in equilibrium, with the average number of electrons in level 2 unchanging in time, then the number of nuts eaten and spit back out must be equal. If you are in the line of fire of the cannon (looking along the line of sight to the star), you will see an intense beam of mixed nuts from the cannon coming at you, minus acorns and macadamia nuts. However, if you stand at a random place looking at the tree but not in the line of fire of the cannon (not along the line of sight to the star), you will not see mixed nuts from the cannon, but looking just at the tree (the gas cloud), you will see acorns and macadamia nuts flying out of the tree. These will be bright *emission lines* at just those wavelengths that were absorbed before. From the acorns and macadamia nuts you see, you can deduce that there are squirrels and chipmunks in the tree. By analogy, the emission lines you see emanating from the gas cloud can enable you to identify some of the elements it contains. The picture of the Rosette Nebula in figure 6.4 shows that the nebula is red. The gas is emitting light in the emission line of hydrogen-alpha (Hα) at a wavelength of 6,563 Ångstroms. So this cloud contains hydrogen. Astronomers can take excellent pictures of emission nebulae like the Rosette Nebula using a filter that only lets in the Hα wavelength. That way, almost all the light from the rest of Earth's atmosphere—the light pollution—is blocked out. The light from the young bright blue stars in the center of the Rosette Nebula (which you can see in the figure) bump its hydrogen atoms up to level $n = 3$, and when they drop back down to $n = 2$, they emit Hα photons in all directions, making the nebula glow in red Hα light, in the same way a neon sign glows orange.

We have been discussing the family of transitions for hydrogen, Hα, Hβ, Hγ, Hδ, and so on, called the *Balmer series.* This series of transitions was discovered in 1885 and is named after Johann Jakob Balmer, who figured them out. It doesn't matter which way you draw the arrowhead in the energy level diagram—it's the same photon coming in or going out. It can be absorbed (up) or emitted (down), but all the transitions in the Balmer series have the first excited state, $n = 2$, as the base, and the relevant photons find themselves in the visible part of the spectrum (see figure 6.2, which shows the emission of photons as electrons fall down). That's why the Balmer series was discovered first, because the Balmer photons are in the visible region of the spectrum. But there are two other common series we can refer to. One of them, the *Paschen series*, is based at the $n = 3$ state. These are shorter jumps on the energy scale, so all the participating photons will have lower energy than visible light (see figure 6.2). This lands the Paschen series entirely in the infrared. Once we invented good detectors to reliably measure infrared light, the Paschen series

FIGURE 6.4. Rosette Nebula, a star-forming gas cloud. The red color is due to emission from hydrogen, specifically the $n = 3$ to $n = 2$ transition (Hα). *Photo credit:* Robert J. Vanderbei

showed up. You should know that these families continue, but I'm only going to mention three: Paschen, Balmer, and one more, the *Lyman series* (as before, our Greek letter nomenclature gives Lyman-alpha, Lyman-beta, etc.). The ground state, $n = 1$, forms its base, and all its transitions are in the ultraviolet. The lowest energy transition of the Lyman series has a higher energy than the highest energy transition of the Balmer series (see figure 6.2 again).

This means that when you look for those transitions in the spectrum, the Balmer series is sitting there distinct from the other series, the Lyman series is distinct, and the Paschen series is also distinct, which made them easy to isolate and understand. I could draw an atom for which that is not the case. I can make up an atom—we've got some strange atoms out there—for which the energy jumps in its Lyman and Balmer and Paschen series might be similar, such that these three families would overlap in a spectrum. When thinking about these lines and how we decode them for yet-to-identified elements, we must account for this possibility.

For thousands of years, all we could do was measure the brightness of a star, its position on the sky, and maybe note its color. This was classical astronomy. It became modern astrophysics when we started obtaining spectra, because spectra allowed us to understand chemical composition, and our accurate interpretation of spectra came from quantum mechanics. I want to impress on you the importance of this. We had no understanding of spectra until quantum mechanics was developed. Planck introduced his constant in 1900, and in 1913 Bohr made his model of the hydrogen atom, with electrons in orbitals based on quantum mechanics, which explained the Balmer series. Modern astrophysics really didn't get under way until after that, in the 1920s. Think about how recent this is. The oldest people alive today were born when astrophysics was starting. For thousands of years, we were essentially clueless about stars, yet in one human lifetime we have come to know them well. I have an astronomy book from 1900, and all it talks about is "here is a constellation," "there is a pretty star," "there are a lot of stars over here," and "fewer stars here." It has an entire chapter on Moon phases, another whole chapter on eclipses—that's all they could talk about. Textbooks written after the 1920s, however, talk about the chemical composition of the Sun, the sources of nuclear energy, the fate of the universe. In 1926, Edwin Hubble discovered that the universe is bigger than anybody had thought, because he revealed that galaxies live far beyond the stars of our own Milky Way. And in 1929, he discovered that the universe is expanding. These leaps of understanding happened in the lifetime of people alive today. Extraordinary. I often ask myself, what revolutions await us in the next several decades? What cosmic discoveries will you witness that you can tell your descendants about?

With these lessons of history, you might just avoid making bonehead predictions like that of the French philosopher Auguste Comte, who, in his 1842 book, *The Positive Philosophy*, declared of the stars: "We can never learn their internal constitution, nor, in regard to some of them, how heat is absorbed by their atmosphere."

7

THE LIVES AND DEATHS
OF STARS (I)

NEIL DeGRASSE TYSON

Two astronomers working independently, Henry Norris Russell and Ejnar Hertz-sprung, decided to take all the known stars and plot their luminosity versus their color (see figure 7.1). This graph is, not surprisingly, called the *Hertzsprung–Russell (HR) diagram*. You can quantify the colors of stars if you know their spectra. We know today, as they knew then, that color is a measure of temperature (through the Planck function). The vertical coordinate on the HR diagram shows luminosity, and the horizontal coordinate shows color or temperature, with the hottest (blue) stars on the left, and the least hot (red) stars on the right.

Henry Norris Russell was chair of the Princeton astrophysics department. By many accounts, he was the first American astrophysicist. Because his early diagram showed temperature increasing toward the left, we follow that tradition today. He had data on thousands and thousands of stars, obtained mostly by women at the Harvard College Observatory, doing what most men considered to be menial work, classifying spectra of all these stars. That was back when the humans who did calculations were called "computers." People were computers. There was one large room—filled with these women. Back then, around the turn of the twentieth century, women weren't professors and had no access to any of the jobs that men coveted. But this room of computers included some smart, motivated women who, in the analyses of these spectra, deduced important features of the universe—features that you will learn about in subsequent chapters. Henrietta Leavitt was among them. Cecilia Payne also

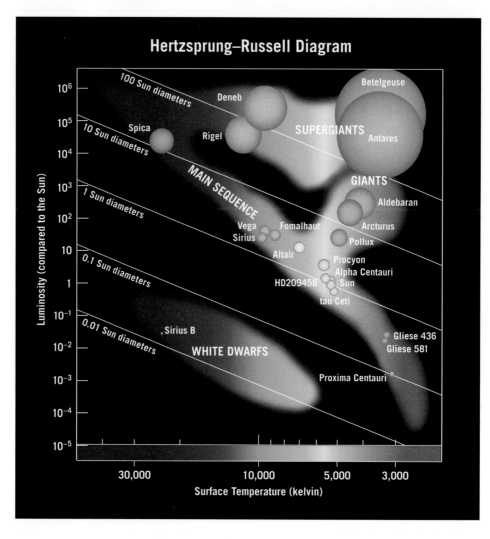

FIGURE 7.1. Hertzsprung–Russell diagram for stars. Luminosities of stars are plotted against their surface temperatures. Note that by convention, surface temperature decreases to the right. Stars with cooler surface temperatures are red, while the hotter ones are blue, as indicated here. The shading indicates where stars are commonly found. Stars lying along a particular labeled diagonal line all have the same radii.

Credit: Adapted from J. Richard Gott, Robert J. Vanderbei (*Sizing Up the Universe*, National Geographic, 2011)

worked on spectra at Harvard for a decade as Harlow Shapley's assistant before eventually being appointed a professor. She was also the one who discovered that the Sun is made mostly of hydrogen. Astronomy, because of that peculiar history, has a fascinating legacy of early contributions by women.

From catalogs of stellar luminosities and temperatures, Hertzsprung and Russell started filling in the diagram. They discovered that stars did not occupy just any place in this diagram. Some regions had no stars—the blank places in this diagram—but diagonally, right down the middle, a prominent sequence of stars emerged. They called it the *main sequence*, as is the way of my field, giving something the simplest possible name.

Ninety percent of the cataloged stars land in that zone. There's a smattering in the upper right corner. These stars have relatively low temperature, yet they are highly luminous. If they have low temperature, what color might they be? Red. How do you have a low-temperature red thing that has extremely high luminosity? What must be true about it? It must be huge. Indeed, these stars are big red things. We call them *red giants*. Armed with our knowledge of the Planck function, we know that they must be red and they must be big. I live for power of deduction like that. Even higher on the upper right are the *red supergiants*. We can now walk into a new astronomical arena and analyze the whole situation armed just with the physics you now carry on your utility belt. In fact, using the *Stefan–Boltzmann law* and the radius r of the star, giving $(L = 4\pi r^2 \sigma T^4)$, we can draw diagonal lines of constant size on the diagram: 0.01 Sun diameters, 0.1 Sun diameters, 1 Sun diameter, 10 Sun diameters, and 100 Sun diameters. Now we know how big these stars are. The Sun lies on the 1-Sun-diameter line, of course. Red supergiants are larger than 100 Sun diameters. Below the main sequence we find another group of stars. These are hot but not too hot; that makes them white. They are extremely low in luminosity, so they must be small. We call them *white dwarfs*. Some people in the U.K. (like J.R.R. Tolkien) might prefer to say *dwarves*. But in America we form the plural of *dwarf* as *dwarfs*. Astronomers are not alone in their preference. Disney's 1937 film is *Snow White and the Seven Dwarfs*, not Seven Dwarves.

At the time the HR diagram was published, classifying stars into zones, we didn't know why they grouped that way. Maybe a star is born with high luminosity and over time, it gets progressively weaker until it dies as a low-luminosity, low-temperature thing. Perhaps it slides downward along the main sequence (simultaneously cooling and losing luminosity) as it ages. A reasonable guess, but that kind of reasoning led to an estimate for the age of the Sun of about a trillion years, which was much larger than the age of Earth. For dozens of years, we proposed educated guesses to answer the question—until we figured out what was really going on. That insight began by taking a look at some different kinds of objects in the sky (figures 7.2 and 7.3).

These images show stars clustered together, officially called *star clusters*. Some have a few hundred stars, others have hundreds of thousands. If the

FIGURE 7.2. The Pleiades, an open star cluster. This is a young star cluster (probably less than 100 million years old).
Photo credit: Robert J. Vanderbei

number of stars is only a few hundred (like the Pleiades in figure 7.2) we call it an *open cluster*; if the cluster has hundreds of thousands of stars, it tends to be spherical or globe-shaped, like M13 (in figure 7.3), and we call it a *globular cluster*.

Globular clusters can have hundreds of thousands of stars, but open clusters, up to a thousand. When you see one of these objects in the sky, it's clear and obvious which kind of cluster you're looking at. There's no argument, because there is no middle ground: either they have a few stars or they have a whole bunch. The stars in a particular cluster have a common birthday—they formed from a gas cloud all at the same time.

The Pleiades is a young star cluster—it's like looking at a kindergarten class. Young, bright, blue stars dominate the picture. But the HR diagram for this cluster shows a complete main sequence and no red giants. The blue stars at the top of the main sequence are so bright that they dominate, but red stars lower down on the main sequence are also in evidence. The Pleiades shows what an ensemble of stars looks like soon after they're born. From it, we can see that

FIGURE 7.3. M13, a globular cluster.

Photo credit: Adapted from J. Richard Gott, Robert J. Vanderbei (*Sizing Up the Universe*, National Geographic, 2011)

some stars are born having high luminosity and high temperature while other stars are born having low luminosity and low temperature—they're just born that way—along the entire main sequence.

Globular clusters like M13 show a main sequence minus an upper end, plus some red giants, which are not part of the main sequence. The picture of M13 is like looking at a fiftieth college reunion—all the stars are old. The red giants are the brightest and dominate the picture. M13's main sequence still has low-luminosity, low-temperature objects, but where did the bright blue ones go? Did they exit the scene? What happened? You can probably guess where they "went": they became red giants. The upper part of the main sequence was peeling away, with luminous blue stars becoming red giants.

We also found middle-aged cases: clusters for which just part of the upper main sequence was gone and only some red giant stars had appeared.

To figure out the masses of different types of stars, we had to be clever. We measured the Doppler shifts in the spectral lines of binary stars as they orbited each other, and applied Newton's law of gravity. From this exercise, we discovered that the main sequence is also a mass sequence, running from massive, luminous, blue stars at the top left to low-mass, low-luminosity red stars at the bottom right. Low-mass stars are born with low luminosity and low temperature, whereas high-mass stars are born with high luminosity and high temperature.

Massive blue stars on the upper main sequence live for perhaps 10 million years. That's actually not much time. Around the middle of the main sequence, a star like the Sun lives for 10 billion years, a thousand times longer. Following the main sequence all the way down to the bottom, the low-luminosity red stars should live for trillions of years. We see 90% of stars on the main sequence. Why? It turns out that stars spend 90% of their lifetimes with luminosity and temperature that land them on the main sequence. Think of it this way: I know, or I am pretty confident, that you brush your teeth in a bathroom every day. But if I take snapshots of you during the day at random times, I'm not likely to catch you in the act, because even though every day you spend some time brushing your teeth, you don't spend much time doing it. We've come to learn that for some regions of the HR diagram that are sparsely populated, stars are actually "passing through" those regions as their luminosity and/or temperature changes, but they do so quickly, not spending much time there. It is rare to catch stars in the act of brushing their teeth.

What's going on down in the center of stars? We agreed that as you raise the temperature, particles move faster and faster. We also agreed that 90% of the atomic nuclei in the universe are hydrogen, the same percentage found in stars. Take a blob of gas that is 90% hydrogen—it's not a star yet. Let it collapse and form a star. As you might suspect, the center becomes the hottest part of the star. If you compress something it becomes hot. The centers of stars are hot enough (as we shall see) to create a nuclear furnace that keeps the center hot. It's much less hot up on the surface. The centers of stars are so hot that all electrons are stripped entirely from their atoms, leaving their nuclei bare.

The hydrogen nucleus has one proton. When another proton approaches it, the two protons repel one another. Protons are positively charged, and like charges repel one another with a $1/r^2$ force. The closer they get, the harder they repel. But increase their temperature. Higher temperature means larger average kinetic energies, and higher velocities for the protons. Higher velocities mean that the protons can approach closer to one another before the electrostatic forces make them turn around. It turns out that there is a magic temperature—about 10 million K—at which these protons are able to get so close together that a whole new short-range, strong nuclear force takes over, attracting them and binding them together, as I mentioned in chapter 1. This attractive nuclear force, unknown a hundred years ago, must be quite strong to overcome the natural electrostatic repulsion of the protons. What else to call it other than the *strong nuclear force*? It's what enables what came to be called *thermonuclear fusion*. (The strong nuclear force is also what holds more massive nuclei together. The helium nucleus has two protons and two neutrons. The two protons repel due to electrostatic repulsion; it is the strong nuclear force that holds them in the nucleus. Similarly with the carbon nucleus [six protons and six neutrons] and the oxygen nucleus [eight protons and eight neutrons]).

The ensuing reaction when two protons come together at 10 million K is kind of fun. You end up with a proton and a neutron stuck together—one of the protons has spontaneously turned into a neutron—and a positively charged electron, called a *positron*, is simultaneously ejected. That's antimatter, exotic stuff. That positron weighs the same as an electron, but when it meets up with an electron, they annihilate, converting all their mass into the energy carried away by two photons. This follows precisely Einstein's mass-energy equation $E = mc^2$, about which Rich will have much more to say in chapter 18. Also ejected is an electron *neutrino*, a neutral (zero-charge) particle that interacts so weakly with other stuff in the universe that it promptly escapes from the Sun. Notice that charge is conserved in this reaction. We start with two positive charges (each proton has one) and end with two positive charges (one on the proton and one on the positron). The reaction creates energy, because the sum of the masses of the original particles is more than the sum of the masses of the particles at the end. Mass is lost, converted to energy via $E = mc^2$. What is a nucleus with a proton and a neutron? It has only one proton in it, so it is still hydrogen, but now it is a heavier version of hydrogen. We often call it "heavy hydrogen," but it also has its own name, *deuterium*.

Now I have some deuterium. Deuterium plus another proton gives me a ppn nucleus (two protons, one neutron) plus more energy. What have I just made? I now have two protons in my nucleus, and when you have two protons, it's called helium. *Helium* derives from Helios—the Greek god of the Sun. We have an element named after the Sun. That's because this element was discovered in the Sun, through spectral analysis, before we discovered it on Earth. This ppn nucleus is a lighter-than-normal version of helium, called *helium-3* because it has three nuclear particles (two protons and one neutron). Now collide two of these helium-3 nuclei: ppn + ppn = ppnn + p + p + more energy. This resulting ppnn is full, red-blooded *helium-4* (the normal helium you find in helium balloons).

All this goes on at 15 million K in the center of the Sun, which converts 4 million tons of matter into energy every second. We came to understand that stars in the main sequence are converting hydrogen into helium. Eventually, the hydrogen in the core runs out, and then, all heaven breaks loose: the star's envelope expands, and it becomes a red giant. About 5 billion years from now, our Sun will become a red giant, throw off its gaseous envelope, and settle down to become a white dwarf. More massive stars will become red giants, and supergiants. They may explode as supernovae, with their cores collapsing to form neutron stars or black holes. We will return to this topic in chapter 8.

For now, let's go back to the HR diagram. We have the main sequence, red giants, and white dwarfs, with temperature increasing to the left and luminosity getting higher as you go upward. Stars are given spectral classification letters. Some are relics from a pre-quantum classification scheme in which they were actually in alphabetical order, but the system is still in use: O B A F G K M L T Y. Each letter designates a class of surface temperature for stars; the Sun is a G star. Their approximate surface temperatures and colors are:

O (>33,000 K, blue),
B (10,000–33,000 K, blue-white),
A (7,500–10,000 K, white to blue-white),
F (6,000–7,500 K, white),
G (5,200–6,000 K, white),
K (3,700–5,200 K, orange), and
M (2,000–3,700 K, red),

all of which are included in figure 7.1. Off to the right, beyond our chart, would be the remaining classes: L (1,300–2,000 K, red), T (700–1,300 K, red), and Y (<700 K, red). If you look at the temperatures on the scale at the bottom of the figure, you can see where these classes go. Spica is a B star, Sirius is an A star, Procyon is an F star, and Gliese 581 is an M star. Each star has both a horizontal position on the chart that shows its temperature (hotter on the left, cooler on the right) and a vertical position that shows its luminosity (increasing from bottom to top) . The Sun has exactly one solar luminosity, of course, by definition, as can be seen by noting its luminosity on the vertical scale. This is a logarithmic scale, allowing us to plot the huge range of observed luminosities, with each tick mark going up representing a star 10 times as luminous.

Along the top edge of figure 7.1 are stars with a million times the Sun's luminosity. At the bottom of the chart are stars with 1/100,000 of the Sun's luminosity. The range in luminosity among the main sequence stars in the universe is staggering. We would eventually figure out that stars at the top end of the main sequence are only about 60 times the mass of the Sun, not a million times more massive. At the bottom end, they are only about a tenth the mass of the Sun, but as indicated, are much, much fainter than the Sun. So the range of masses is large but not nearly as large as the range we find in luminosities. In fact, we can give a formal relationship describing how the luminosity depends on a star's mass on the main sequence, but it's nonlinear: the luminosity is proportional to mass raised to the 3.5 power. Which tells us that two stars of slightly different mass will have very different luminosities.

Here is a cool calculation. Start with $E = mc^2$. That's one of the first equations anybody ever learns in school. You know this equation before you even know what it means. You learn it in third grade perhaps, and find out Einstein came up with it. Good old Albert, from work he did in 1905. The equation says, as we have discussed, that a certain amount of mass can be converted into energy through this relationship, in which c, the enormous speed of light, gets squared, becoming very large indeed. Nuclear bombs owe their power to what goes on in this equation. Rich will explore the origins of this equation in Einstein's theory of Special Relativity in chapter 18.

If a star has a certain amount of mass and a certain amount of luminosity, for how long will it stay alive? Of course, you can ask the same question of your gas-driven car: you know the capacity of its gas tank when you fill it up, and you also know its gas mileage, in miles per gallon. From these facts, you can predict how far you can go before your car runs out of gas. A star's luminosity is its energy emitted per unit time. If you multiply the lifetime ℓ of the star by its luminosity L, you will get the total energy it emits over its lifetime, ℓL. We know the luminosity of a star, the rate at which a star is consuming fuel, and we know how much hydrogen fuel it has, so what is the lifetime of a main-sequence star? That is, how long will it stay on the main sequence? The total energy you can get out of a star by fusing its hydrogen fuel is proportional to its mass M. Remember $E = mc^2$. Total energy emitted is proportional to M and is also proportional to ℓL, so M is proportional to ℓL. That means ℓ is proportional to M/L. If L is proportional to $M^{3.5}$, as I have said, then ℓ is proportional to $M/M^{3.5}$, which is the same as $1/M^{2.5}$. The more massive a star is, the shorter will be its main sequence lifetime will be!

Let's see what that means. If the lifetime of a star is proportional to $1/M^{2.5}$, then if I have a star that's 4 times the mass of the Sun, its lifetime should be $1/4^{2.5}$ times as long as the Sun's. Now $1/4^{2.5}$ is one divided by 4 squared times the square root of 4. The square root of 4 is 2, and 4 squared is 16. Thus, this 4-solar-mass star has a lifetime that is $1/32$ of the lifetime of the Sun. The main-sequence lifetime of the Sun is about 10 billion years. So, this 4-solar-mass star will have a main-sequence lifetime only $1/32$ of 10 billion years, or about 300 million years. That's short.

Another example: $1/40^{2.5}$ is about $1/10,000$, so if you have a 40-solar-mass star, it will live only 1 million years—that's tiny compared to a billion years. Let's take a step in the other direction. Consider a star that has $1/10$ the mass of the Sun. One over $1/10$ is 10, and 10 raised to the 2.5 power is about 300. That star will live 300 times as long as the Sun. What is 300 times 10 billion? It's 3,000 billion, or 3 trillion years, much longer than the current age of the universe—making that star very efficient in its fuel consumption. A 10-solar-mass star lives $1/300$ as long as the Sun, whereas a $1/10$-solar-mass star lives 300 times longer.

Hydrogen fuses to form helium inside a star on the main sequence. Stars do other stuff in their cores during their red-giant phase. More fusing occurs, producing such elements as carbon and oxygen, and others down the periodic table to iron (which has 26 protons and 30 neutrons). A star spends 90% of its life on the main sequence, before it starts cranking out these additional elements as a red giant. That last phase happens fast, occupying a mere 10% of a star's life. Every time you bring together light elements (lighter than iron, number 26 on the periodic table) to make a heavier one, all these reactions lose mass, and the fusion reaction progresses via $E = mc^2$ and emits energy. This fusion process is called *exothermic*, because it gives off energy. But we know other nuclear processes that give off energy as well. Take uranium (number 92), split its nucleus into smaller ones, and that will be exothermic too. This was done in World War II—the Hiroshima bomb was a uranium bomb; the Nagasaki bomb used plutonium (number 94). Each of these elements has a huge nucleus, and has *isotopes* (versions having the same number of protons but different numbers of neutrons) that are unstable. If you split them into parts, creating lighter elements, energy is released. This is also exothermic and is called *fission*. Most of the world's nuclear arsenal going into the Cold War consisted of fission bombs, whereas today, most of the power of our nuclear arsenal resides in bombs that fuse hydrogen into helium. Just to put their relative destructive energy in perspective, fusion bombs use fission bombs as their trigger, giving a sense of how devastating these fusion-based weapons really are. We know how efficiently they convert matter into energy, and that's exactly what stars do. The Sun is one big thermonuclear fusion bomb, except its awesome energy is contained by all that mass pressing down on the core. We have not yet been able to make a contained nuclear fusion power plant. All nuclear power plants in America, France, and other countries are contained fission power plants.

You just can't just split atoms and keep getting energy forever; you can't fuse atoms and get energy forever either. Figure 7.4 explains why. The horizontal axis shows the *atomic number*, the number of *nucleons* (i.e., protons or neutrons) that each naturally occurring element contains, and it starts at 1, hydrogen. Hydrogen's nucleus has one proton. The chart goes all the way out to 238, uranium; its nucleus has 92 protons and 146 neutrons. Some elements, like uranium, have different isotopes; uranium-235, which has 92 protons and only 143 neutrons, is radioactive and highly fissionable (it was the isotope used in the atomic bomb dropped on Hiroshima). All the other elements lie between hydrogen and uranium on the chart. Plotted vertically is the *binding energy*—binding energy per nucleon. The larger the binding energy is, the lower on the chart the element is placed.

To appreciate binding energy, imagine taking two magnets stuck together with the north pole of one matched up with the south pole of the other. In this configuration, you will need to invest energy if you want to pull the magnets apart. Binding energy is what keeps the two magnets together. Figure 7.4 shows hydrogen at the top of the chart—zero binding energy. Hydrogen fusing into helium falls down the hill, releasing energy. Helium has a larger binding energy relative to hydrogen—it's like being down in a valley relative to hydrogen. Note the scale: these binding energies are large, (measured in millions of electron volts per nucleon). Recall that we introduced the electron volt (eV) in chapter 6. You have to add energy to helium (over 7 million electron volts times 4 nucleons, or more than 28 million electron volts) to break it apart into hydrogen. This curve dips in the middle to its lowest point. Uranium at the right-hand edge of the diagram is higher than this lowest point in

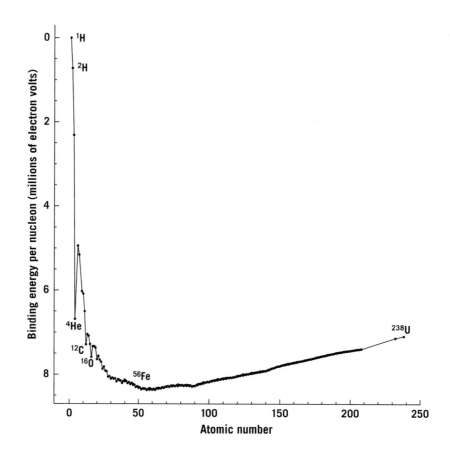

FIGURE 7.4. Binding energy per nucleon of atomic nuclei. Only stable isotopes for each element are shown. Binding energy is shown in millions of electron volts per nucleon (i.e., proton or neutron). This represents the energy per nucleon that would be released in creating this nucleus from free protons. The greater the binding energy per nucleon (lower in the diagram), the less mass there will be per nucleon in the nucleus (according to Einstein's relation $E = mc^2$). *Credit:* Michael A. Strauss, using data from: http://www.nndc.bnl.gov/amdc/nubase/nubtab03.asc; G. Audia, O. Bersillon, J. Blachot, and A. H. Wapstra, *Nuclear Physics* A 729 (2003): 3–128

the middle. If you are an element, you can undergo fission exothermically, or fusion exothermically, until you land all the way down at the bottom. Iron, with its 26 protons and 30 neutrons (i.e., 56 nucleons), occupies that bottom spot. If I try to fuse iron, it goes *endothermic* and absorbs energy. If I try to fission iron, it's endothermic again. The buck stops on iron: there is no more energy to be released when you get to iron.

Stars are in the business of making energy. If a star is cranking along, fusing its elements down the line, and if it's getting energy for doing so, you have a happy star. The energy being generated keeps the center of the star hot, and the thermal pressure of that hot gas keeps gravity from collapsing the star under its own weight. Let's say I have a main sequence star ten times as massive as the Sun: it's mostly hydrogen and helium, and in its core it is still converting hydrogen to helium; that's Scene 1. By Scene 2, the core is now pure helium, but it still has hydrogen and helium in the surrounding envelope. Fusion stops in the center, and the center can't hold the star up anymore, so what does the star do? The star's core collapses, the pressure builds, and the temperature increases, becoming hot enough to fuse helium. It takes a higher temperature to bring helium nuclei together (ppnn + ppnn) than it does to bring hydrogen nuclei together (p + p), because each helium nucleus (ppnn) has two protons—doubling the number of positive charges repelling one another. Continuing with Scene 2, helium fusion kicks in (at 100 million K), keeping the star stable. In the middle of that very hot core, helium is becoming carbon; outside the core, our envelope has hydrogen fusion in a shell. Eventually, I get a ball of carbon in the center, and the center is not hot enough to fuse carbon, so the fusion stops. The core collapses further, the temperature rises again, and carbon fusion begins. That's Scene 3. We now have carbon fusing to make oxygen in the center of the carbon core in the center of the helium core, in the center of the star's envelope, which still has hydrogen and helium. We're creating an onion of elements, layer upon layer, because it's always hottest in the middle. Each reaction releases energy. Eventually, you get iron in the middle, surrounded by successive shells of all the other lighter elements. Therein sits the future chemical enrichment of the galaxy.

But these elements are still locked inside a star, and they have to get out of the star somehow, because we're made of these elements! We now know that since iron is the end of the road, once iron accumulates in the core, the fusion stops, and the star collapses. If it tries to fuse iron, doing so sucks energy out of the star, collapsing the star even faster. Stars are in the business of making energy, not absorbing it. As the core collapses faster and faster, the star implodes, leaving a tiny, superdense neutron star in the center, whose formation generates enough kinetic energy to blow off the entire envelope and outer core of the star and causes a titanic explosion, for several weeks shining billions of times brighter than the Sun. The guts of this star are now released into the galaxy, into what we call the *interstellar medium*, chemically enriching gas clouds with heavy elements, enabling gas clouds to become something more interesting than clouds of pure hydrogen and helium.

Figure 7.5 shows a beautiful spiral galaxy M51, containing a hundred billion stars, sitting there nicely (top view) until a star explodes (in the bottom view).

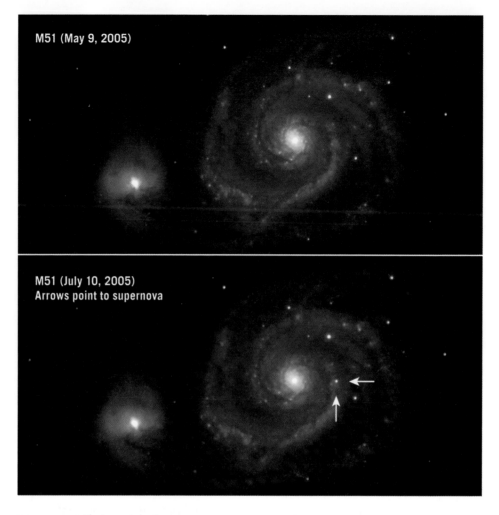

FIGURE 7.5. Spiral Galaxy M51 and supernova.

Photo credit: J. Richard Gott, Robert J. Vanderbei (*Sizing Up the Universe*, National Geographic, 2011)

As we'll see in chapter 12, we live in a spiral galaxy, not unlike M51. Before the explosion (top) you can see the galaxy and some foreground stars in the Milky Way, which are much closer to us (and of course far less luminous) than the galaxy. When one of these explosions goes off, we see a new star in the galaxy (bottom), one that was not visible before and is by far the brightest thing in the galaxy. It's a single star. If you're a planet in orbit around that star, you're toast. Quite simply, and quite literally. We call these things *supernovae. Nova* means "new" in Latin, and it meant a new star in the sky. We later learned that in a supernova, we were seeing the death throes of a star. Not all stars can do this; only relatively high-mass stars become supernovae, leaving tiny, incredibly dense neutron stars at the center when they blow off their outer parts. Even higher-mass stars exist. And they explode too. But when one of them collapses, the increase in gravity near the center warps space so severely that it closes itself off from the rest of the universe, and guess what you get: a black hole. A black hole may sometimes form at the center while the envelope of the star is being thrown off, also creating a supernova explosion.

Stephen Hawking works on black holes; he has made major discoveries about their strange behavior, and Rich will have much more to say about black holes and Hawking's discoveries in chapter 20. The animated TV sitcom *The Simpsons* has given Stephen Hawking the reputation of being the smartest person alive. Most of us agree.

Now, let me tell you about star births. The Orion Nebula is a stellar nursery; a gas cloud that has already been enriched with heavier elements forged in the cores of a previous generation of dying stars.

In the center of the nebula are bright, newly born massive O and B stars. These O and B stars are radiating intensely in the ultraviolet. This hot UV radiation has photons with enough energy to ionize (strip the electrons off) the hydrogen gas near the center. The gas is trying to form stars, but it's being thwarted by the intense luminosity from the high-mass stars in the center. Meanwhile, some of this enriched gas is ready to make something more interesting than just smaller balls of gas. It can also make balls of solid stuff that contain oxygen, silicon, iron—things like terrestrial planets. Some nascent stars are also forming planetary systems from the gas that swaddles them. These are new solar systems being born from rotating disks of material (see figure 7.6). And in the Orion Nebula it's still happening now. Some stellar nurseries are birthing thousands upon thousands of solar systems. Our galaxy has 300 billion stars, many of them likely surrounded by planets of their own.

How important are we in this picture? We're quite small—cosmically insignificant. A depressing revelation for some, who would prefer to feel large. The problem is history. Every time we make an argument that we're special in the cosmos, either that we are in the center or that the whole universe revolves around us, or that we are made of special ingredients, or that we've been around since the beginning, we learn that the opposite is true. In fact, we occupy a humble corner of the galaxy, which occupies its own humble corner in the universe. Every astrophysicist lives with that reality.

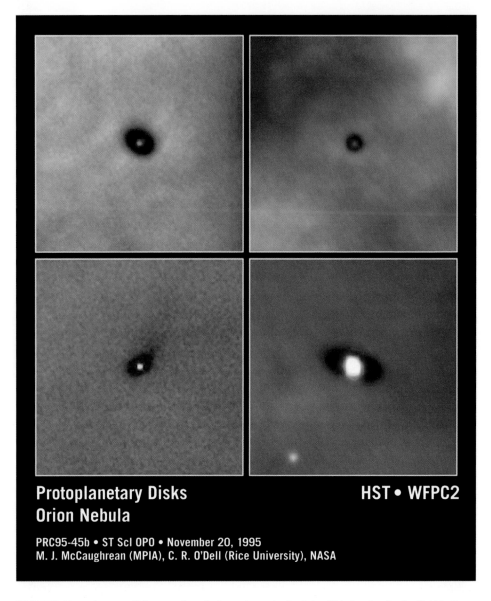

Protoplanetary Disks
Orion Nebula

HST • WFPC2

PRC95-45b • ST ScI OPO • November 20, 1995
M. J. McCaughrean (MPIA), C. R. O'Dell (Rice University), NASA

FIGURE 7.6. Protoplanetary disks around newly formed stars in the Orion Nebula taken by the Hubble Space Telescope. *Photo credit:* M. J. McCaughrean (MPIA), C. R. O'Dell (Rice University), NASA

Let me make you feel even smaller. In figure 7.7, taken by the Hubble Space Telescope, all the smudges in the picture are entire galaxies, so far away that each of them occupies only a small part of the image. Each one of these smudges harbors more than 100 billion stars unto itself. And this is just another small pocket of the universe. This Hubble Ultra-Deep Field, as it is known, is the deepest image of the universe ever acquired. It shows about 10,000 galaxies. This whole picture

Hubble Ultra Deep Field
Hubble Space Telescope • Advanced Camera for Surveys

NASA, ESA, S. Beckwith (STScI) and the HUDF Team
Color representation by Wherry, Blanton, Hogg (NYU), Lupton (Princeton)

FIGURE 7.7. Hubble Ultra Deep Field. This long-exposure photograph taken by the Hubble Space Telescope shows about 10,000 galaxies. But it covers only about 1/13 millionth of the sky. Therefore, there are about 130 billion galaxies within the range of this telescope over the whole sky. *Photo credit:* NASA/ESA/S. Beckwith(STScI) and The HUDF Team. Color representation by Nic Wherry, David W. Hogg, Michael Blanton (New York University), Robert Lupton (Princeton)

covers a patch of sky 1/65 of the area of the full Moon, about 1/13 millionth of the whole sky. Since this spot on the sky is not unusual, the number of galaxies we can see in the whole sky is 13 million times as many as we can see in this picture. That means 130 billion galaxies are within the reach of the Hubble Space Telescope.

In his book *Pale Blue Dot*, Carl Sagan noted that everyone we ever knew, everyone we have ever read about in history, lived on Earth, this one tiny dot in the universe—something I think about often. I think about it, because your mind says "I feel small," your heart says "I feel small," but now you're empowered, and you'll continue to be empowered as this book unfolds, not to think small, but to think big. Why? Because you're now enlightened by the laws of physics, the machinery by which the universe operates. In effect, understanding astrophysics emboldens and empowers you to look up in the sky and say, *No, I don't feel small, I feel large, because the human brain, our three pounds of gray matter, figured this stuff out. And yet, even more mysteries await me.*

8

THE LIVES AND DEATHS
OF STARS (II)

MICHAEL A. STRAUSS

In this chapter, we're going to explore the nature of stars in a bit more detail, building off what we learned in the last chapter. What makes an object qualify as a star? An astronomer defines a star to be a self-gravitating object that is undergoing nuclear fusion in its core. *Self-gravitating* means that it holds itself together by gravity. Earth also holds itself together by gravity. In fact, for an object as massive as Earth, the strength of gravity is actually much greater than the internal strength of rocks. We can see that by noticing that the shape of Earth is spherical, just like it is for stars. Gravity pulls everything together equally in all directions; the mark of an object held together by gravity is its tendency to be spherical. Smaller objects, such as asteroids, whose gravitational force is not as great, are held together by the tensile strength of their rocks or they are irregular rubble piles, often quite lumpy and elongated (figure 8.1).

But for a large massive object like the Sun, gravity is so strong relative to other forces that it compresses the mass into a spherical shape—the most compact configuration. If a large self-gravitating object is spinning rapidly, however, it will not be spherical; the spinning causes it to flatten. Isaac Newton himself understood this. Jupiter is spinning rather rapidly, and it is slightly elliptical as a result; its equatorial radius is about 7% larger than its polar radius. The most dramatic examples of flattened spinning objects are spiral galaxies, which we discuss in chapter 13.

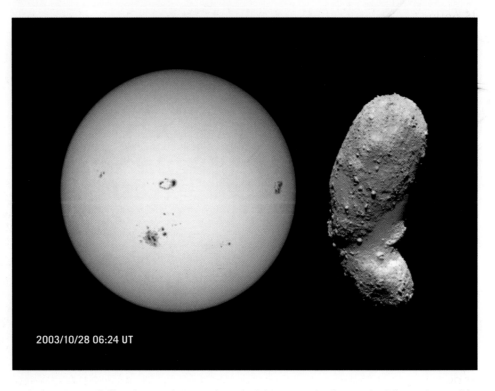

2003/10/28 06:24 UT

FIGURE 8.1. The Sun (left) and Asteroid 25143 Itokawa (right), not to scale, showing the different shapes of the two. The Sun, with a diameter of 1.4 million kilometers, is pulled into a spherical shape by its own gravity. Note the dramatic sunspots. The asteroid is only half a kilometer in diameter; its self-gravity is not adequate to make it spherical; it is thought to be a loose agglomeration of material that has accreted over time. The image of the Sun was taken by the Solar and Heliospheric Observatory (SOHO), a spacecraft dedicated to observing the Sun. The image of the asteroid was taken by the Hayabusa spacecraft launched by the Japan Aerospace Exploration Agency (JAXA). *Photo credits:* Sun: NASA, from http://sohowww.nascom.nasa.gov/gallery/images/large/mdi20031028_prev.jpg; Asteroid Itokawa: JAXA, from http://apod.nasa.gov/apod/ap051228.html

If the gas in a star is held together by gravity, what keeps all that gas from collapsing to a point? It is the internal pressure of the gas. Every parcel of gas feels gravity pulling it inward, and pressure pushing outward, with the two forces balanced in equilibrium.

An analogy is a balloon: it is not held together by gravity, but rather by the tension in the latex of the balloon. The balloon wants to shrink like a rubber band, but as in a star, the pressure of the air inside a balloon prevents it from shrinking. The air pressure and the tension are in equilibrium, and the balloon holds its spherical shape.

The gas pressure inside a star increases as you go to the center and decreases as you go out to larger radii. Gas pressure that decreases with larger radius is familiar here on Earth. Here at sea level, the atmospheric pressure is about

15 pounds per square inch; this represents the weight of the total column of air sitting above every square inch of Earth's surface and extending to the top of the atmosphere. As you go up in Earth's atmosphere, more and more of the atmosphere is below you, and the remaining column of atmosphere above you pressing down on you weighs less. The air pressure therefore decreases with altitude.

The pressure of the gas in a star is a reflection of its temperature and density, both of which increase dramatically as you go toward the center.

Now let's go to the core. We can't observe the core directly, but we can infer its properties by writing down the equations of stellar structure, which include the effects of pressure and gravity. These equations incorporate the observation that the Sun is in equilibrium, with gravity and pressure balanced throughout the star. These calculations show that in the very center of the Sun, the temperature is 15 million K, as we have discussed. This calculation also reveals the density at the center to be about 160 grams per cubic centimeter, 160 times denser than water. For comparison, the densest naturally occurring element on Earth is osmium, at 22.6 grams per cubic centimeter (about twice as dense as lead). With its enormously high temperature, the gas in the core of the Sun is ionized, meaning that the electrons are stripped from atoms, and the nuclei and electrons are zipping around at high speed—this is called a *plasma*. It is the pressure of these particles moving at high speed that resists gravity, keeps the Sun from collapsing, and holds it in equilibrium.

We've already seen that a basic property of material at a given temperature is that it emits photons. This is true for the center of the Sun as well, at 15 million K. The blackbody spectrum of an object at this temperature peaks at X-ray wavelengths. Does this mean that the Sun is shining brightly in X-rays? No. Consider an X-ray photon emitted from the center of the Sun. Can it make its way unimpeded from the center? When you go to the doctor's office to get an X-ray; they shield those parts of your body they don't want to X-ray with a blanket containing lead. Thus, a fraction of an inch of lead, with a density of a measly 11.34 grams per cubic centimeter, will absorb any X-rays that hit it. If that's all it takes to absorb X-rays, you can imagine that the X-rays coming from the center of the Sun are not going to get very far. In fact, they travel only a fraction of a centimeter before they are absorbed.

The energy of an absorbed photon has to go somewhere, however. It heats up the material that absorbed it, which will then radiate blackbody radiation—

more X-rays to be re-emitted. So you can think of our little photon as getting absorbed and getting reradiated over and over again. When you work through all the numbers, the time it takes for energy generated in the center in the Sun to propagate to the surface is about 170,000 years. The distance from the center to the surface is just 2.3 light-seconds, so if that photon could travel unimpeded, it would take only 2.3 seconds to go from the center of the Sun to its surface. But because the photon is getting jostled around, it makes a random, drunken walk, being absorbed and re-emitted as it slowly wanders out from the center of the Sun.

The original photon in the center is an X-ray photon, emitted by gas at 15 million K. Will it still be an X-ray photon when it reaches the surface? No; each time the energy is re-emitted, it is in the form of photons appropriate for the temperature at that position in the star. As the energy makes its way from the center to the surface, the temperature drops, and individual photons lose their identity. The energy becomes distributed among photons of lower energies, corresponding to a lower temperature. So even though X-rays are produced in the center, we don't see X-rays from the surface. X-rays are slowly degraded down to mere visible light photons, the kind that we see coming from the surface of the Sun.

If there were no nuclear furnace in the center of the Sun to keep the center hot and the pressure up, the Sun would slowly start to shrink under the influence of gravity as it lost energy radiated from its surface. This gravitational shrinkage, as the envelope of the star falls toward the center, would generate energy, in exact analogy to the way a piece of chalk gathers speed, and thus kinetic energy, as it falls to the floor. That gravitational energy of contraction would, by itself, be enough to keep the Sun shining at its current luminosity for about 20 million years. Before Einstein, Hermann von Helmholtz (in 1856) hypothesized that this slow gravitational contraction was actually the source of energy powering the Sun. This was plausible at the time, since nuclear fusion was unknown and would not be discovered for another 82 years. This mechanism implied that the Sun had been shining as it is today for 20 million years at most. But we now know, from dating via radioactive isotopes (e.g., noting how much uranium in particular rocks has decayed into lead), that Earth is several billion years old. Moreover, fossils demonstrate that Earth's surface temperature has remained at least approximately constant for a significant portion of that time. Thus, the Sun has been shining more or less as it does at present for much longer than 20 million years, and the gravitational contraction hypothesis for the Sun's energy generation cannot be true.

With the understanding of the significance of $E = mc^2$, all was answered. The Sun burns nuclear fuel in its center, providing energy. This nuclear energy generation balances the luminosity given off by the Sun and maintains its internal pressure. Thus the Sun is stable and does not contract. Nuclear fusion is so efficient in generating energy that the Sun has shone steadily for the past 4.6 billion years, giving life on Earth a long period of stable conditions over which to evolve. The Sun is now about halfway through its main sequence lifetime.

By the way, how do we measure the basic properties of the Sun: its radius, mass, and luminosity? To measure the Sun's radius, we go through a series of steps. We have known the radius of *Earth* since the time of the Greek mathematician and geographer Eratosthenes, about 240 BC. Every year at noon on June 21, the Sun passed directly overhead at Syene, Egypt. Eratosthenes knew this fact. At that same moment, he measured the Sun to be 7.2° off vertical in Alexandria, which is directly north of Syene. Aristotle had argued that Earth, no matter what its orientation, always casts a circular shadow on the Moon during an eclipse of the Moon. The only object that always casts a circular shadow is a sphere; thus, Eratosthenes knew that Earth must be a sphere. He also understood that the 7.2° shift in the altitude of the Sun, as measured from the two cities at the same time, was due to the curvature of Earth's surface, meaning that the two cities were separated by 7.2° of latitude, or about 1/50 of Earth's entire 360° circumference. Hire someone to pace off the distance from Alexandria to Syene, multiply by 50, and you have the circumference of Earth—about 25,000 miles. Divide by 2π and you have the radius. It was easy, once someone figured out how to do it!

From different observatories widely separated on Earth's surface, we get slightly different parallax views of Mars against the distant stars. Knowing the radius of Earth and measuring that parallax shift allows us to measure the distance to Mars. Giovanni Cassini was the first to do this. Kepler's work allowed us to plot the orbits of planets to scale—to make a scale model of the solar system. Once you had that one Earth–Mars distance, you could deduce the size of all the orbits, including the radius of Earth's orbit, the Astronomical Unit. Thus, Cassini in 1672 determined that the distance from Earth to the Sun was about 140 million kilometers—not far from the true value of 150 million kilometers.

We know the angular size of the Sun (about half a degree across) as seen from Earth, and knowing the distance to the Sun, we can determine the Sun's radius. It is equal to the angular half-diameter of the Sun in degrees ($1/4°$) divided by $360°$, multiplied by 2π times the distance to the Sun. The radius of the Sun is about 700,000 kilometers, about 109 times larger than the radius of Earth. The luminosity of the Sun is also straightforward: we can measure its brightness as seen from Earth, and now that we know its distance r, the inverse-square law allows us to determine its luminosity—about 4×10^{26} watts.

We can also determine the mass of the Sun. Newton's laws allow us to figure out the ratio between the mass of Earth and that of the Sun. We know the acceleration Earth produces at a distance of one Earth radius (i.e., here on Earth's surface), $GM_{Earth}/r_{Earth}^2 = 9.8$ meters per second per second, which we can determine by watching apples drop. We also know the acceleration the Sun produces at a distance of 1 AU: $GM_{Sun}/(1\,AU)^2 = 0.006$ meters per second per second, which we have already calculated in chapter 3. Take the ratio of these two accelerations: 0.006 meters per second per second/9.8 meters per second per second = $0.0006 = [GM_{Sun}/(1\,AU)^2]/[GM_{Earth}/r_{Earth}^2] = (M_{Sun}/M_{Earth})$ $(r_{Earth}/1\,AU)^2$. Plugging in the known values for the radius of Earth and 1 AU, and solving, we find that the Sun has a mass of about 330,000 times that of Earth. Because the constant G factors out of the ratio, we don't have to know it to find the ratio of the mass of the Sun to the mass of Earth.

But what is the mass of Earth in kilograms? We could solve for its mass using the equation for the acceleration of gravity at the surface of Earth, 9.8 meters per second per second = GM_{Earth}/r_{Earth}^2, if we only knew the numerical value of Newton's constant G. Henry Cavendish —who discovered hydrogen, the most abundant element in the universe—did a clever experiment to find the value of G. He used a torsion pendulum to measure the ratio of the force exerted on a test ball by Earth and the force exerted on it by a nearby 159 kilogram lead ball. Earth pulled the test ball downward, and the nearby heavy lead ball pulled it to the side, and he could compare the forces they exerted by measuring the angle of deflection produced in the pendulum. Knowing the distances to the nearby lead ball and to the center of Earth, and using Newton's laws, he determined the ratio of the mass of Earth and that of the lead ball. This enabled Cavendish in 1798 to determine the value of Newton's constant G and find the mass of Earth in kilograms. Multiply by 330,000, and you have the mass of the Sun. It turns out to be 2×10^{30} kilograms. That's a lot!

We've been focusing on the Sun here, but we'd like to understand the nature of other stars as well. Just as we use the orbit of Earth around the Sun to determine the Sun's mass using Newton's laws, we can use observations of pairs of stars ("binary stars") orbiting each other to determine their masses.

The lowest-mass stars on the main sequence (which are M stars) have a mass about 1/12 of that of the Sun. What about stars that are even lower in mass? With lower gravity, they will have lower temperature and lower density in the core. What happens when you get a gaseous mass held together by gravity that is simply not hot enough in its center for nuclear fusion of hydrogen to take place? Such a star we call a *brown dwarf* (they are not really brown, but actually appear very red and glow mainly in the infrared; sometimes astronomical nomenclature can be a bit misleading). They exist but are hard to find. Such stars are glowing feebly by the residual heat from their gravitational collapse (just as Helmholtz had imagined for the Sun); they have no significant internal nuclear furnace and are of low luminosity. They are also cool, with surface temperatures ranging from 600 K to 2,000 K, and thus their radiation is coming out mostly in the infrared rather than the visible part of the spectrum. By comparison, your oven at home gets up to 500 K. (A little known fact: 574° F is also 574 K—the crossing point of these two temperature scales.)

Most of our most powerful telescopes are sensitive to visible light, and only in the past few decades have we built telescopes that can survey the sky in infrared light (which turns out, for all kinds of technical reasons, to be quite a bit more difficult). It's only been with the advent of powerful telescopes that are sensitive to infrared light that astronomers have even been able to find such objects.

The O, B, A, F, G, K, and M classes of stars have been around for about 100 years, but since 1999, as we have discovered brown dwarfs, we have added two new classes: L and T stars. Even more recently, an infrared satellite called the Wide-Field Infrared Survey Explorer has discovered still cooler stars, classified as Y stars, with surface temperatures as low as 400 K, just a bit above the boiling point of water. Brown dwarfs with masses between 1/80 and 1/12 of the Sun (i.e., between about 13 and 80 times the mass of Jupiter) feebly burn the trace amount of deuterium that exists in their cores. Thus, since they do have some nuclear burning in their cores, they are still are called stars. Objects of still lower mass, less than 13 times the mass of Jupiter, will have absolutely no nuclear fusion of any sort in their cores. We call such objects planets!

Let's consider the death of stars in greater detail than we did in chapter 7. Even during its late main sequence phase, the Sun will gradually increase in luminosity, and Earth's oceans will boil away about a billion years from now. This will represent the end of life as we know it on Earth. Roughly 5 billion years from now, when no more hydrogen remains in the Sun's core (having all been turned into helium), the Sun's nuclear furnace turns off, and the pressure that's been holding up the star against gravity drops. Gravity wins, and the star will start to collapse. But recall that it takes a couple of hundred thousand years for energy generated in the core to make its way out to the surface. The inner parts of the star will start to collapse, even as energy is still flowing through the outer parts of the star, holding them up. The outer parts of the star have a couple of hundred thousand years before they get the bad news that the energy source at the center of the Sun is gone.

Consider the hydrogen shell immediately surrounding the (now pure helium) core. Outside the core, there is still plenty of hydrogen, but that region has so far been uninvolved in nuclear fusion, as its density and temperature are simply too low. As this shell of hydrogen collapses, however, it becomes hotter and denser. Very quickly, its density and temperature get high enough to trigger the fusion of hydrogen to helium in the shell. We have a new source of fuel to run the nuclear furnace: hydrogen burning in a shell.

Suddenly, the star has a new lease on life. The rate of energy production in the hydrogen-burning shell is enormous—much higher than that of the core while the star was still on the main sequence. Furthermore, the volume of the hydrogen-burning shell is much larger than that of the core.

And so, for a brief period at least, the star produces a huge luminosity, but it takes a long time for that radiation to get out, and the increased pressure starts winning the tug-of-war with gravity. As a consequence, the outer parts of the star expand (and cool somewhat), even while the inner parts contract. The Sun becomes a *red giant*, as discussed in chapter 7. Outside the hydrogen-burning shell, the outer parts of the star have expanded to an enormous radius, about 1 AU (or 200 times the current radius of the Sun). About 8 billion years from now, tidal interactions with the Sun during its red-giant phase will probably cause Earth to spiral into the envelope of the Sun and burn up.

While the star's hydrogen shell is burning, its helium core has no internal energy source; gravity causes it to continue to contract and thus heat up. When the temperature reaches about 100 million K in the core of the star, the helium nuclei start fusing to make carbon and oxygen nuclei. That helium-burning phase will last for about 2 billion years for the Sun, but eventually, the helium in the core will all be depleted, and the core begins to collapse again.

For stars of the mass of the Sun, we're near the end of our story. The outer parts of the star are far away from the core, and thus feel only a weak gravitational pull. It takes only a bit more energy to eject the outer parts of the star, which gently expand as a diffuse gaseous envelope, revealing the hot dense carbon-oxygen core of the star left behind. This ejected gas is excited by the ultraviolet light of the central star, causing it to fluoresce, and making a nebula like the Dumbbell Nebula pictured in figure 8.2. Such objects are called, confusingly, *planetary nebulae*, because the first astronomers to spy them through telescopes thought they looked something like planets, and the name has stuck ever since. Astronomers have a nostalgic tendency to keep names for things even when they are outmoded and misleading.

This extended envelope of material, which used to be part of the star itself, is now gently expanding outward. Stars sometimes blow off their outer layers in complex ways, giving rise to planetary nebulae with multiple shells of gas around them. Different layers are coming from different depths inside the star, which may be enriched in different elements. Rotation of the original star can cause the layers to be blown out preferentially along the spin axis, as occurs in the Dumbbell Nebula (see figure 8.2).

The now-exposed glowing core of the star is visible in the very center of the nebula. It is small (about the size of Earth) and hot enough to appear white; we thus call it a white dwarf. The white dwarf has no internal source of energy, and thus slowly cools off over billions of years. We still call a white dwarf a star, even though it is not burning nuclear fuel. (I admit it: this nomenclature is not totally consistent!)

What holds up the white dwarf against gravitational collapse? The *Pauli exclusion principle*, named for physicist Wolfgang Pauli, states that no two electrons can occupy the same quantum state. This is key for understanding the structure of atoms. In atoms having many electrons, the electrons must stack up into higher energy levels when the lower energy levels are filled. For white dwarfs, the Pauli exclusion principle means that the electrons don't like to be squeezed too close together; this gives rise to a pressure that holds the white dwarf up against gravity. Our Sun will end its life as a white dwarf.

FIGURE 8.2. The Dumbbell Nebula. This is a red giant star that has ejected its outer layers, revealing its hot dense core. The core is a white dwarf star, seen at the center, while the outer layers are fluorescing as a planetary nebula from the ultraviolet light that the white dwarf emits. *Photo credit:* J. Richard Gott, Robert J. Vanderbei (*Sizing Up the Universe*, National Geographic, 2011)

As described in chapter 7, stars more than 8 times as massive as the Sun go through a much more dramatic series of reactions. There's enough mass in their cores for the carbon and oxygen, which would otherwise sit inertly while the star quietly becomes a white dwarf, to instead heat up enough to fuse into such elements as neon, silicon, and the others all the way up the periodic table to iron.

The outer layers of these more massive stars grow appreciably larger than mere red giants. They become red supergiants, with radii of several AUs.

In the night sky, some bright stars are clearly red to the naked eye. Red stars that are on the main sequence have a low luminosity; none are visible to the naked eye. In contrast, a red giant is large, has a huge luminosity, and can be seen out to a great distance. All the bright red stars in the sky are either red giants (like Arcturus in the constellation Boötes and Aldebaran in Taurus) or red supergiants (like Betelgeuse in Orion).

Scientists overuse the prefix *super-*. We stick it in front of just about everything, because we keep discovering or making things that are even bigger or more spectacular than what we knew before: supernovae, supermassive black holes, and, of course, the never-completed particle accelerator known as the superconducting supercollider! The most famous red supergiant in the sky is Betelgeuse (pronounced "Beetlejuice"). It has a radius about 1,000 times as large as the Sun's and has a mass of at least 10 solar masses. In its core, helium is burning into carbon, oxygen, and heavier elements. Outside the core is a thin shell of essentially pure helium that is not hot or dense enough to burn yet, so it's sitting there more or less quiescent. Outside that is a shell of hydrogen burning into helium, and outside *that*, the vast majority of the volume of the star, is a huge extended envelope of hydrogen and helium.

This story of the evolution of stars after the main sequence was worked out in detail in the 1940s and 1950s, as people came to understand the detailed nuclear physics that takes place in the cores of stars and were able to use the first computers to solve the relevant equations of stellar structure. Much of this work was done at Princeton University, led by Professor Martin Schwarzschild. Neil, Rich, and I had the opportunity to interact with him in his later years; he was a wonderful man.

Figure 8.3 shows Schwarzschild with Lyman Spitzer and Rich Gott. When Henry Norris Russell (of HR diagram fame) retired as chair of Princeton University Observatory in 1947, he brought in two young astronomers, Martin Schwarzschild and Lyman Spitzer, both in their early thirties. Spitzer, who became chair of the department, went on to develop much of our modern understanding of the interstellar medium (the gas and dust between the stars) and founded the Princeton Plasma Physics Laboratory, where scientists are working to harness nuclear fusion as a source of energy. Spitzer will always be known to the public as the father of the Hubble Space Telescope, having developed the initial concept and working for decades to convince the astronomical community and the US Congress that it should be built. Spitzer and Schwarzschild were the core of the Princeton Astrophysics Department for the next 48 years. They passed away within 11 days of each other in 1997, quite a shock to all of us.

In the 1950s, Schwarzschild and his students worked out all the details of the story I am describing. He was one of the first people who understood the whole story of the evolution of stars. Martin's father, Karl Schwarzschild, played a key role in the study of black holes; his name will come up again in chapter 20.

FIGURE 8.3. Left to right: Lyman Spitzer, Martin Schwarzschild, and Rich Gott in the 1990s.
Photo credit: Collection of J. Richard Gott

Continuing our story of stars, the pressure of electrons prevents a white dwarf from collapsing. However, if the mass of the star's core is more than 1.4 solar masses, even this pressure is not large enough to hold it up against gravity. Compressed by gravity, the electrons and protons are pushed together to form neutrons (releasing electron neutrinos in the process). This leaves us with a *neutron star*—actually a giant atomic nucleus of mostly pure neutrons. The Pauli exclusion principle holds for neutrons as it does for electrons, and the resulting neutron pressure now supports the star against gravity. However, as neutrons are much more massive than electrons, the equilibrium size of a neutron star (about 25 kilometers) is much smaller than that of a white dwarf. Imagine more than a solar mass of material squeezed into a volume the size of Manhattan Island (or 100 million elephants in a thimble, from chapter 1)! Neutron-star matter is the densest material we know of. The center of a neutron star can have a density of almost 10^{15} gm/cm^3.

If the core of a massive star is more massive than about 2 solar masses, the neutron star that tries to form is unstable to further collapse; the neutron pressure is inadequate to hold the star up against gravity, and a black hole forms. Whether the core collapses to a neutron star or a black hole, the infalling material compresses violently, triggering further nuclear reactions (remember that the material outside the core is still made up of elements lighter than iron). The energy that is suddenly released can eject the entire exterior of the star above the core, causing a supernova explosion. Stars with initial masses on the main sequence greater than about 8 solar masses die by exploding as supernovae, forming either neutron stars or black holes in the process. Massive exploding stars are called *Type II* supernovae, as there is another type of stellar explosion. Consider three stars orbiting each other, two of which are white dwarfs. Gravitational interactions between them can cause two white dwarf stars to collide. The heating due to the collision detonates their nuclear fuel and produces a supernova. Alternatively, a red giant star in a binary system can transfer mass onto a white dwarf star, pushing it over the limit of 1.4 solar masses and causing it to collapse, and a supernova ensues. Such explosions are called *Type Ia* supernovae, to distinguish them from the explosions of massive collapsing stars: we'll return to them briefly in chapter 23, as they become important tools to help us measure the accelerating expansion of the universe.

In any case, in a supernova explosion, gas flies outward in all directions. This is not a gentle process like the slow wafting away of the outer parts of a planetary nebula. Rather, it is an extremely violent explosion. Most or all of the star is destroyed in the explosion, and material is sent out at speeds approaching 10% of the speed of light. Heavy elements produced in the core of the star are now returned to the interstellar medium, ready to be included in the next generation of stars and planets.

In 1054, Chinese astronomers noticed a new star in the constellation we call Taurus. The ancient Chinese were careful observers of the sky, looking for portents of future events, so they were particularly impressed by this "guest star," which was visible for many weeks and was initially bright enough to be seen during the day. Interestingly, there are no records in any European manuscripts whatsoever that anyone living there had seen this thing, despite it being the brightest object in the sky for weeks on end. Perhaps it was cloudy for that entire period in Europe, or any written European accounts were lost, or maybe the Chinese astronomers were just paying much more attention to what was going on in the sky.

Images of the Crab Nebula in Taurus (figure 8.4) taken a few decades apart clearly show that it is expanding. Given the observed rate of expansion and its current size, we can work out when the expansion must have started; the

FIGURE 8.4. The Crab Nebula. This is the expanding remnant from a supernova explosion (seen on Earth in 1054 AD).
Photo credit: Hubble Space Telescope, NASA

answer is about a thousand years ago, just about the time the Chinese observed their "guest star." Thus the Crab Nebula, located in exactly the same region of the sky that the Chinese records describe, is certainly the remnant of the supernova they discovered. In a few hundred thousand years more, this gas will have become so diffuse as to be essentially invisible, its enriched gases having mixed with the interstellar medium.

In the center of the Crab Nebula, a rapidly spinning neutron star was discovered, rotating 30 times a second. When a star collapses, it retains its angular momentum and begins spinning more rapidly, like an ice skater pulling in her arms. Its magnetic fields become compressed and intensified as well. The magnetic field at the surface of the Crab Nebula's neutron star is about 10^{12} times stronger than the magnetic field at the surface of Earth. As the neutron star rotates, its north and south magnetic poles swing around, and the neutron star emits radio waves in two beams like a lighthouse. Every time the lighthouse beam swings past Earth, we see a pulse of radio radiation. Thus, the neutron star is called a *radio pulsar*. The first radio pulsar was discovered by graduate student Jocelyn Bell in 1967. It had a rotation period of 1.33 seconds. Her thesis advisor Antony Hewish was awarded the Nobel Prize in Physics for the discovery. I find it outrageous that she did not share in the prize.

The Crab Nebula pulsar emits electromagnetic radiation over the full electromagnetic spectrum, from radio wavelengths all the way to gamma-ray wavelengths. The pulsar can be seen blinking rapidly 60 times a second (as each of the two lighthouse beams sweep by us) in visible light as well, but astronomers never noticed that until after its radio pulses were discovered. It just looked like a faint star at the center of the Crab Nebula. The Crab Nebula is about 6,500 light-years away, which means that the explosion really occurred about 5445 BC, but it took the light until 1054 AD to reach us.

Recall the inverse-square law. The nearest star system is Alpha Centauri, 4 light-years away. The Crab Nebula is very much farther away, yet the supernova was much brighter than any star we see in the night sky, being easily visible in the daytime. At its peak luminosity, it was about 2.5 billion times as luminous as the Sun.

Supernovae are rare. The last time a supernova is known to have gone off in the Milky Way was about 400 years ago, before Galileo first pointed a telescope at the heavens. Thus, in 1987, astronomers were particularly excited when they saw a supernova explode in the Large Magellanic Cloud, a small satellite galaxy of the Milky Way. This was the closest supernova to go off in modern history. It was bright enough to be seen with the naked eye, even though it was 150,000 light-years away. I was lucky enough to travel to Chile to use telescopes there for my PhD research in May 1987; it was very exciting (and quite easy) for me to see this "new" star in the Large Magellanic Cloud.

9

WHY PLUTO IS NOT
A PLANET

NEIL DeGRASSE TYSON

Here's the story of how Pluto lost its planetary status and was demoted to an ice ball in the outer solar system. It's also about my role in this at the Rose Center for Earth and Space at the American Museum of Natural History.

In building the Rose Center, we decided to create a facility that would do more than show pretty pictures of the cosmos—you can get those on the internet. We created an 87-foot-diameter sphere in a glass cube, in which the architecture and the exhibits combine to make you feel like you're part of the universe—that you're walking through the universe. Our sphere is whole. Most planetariums have just a dome, with the planetarium sky projector housed within, and corridors surrounding it, displaying pictures of the universe. That's the way most planetariums in the country are designed. Pretty pictures are beautiful, but we thought it was time to learn something more about *how* the universe works, so instead we assembled the deepest concepts in the cosmos and made them our exhibits.

In our collaboration with the architects, Jim Polshek and partners, and the exhibit designer, Ralph Appelbaum and associates (perhaps best known for the Holocaust Museum, in Washington, DC), we proceeded. The universe loves spheres. You gain significant doses of insight into how the universe works by recognizing that the laws of physics conspire to make things round, from stars to planets to atoms. And in most cases where things are not round, something interesting is going on to prevent it, such as the object is rotating quickly. If we begin with an architectural structure that is round, we can put it to work as an exhibit element, allowing us to compare the sizes of things in the universe. By taking the dome that houses the space theater of the Hayden Planetarium in its upper half and completing it to form a sphere, we also gained a whole new exhibit space in the sphere's belly. That became the Big Bang theater, where visitors could look down and observe a simulation of the beginning of the universe.

Around the 87-foot-diameter sphere we built a walkway where we invite you to envision the "Scales of the Universe." As you start out, imagine first that the planetarium sphere is the entire observable universe. On the railing sits a model, about 4 inches across, showing the extent of our supercluster, containing thousands of galaxies, including the Milky Way. You realize that the universe is much, much bigger than our piece of it, the piece for which we have a word, a line in an address: the Virgo Supercluster. Then you take a few more steps, and we ask you to change scales: to imagine that the planetarium sphere now represents the Virgo Supercluster—87 feet across. On the railing, you next see a model about 2 feet across, which includes the Milky Way, the Andromeda galaxy, and some satellite galaxies—that's our Local Group of galaxies. Next, the planetarium sphere becomes the extent of the Local Group, and on the railing we have a model of the Milky Way, a couple of feet across, looking like a large fried egg—flat, with a bulge in the center. Take a few more steps, and the Milky Way itself becomes the planetarium sphere, where on the railing, we have a Plexiglas sphere a couple of inches across with a hundred thousand specks in it, representing a globular cluster of stars in the Milky Way. Continuing, that model of a globular cluster then becomes the planetarium sphere, and on the railing, we have a sphere about 6 inches across, showing the entire extent of the sphere of comets that surrounds our solar system: the Oort Cloud.

These countless comets from the Oort Cloud, raining down on the inner solar system, are the most dangerous class of objects to hit Earth. Each one packs extreme kinetic energy as it comes in from the outer solar system, gaining speed as it descends toward the Sun. The last time that comet coming in from the Oort Cloud visited the inner solar system was probably more than 40,000 years ago, so we don't have any historical information for it. If one of the comets coming in is headed for us, we don't have much time to do anything about it. When a normal asteroid swings around, we can typically predict its trajectory a hundred orbits in advance. We can chart the asteroid's orbit, along with Earth's orbit, and determine whether a hundred orbits later, it will collide with us or not. That might give us a hundred years to prepare a space mission to deflect it. But if a comet comes in from beyond the orbit of Neptune, and it's headed straight for us, we will have little advance warning.[1]

At a subsequent stop along the "Scales of the Universe" walkway, the big sphere represents the Sun with models of the planets placed next to it, in correct size relative to the Sun as the big sphere. This exercise continues, comparing smaller and smaller scales, until we reach the center of the atom. When the planetarium sphere is the hydrogen atom, we show a dot the size of its nucleus—1/130 of an inch across, revealing that most of the atom's volume is empty space.

The planetarium sphere has become a potent tool to explore the relative sizes of things in the universe.

Today, the Rose Center is a gorgeous place at night (figure 9.1). On the left you can see the walkway where you would stand to compare the big sphere,

FIGURE 9.1. The Rose Center for Earth and Space at night. At night the 87-foot-diameter sphere is bathed in blue light and can be seen inside its glass cube. Models of Jupiter and Saturn, shown to scale, can be seen hanging near the big sphere, which stands in for the Sun. It was the lack of a model for Pluto in this section of the exhibit that started all the controversy. *Photo credit:* Alfredo Gracombe

when representing the Sun, to the sizes of the planets. In the picture you can spot Saturn (with its rings) and Jupiter, next to each other, and of course, Uranus and Neptune are there too. As for Mercury, Venus, Earth, and Mars—they are too small to see in the picture. Their models, ranging from baseball to grapefruit size, are displayed down on the walkway railing, not suspended from the ceiling on cords. That's where all the trouble with Pluto started. We did not include a scale model of Pluto on that railing next to Mercury, Venus, Earth, and Mars. And we had good reasons.

We became the center of a controversy that we did not start. A reporter visited a year after we opened our exhibit, noticed that Pluto was missing from the display of the relative sizes of planets and decided to make a big deal of it, writing a front-page story about it in the *New York Times*, and that's when all hell broke loose. Here's the background of what we did and why we did it.

The story of Pluto starts with Percival Lowell, a very tweedy gentleman from New England. He liked astronomy, and he was wealthy, so he built his own observatory, called, as you might suspect, the Lowell Observatory. It is located in Arizona at an altitude of 7,250 feet. It's still there, on a site called "Mars Hill." Lowell was a Mars fanatic; he loved Mars so much and wanted so badly for it to harbor life that he wrote three books on the subject. It's one thing to write books on the *possibility* of life on Mars, but according to him, and him alone, he looked through his eyepiece and actually claimed to see evidence of life on Mars. He saw seasonally changing vegetation and canals and where the canals crossed, he thought there were oases. As far as he was concerned, the Martians were running out of water, because the canals he saw connected the poles to the vegetation areas. Mars has polar ice caps. He imagined them melting the ice and channeling the water to all the places it was needed. Without this massive public works project, Martian life would run out of water, dooming it to destruction. The human mind is powerfully imaginative; that's why we have the scientific method to check our hypotheses. During the close approach of Mars to Earth in 1877, Giovanni Schiaparelli had seen lines or channels on Mars, which he called *canali*, a word easily mistranslated into English as "canals." Channels are natural formations in a planetary landscape. Canals are made by an intelligent civilization. The words mean two different things. But it was too late. Lowell took up the idea, drawing an elaborate system of canals. Eventually, when others failed to see them with their telescopes, it was realized that they could be the result of an optical illusion, wherein the eye connects up random features to create lines. Modern photographs show no network of canals. The "vegetation" areas turned out to be dark areas of basalt-type rock, which appeared green in contrast with Mars' red deserts, and which were seasonally covered and uncovered by wind-blown dust.

In addition to his interest in Mars, Percival Lowell initiated the search for *Planet X*. At the turn of the century, there were eight known planets: Mercury, Venus, Earth, Mars, Jupiter, Saturn, Uranus, and Neptune. Turns out, Newton's laws accounted beautifully for all the planetary motions in the solar system except Neptune's. Maybe there was an unknown and undetected source of gravity out there affecting its path—a planet yet to be discovered. Lowell was convinced that such a planet existed, and he called it Planet X. Clyde Tombaugh was hired to look for it: to launch a search near the *ecliptic*, the plane in which the known planets orbit the Sun. He was looking for an object that moved a little bit between one photo and another taken of the same region of the sky days or weeks later, indicating that the moving object was a distant planet in orbit around the Sun. Tombaugh used an instrument called a *blink comparator*—an important instrument in the history of astronomy, although we make these comparisons by computer today. One photograph is mounted to one side of the instrument. A second photograph mounts to the other side, and it has a single viewer with two lenses, which observes the two images lit in rapid sequence. As the light flashes back and forth, the viewer's brain fuses the two images into one image, except for the object whose position shifts back and forth from one image to another. Anything that moves stands out quite readily, and it's by that method that Clyde Tombaugh discovered Pluto in 1930.

Pluto was named by an 11-year-old girl, Venetia Burney, who had just been studying Roman mythology in school. Planets were named after Roman gods, and Pluto was the god of the underworld. The official symbol for Pluto combines a *P* and an *L*, which coincidentally are the initials of Percival Lowell. Almost a half century later, a moon of Pluto was discovered. The first photographic evidence, obtained in 1978, was just a little bump on Pluto's blob-like image on the photograph. Years later, as angles of view to the Pluto system became favorable, we could detect eclipses and transits by the resulting diminutions in the light of the image when Pluto and its moon passed in front of each other as seen from our line of sight as they orbited. When we obtained higher-resolution images, from the Hubble Space Telescope, were we able to get direct images of Pluto's moon—named Charon, after the ferryboat driver that carries souls across the river Styx into Hades. Pluto has a moon—that's good. If you want to be in the planet club, that's a good start. No problem, we thought.

But there was a problem. First of all, when Pluto was discovered, we thought we had found the missing Planet X perturbing Neptune. To do this, Planet X must be massive, not insignificant compared to Neptune or Uranus. However, the more data we got on Pluto, and the better our measurements became, the smaller its dimensions and mass were revealed to be. Decade by decade: estimates of Pluto's size got smaller and smaller. Only after Charon was discovered could we measure Pluto's mass accurately by its gravitational attraction on Charon. The result? Pluto's mass is a mere 1/500 that of Earth, tiny relative to what would be needed to perceptibly perturb Neptune's orbit. We could no longer appeal to Pluto to explain Neptune. What was perturbing Neptune? Was there yet another Planet X? So people kept looking. They kept looking, until 1992, when a chap named Myles Standish, the twelfth direct descendant of another Myles Standish (one of the original pilgrims), analyzed the historical data, indicating that Neptune had a variant orbit. The modern Myles Standish is an astrophysicist at the Jet Propulsion Labs in Pasadena, California. He used better estimates for the masses of the planets Jupiter, Saturn, Uranus, and Neptune from the Voyager flybys of the 1980s and excluded one set of suspicious data from the U.S. Naval Observatory taken between 1895 and 1905. After doing so, he determined that the Neptune's orbit matched precisely with what Newton's laws predict, without requiring any mysterious gravity in addition to that exerted by the previously known objects. Planet X was dead and buried overnight.

So what do we do about Pluto? Pluto is the smallest planet, by far. There are seven *moons* in the solar system bigger than Pluto, including Earth's moon. Pluto is the only planet whose orbit crosses the orbit of another planet, because its path is so elliptical. Pluto is made mostly of ice—it's 55% ice by volume. We have a word for icy things in the solar system. They might have been called "ice balls," but they were named before we knew they were made of ice: comets. People back then tended to be poetically descriptive of cosmic objects; describing these things as "hairy objects in the sky," because, if you have long flowing hair, and you're running, your hair will naturally flow backward. They called these things "hair," which in Greek translates to "comet." Comets. That's the other word we already have for icy bodies orbiting the Sun. Pluto has a lot of features in common with comets. But it was alone out there. It wasn't zooming in close to the Sun and then swinging back out, as most comets do. When an icy comet comes close to the Sun, the comet outgasses vapor, producing a long tail. Pluto never gets that close to the Sun, so it doesn't do that. Despite its atypical features, people were happy to keep Pluto within our definition of planet.

In the Rose Center, however, we wanted to future-proof our exhibits as much as possible. So trend lines in planetary exploration mattered to us greatly. Pluto is more different from Mercury, Venus, Earth, and Mars than any of them are from one another. Mercury, Venus, Earth, and Mars are all small and rocky (figure 9.2.) That's one family.

Mercury, the planet, nearest the Sun, has a large iron core, only a trace atmosphere, and a cratered surface. Venus is cloud covered. In figure 9.2, we have removed the cloud cover to show the surface features, dramatic mountain ranges, and a few craters. Venus has a thick atmosphere of carbon dioxide (CO_2), an enormous greenhouse effect, and an intolerably high surface temperature. Mars is smaller than Earth or Venus but larger than Mercury. It retains a thin atmosphere of CO_2 that produces very little greenhouse effect. This coupled with its larger distance from the Sun makes Mars considerably colder than Earth. The atmospheric pressure on the surface of Mars is about 1/100 that of Earth. The dark areas in the picture are areas of darker basalt rock

FIGURE 9.2. Terrestrial/rocky planets to scale (with Earth's moon for comparison). We show Venus here without its cloud-covered atmosphere, so that you can see its surface features as revealed by radar imaging from the Magellan spacecraft. *Photo credit:* Adapted from J. Richard Gott, Robert J. Vanderbei (*Sizing Up the Universe*, National Geographic, 2011)

not covered by sand. The red areas, making Mars the "red planet," are sandy deserts. Mars has a large, long rift valley that could span the United States from coast to coast. It has an extinct volcano, Olympus Mons, that is 70,000 feet high. Mars has two polar caps composed mostly of water ice, with a frosting of dry ice (frozen CO_2) on top. Mars is the most habitable of the planets other than Earth.

What else is out there? We've got Jupiter, Saturn, Uranus, and Neptune. They are all big and gaseous (figure 9.3). That's another family. Once again they have more in common with one another than any one of them has with Pluto.

Jupiter orbits beyond Mars. It is composed mostly of hydrogen and helium. It's outer atmosphere contains methane and ammonia clouds. The bands on Jupiter are cloud belts, and the Great Red Spot, which can easily be seen in the picture, is a storm that has raged for more than 300 years. Saturn is similar to Jupiter but is surrounded by a magnificent set of rings. These rings are composed of icy particles that orbit the planet. Uranus and Neptune are smaller versions. Uranus has thin rings (as does Jupiter, although our picture doesn't show Jupiter's). In 1989, the Voyager 2 spacecraft found that Neptune also had a storm, a Great Dark Spot, shown in the picture, with winds just outside it reaching 1,500 miles per hour. Observations 5 years later, by the Hubble Space Telescope revealed that the Great Dark Spot had vanished.

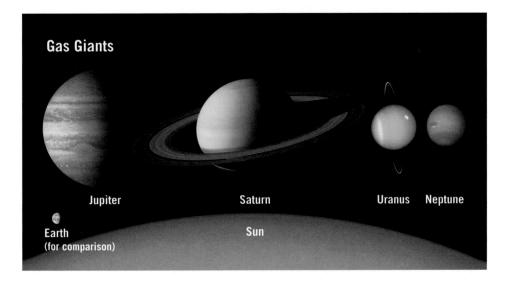

FIGURE 9.3. Gas giant planets to scale (with Earth and Sun for comparison).

Photo credit: Adapted from J. Richard Gott, Robert J. Vanderbei (*Sizing Up the Universe*, National Geographic, 2011)

The terrestrial planets form in the inner solar system, where, warmed by the Sun, light elements such as hydrogen and helium are heated to high enough temperatures that they can escape the planet's gravity. The gas giant planets, formed in the outer solar system, are colder and can retain hydrogen and helium, growing very massive. The terrestrial planets and the gas giant planets form two families. See table 9.1 for their properties.

Pluto doesn't fit in. Over these past decades, we've just been kind to Pluto, keeping it in the family of planets, even though we knew in our hearts that it didn't fit anywhere. A look at textbooks from the late 1970s (when we finally settled on Pluto's size and mass) and the 1980s shows Pluto was beginning to get lumped together with the comets, the asteroids, and other solar system "debris." Those were the first seeds of the unraveling of Pluto's full-blooded planetary status.

Pluto's orbit also has a few problems. First, as already noted, it crosses the orbit of Neptune. That's no kind of behavior for a planet. No excuse. Second, its orbit is significantly tilted relative to the plane of all the other planets. That's embarrassing too. It has orbital properties that just don't belong with the other planets. Then in 1992, in one of those blink comparison pictures, we found another object in the outer solar system whose position shifts over time, another icy body orbiting in the solar system beyond Neptune. Since then, we have discovered more than a thousand of these objects. What are their orbits like? They're all beyond Neptune, and many have orbital tilts and eccentricities that resemble Pluto's orbit. (*Eccentricity* measures how noncircular an elliptical orbit is.) These newly discovered icy bodies constitute a whole new swath of real estate in our solar system. Since they are all small, icy bodies, as predicted by Gerard Kuiper, we call it the *Kuiper Belt*. Pluto's orbit visits the inner edge of that Kuiper Belt, as do most of these other icy bodies. Pluto's existence now makes sense. It has brethren. It has a home. Pluto is a Kuiper Belt object.

Given that Pluto was the biggest known Kuiper Belt object, doesn't it make sense that the first object of a species you are going to find would be the biggest and the brightest? Ceres, which was the first asteroid discovered, is still the biggest known asteroid. The Pluto supporters claimed at first that Pluto is so big it can't be a Kuiper Belt object. But it's out there with them, it's made of the same stuff, and it has similar orbital properties. We look out into the Kuiper Belt and plot the average distance from the Sun of each object, versus eccentricity, and we find a cluster of Kuiper Belt objects with a 3:2 period

TABLE 9.1. PLANETS IN THE SOLAR SYSTEM

	TERRESTRIAL/ROCKY PLANETS				GAS GIANTS			
	MERCURY	VENUS	EARTH	MARS	JUPITER	SATURN	URANUS	NEPTUNE
Semi-major axis (AU)	0.39	0.72	1.00	1.52	5.20	9.55	19.2	30.1
Period of orbit (years)	0.24	0.62	1.00	1.88	11.9	29.5	84.0	165
Diameter/D_{Earth}	0.38	0.95	1.00	0.53	11.4	9.0	3.96	3.86
Mass/M_{Earth}	0.055	0.82	1.00	0.11	318	95.2	14.5	17.1
Principal elements	Fe, Si, O	(Fe, Si, O)?	Fe, Si, O, Mg	Fe, Ni, S, Si, O	H, He	H, He	H, He, CH_4	H, He, CH_4
Atmospheric composition	Trace O, Si, H, He	Thick CO_2, N_2	O_2, N_2	Thin CO_2	H_2, He	H_2, He	H_2, He, CH_4	H_2, He, CH_4
Temperature (°F)	−270 to +800	+820 to +860	−128 to +134	−220 to +95	−256	−310	−364	−368

Note: Temperatures (in degrees Fahrenheit) are at the surface for the rocky planets (giving the full observed range), and are near the top of the atmosphere for the gas giants.

resonance with Neptune; that's an orbital match in which, for every three orbits Neptune makes, the Kuiper Belt object makes two orbits—exactly like Pluto. Kuiper Belt objects that share this pattern are called *plutinos*. They are even more similar to Pluto within the Kuiper belt than they are to the rest of the Kuiper Belt objects.

So back at the Rose Center, we simply grouped Pluto with an exhibit on the Kuiper Belt. Didn't even say it wasn't a planet. Physical properties mattered more to our design than labels.

And so it was—for a year, until the fateful *New York Times* article of January 21, 2001, titled "Pluto's Not a Planet? Only in New York," written by science journalist Kenneth Chang:

As she walked past a display of planets at the Rose Center, Pamela Curtice, of Atlanta, scrunched her brow, perplexed. There didn't seem to be enough planets. She started counting on her fingers, trying to remember the mnemonic her son had learned in school years ago. My Very Educated Mother Just Served Us Nine Pizzas. Mercury, Venus, Earth, Mars, Jupiter, Saturn, Uranus, Neptune. "I had to go through the whole thing to figure out which one was missing," she said. Pluto. Pluto was not there. "Now I know my mother just served us nine," Mrs. Curtice said. "Nine nothings." Quietly, and apparently uniquely among major scientific institutions, the American Museum of Natural History cast Pluto out of the pantheon of planets. . . . "We are not that confrontational about it," said Dr. Neil deGrasse Tyson, director of the museum's Hayden Planetarium. "You actually have to pay attention to make note of this."

I was trying to be diplomatic. We didn't say, "there are only eight planets," or "we kicked Pluto out of the solar system," or "Pluto is not big enough to make it in New York." No. We simply organized the information differently—that's all we did. And the *New York Times* was making a federal case out of it.

The article continued: "Still, the move is surprising, because the museum appears to have unilaterally demoted Pluto, reassigning it as one of more than 300 icy bodies orbiting beyond Neptune, in a region called the Kuiper Belt (pronounced KY-per)."

Pluto orbits with these other similar icy bodies. That's where it lives. We learned this in the 1990s as many new icy bodies like Pluto were discovered, giving us new information and further insight into how the solar system is constructed.

The article quotes a colleague of mine, Dr. Richard Binzel, a professor at the Massachusetts Institute of Technology—we were in graduate school together—and he was upset, because he had devoted part of his career to studying Pluto. Binzel said in the article: "They went too far in demoting Pluto, way beyond what the mainstream astronomers think." Then the head of the planetary division of the American Astronomical Society, Dr. Mark Sykes, called the *New York Times* and said he was going to be in New York and planned to debate me, and invited them to come. They agreed. So they sent another reporter and another photographer to capture and record a private debate that Stykes and I had in my office, and it's quoted verbatim in the February 13, 2001, article that ensued. Meanwhile, with the photographer following along, we went on the scaling walk near the suspended gas giant planets, and Dr. Sykes reached over to me, playfully grabbing my neck. The picture caption said: "Dr. Mark Sykes challenges Dr. Tyson to explain the treatment of Pluto in the planet display in the Hayden Planetarium."

It hit the internet, wired news, boston.com. "Center puts Pluto's planetary status in doubt." It was all the buzz. And I spent three months of my life just fielding media inquiries, getting nothing else done. Let's look at some of the comments from a chat room online.

"Pluto is a true-blue American planet, discovered by an American." This was from a NASA scientist. Someone else in the chat room said, "Such romanticism has no place in science, a system which must never cease trying to determine the objective truth. Neither does nationalism." Here is another one on our side: "I confess I am disappointed in the learned community that joins with astrologers in holding onto an outdated classification scheme." You want to get an astronomer angry, call him an astrologer. Those are fighting words.

This next person couldn't commit: "My personal view is that Pluto should probably have dual citizenship." This was from the president of the International Astronomical Union's planet-naming commission—he didn't want to upset anybody. Want more? "I do not agree with the dual status because it complicates matters too much in the public perception." That was none other than David Levy. Patron saint of comet hunters. More than 20 comets he discovered are named after him. Even the famous comet that hit Jupiter in 1994 was co-discovered by him, comet Shoemaker–Levy 9. David Levy was worried that if we did something that confused the public, that would be bad. My thoughts were that there's a lot of our research that is confusing, but we shouldn't reshape our science just to avoid confusing the public. Another person said: "First, it's amazing that Tyson, an astrophysicist, would even venture into such waters. I feel, as a planetary geologist, equally qualified to demote the Magellanic Clouds from its current status, as a satellite galaxy of the Milky Way, to just a star cluster . . . so in that spirit, I think he's full of baloney." This was someone else from NASA.

Here's another comment: "It's not too hard to imagine the same type of people back in Galileo's age, saying, 'I've been taught that the Earth is the center of the universe since I was a child. Why change it? I like things the way they are.'"

As a scientist, you must embrace the inconstancy of knowledge. You learn to love the questions themselves. Charon, Pluto's moon, is over half the diameter of Pluto. You can easily argue that Pluto is not a planet with a moon, but more like a double planet. In fact, their center of mass is not even within Pluto, but falls in the space between them. Just to be clear, in contrast, the center of mass between Earth and the Moon lies within Earth, about a thousand miles beneath Earth's crust. It's not that we are stationary and the Moon goes around us. We both go around our mutual center of mass; Earth simply jiggles a little, whereas the Moon circles it in a wide orbit. Pluto is massive enough to be round. Charon is also big enough to be round. If you counted Pluto as a planet, Charon would qualify also, as would many other small, but large-enough-to-be-round objects.

Walt Disney's cartoon dog, Pluto, was first sketched in 1930, the same year that Clyde Tombaugh discovered the cosmic object. They have the same age in the American psyche. Disney is a major force in our culture, so I believe that had we demoted Mercury, no one would have cared. But we demoted Pluto. Who is Pluto? Pluto is Mickey Mouse's dog. That's important stuff to us in America. That's our culture. By the way, why is Pluto Mickey's dog, but Mickey is not Pluto's mouse? Ever thought about that? I would come to learn of the Disney pantheon that if you wear clothes, you can own other animals who do not. Goofy is a dog but he wears clothes and can talk, so he is not anyone's pet. Mickey Mouse wears pants. Pluto is naked, except for his collar, doesn't usually talk, and therefore can be owned by a mouse. It's a Disney world.

Let me complete the arguments. I own lots of books, some of which go back for centuries, tracking the evolution of our thought concerning our place in the cosmos. One is from 1802. Know what happened in 1801? Reflecting on the big gap between the orbit of Mars and Jupiter in our solar system, people felt there ought to be a planet out there. That gap is too big not to have a planet. After some effort, Italian astronomer Giuseppe Piazzi, in 1801, found a planet in that gap. They named it Ceres, after the Roman goddess of the harvest. Indeed, the word *cereal* derives from Ceres. Everyone was excited, because a new planet had been discovered. Have you heard of that planet? No. One book from that time has the orbits of the planets in it: Mercury, Venus, Earth, Mars, Ceres, Jupiter, Saturn, and planet Herschel (not yet renamed Uranus). Ceres is on the list.

In 1781, when William Herschel discovered what later became known as Uranus, it was a puzzle as to what to name it, because no one since ancient times had ever discovered a new planet. (In his book on Herschel, Michael Lemonick has argued that Copernicus could also be credited with discovering a new planet—Earth—for he showed that Earth was indeed a planet.) Being a good English subject, Herschel sought to name his new planet after King George III. So he called it *Georgium Sidus* (George's star). That's the same King George to whom the U.S. Declaration of Independence is addressed. King George honored Herschel by giving him a stipend of £200 a year if he would give stargazing parties with his telescopes for the king's guests at Windsor Palace. Then the list of planets would have read Mercury, Venus, Earth, Mars, Jupiter, Saturn, and George.

Fortunately, clearer heads prevailed, and they looked for a suitable Roman god to replace George as the planet's name. Johann Bode suggested "Uranus," after Ouranos, the Greek god of the sky, and that name stuck. Martin Klaproth, a German chemist, was so excited by this he named his newly discovered element "uranium" after the new planet. By the usual scheme, planets are named after Roman gods, whereas satellites are named after Greek characters in the life of the Greek counterpart of that Roman god. Take Jupiter—its biggest moons are Io, Europa, Ganymede, and Callisto. In Greek legend they are characters in the life of Zeus, the Greek counterpart to Jupiter. By this scheme, the names pay homage to both Roman and Greek mythology. For Uranus, however, to mollify the Brits, after dissing the king, we broke tradition and named all the moons of Uranus for fictional characters in English literature. Nearly all of them come from Shakespeare. One of them, Miranda, I chose for the name of my daughter, except at the time I knew of the name only from the moons of Uranus. When I told my wife, "I like this name 'Miranda,'" she said, "Oh, you mean the heroine in Shakespeare's *The Tempest*." I said, "Ah, yeahh . . . that's what I was thinking too."

Back to this planet Ceres. Let's go to another book—30 years later. *The Elements in the Theory of Astronomy*—an advanced textbook. Math intensive. Now it lists ten known planets, Mercury, Venus, Mars, Vesta, Juno, Ceres, Pallas, Jupiter, Saturn, and Uranus, which are denoted by their symbols (♀ for Venus, ⊕ for Earth, ♂ for Mars, etc.). Neptune wasn't discovered yet. Four new planets had cropped up, all needing their own new symbols, making a total of ten planets. What's the problem? The word "planet" was not formally defined. The last time the word had an unambiguous definition was in ancient Greece—"planet" is Greek for "wanderer." You look up at the night sky, and if an object moves against the background of stars, it is a planet. What do we see moving against the background sky of stars? Mercury, Venus, Mars, Jupiter, Saturn—and two more—the Moon and the Sun. The seven planets of the universe. That is an unambiguous definition. But Copernicus put the Sun in the middle and described Earth as going around the Sun—is the Sun still a planet? What about Earth? Is Earth a planet? Planets then became things that went around the Sun. Comets also go around the Sun but were fuzzy and had tails ("hair"), so we didn't call them planets. But that decision was arbitrary. When we found these new non-comet-like objects between Mars and Jupiter—Vesta, Juno, Ceres, and Pallas—we called them planets too. A few years later, we had found 70 more of these things. And you know what we found? They had more in common with one another than any one of them had with anything else in the solar system, and they were all orbiting in the same belt. We hadn't discovered new planets. We had discovered a new swath of real estate in the solar system occupied by a new species of objects. Today we call them *asteroids*, a name invented by William Herschel. He found that they were tiny relative to the established planets and argued that they constituted a new class of objects. Things that started out being called "planet" were later reclassified with a new name, and more importantly, we learned something new about the structure of the solar system. Our knowledge base broadened, and our understanding advanced. That all happened about 10 years after *The Elements in the Theory of Astronomy* was published. Plus, I bet they ran out of ways to make new symbols.

Pluto is about 1/5 of the diameter of Earth—it's small, like the other Kuiper Belt objects (figure 9.4.) We show to scale (in comparison with Earth) the rest of the objects in the solar system (other than the Sun and planets) which are larger than 254 km in diameter. Earth's moon is shown, as well as the large moons of other planets. The four largest moons of Jupiter (discovered by Galileo when he first turned his telescope to the sky) are shown. Ganymede, Jupiter's largest moon, is slightly larger than the planet Mercury, but less than half as massive. Io and Europa are jostled by the other moons gravitationally and are heated by kneading due to Jupiter's tides. Io is covered with active volcanoes. Europa has an 80 km deep water ocean beneath a 10 km deep crust of ice. There is more water in the oceans of Europa than in all the oceans of Earth. Saturn's small moon Enceladus, for similar reasons, has a southern ocean below an icecap, and spectacular water geysers. Saturn's largest moon, Titan, has methane lakes and a mostly nitrogen atmosphere. It rains methane on Titan, and there are frozen methane riverbeds. The dark features are frozen methane-ethane regions, while the white areas are frozen water ice. Neptune's large icy moon Triton has spectacular geysers (perhaps gushing nitrogen). Triton orbits Neptune in a backward direction and may be a captured Kuiper belt object. The largest asteroids and the largest Kuiper Belt objects known as of 2010 are also shown in the figure. Ceres, the largest asteroid, was the first discovered. Vesta is next largest; it is rich in iron and may have had its surface blasted off by an ancient collision with another asteroid. The asteroids are all rocky bodies. The Kuiper belt objects are icy bodies. Pluto and Charon are shown as they were thought to appear in 2010, deduced from mapping their brightness fluctuations as they eclipsed each other. Eris is shown slightly bigger than Pluto, as thought at the time, but improved measurements in 2015 have since shown Eris (diameter: 2,326 ± 12 km) to be slightly smaller than Pluto (diameter: 2374 ± 8 km). The Kuiper belt objects are all smaller than our moon.

I've been giving you some of the background on the science of Pluto, but now let's get back to the story. What happened next? Letters from the public came pouring in. "If Pluto remains a planet, it will cost the museum money to build a model of it. People may complain that they have to buy new posters, but who cares? It will cost them three dollars." That was from a seventh grader. "What's wrong with Pluto? Is it because it is different? Is that why you don't consider it a planet? If you do, then that is racism." Racism?

Here's another theme. "Now the teachers are going to have to teach them that there are eight planets when last year's teacher taught them that there were nine. Students, young students are going to get confused, and I always remember my planets by saying My Very Educated Mother Just Served Us Nine Pizzas. That has always helped me and a lot of kids remember the planets. Now what are they going to teach them? And even though I am a kid I still know what's happening."

In the Rose Center, we don't count planets. Exam questions, such as "What's the fourth planet from the Sun?" There's no science in that. In the Rose Center, we didn't say Pluto is not a planet; we don't even emphasize the word "planet." What we say is, the solar system has families, and one of those families—the terrestrial planets (Mercury, Venus, Earth, Mars)—has properties in common that distinguish their members. The Asteroid Belt is another family—small rocky bodies. The gas giants make a family. The Kuiper Belt objects, including Pluto orbiting near their inner edge, all have similar properties. They constitute yet another family. Then we have a cloud of icy bodies that completely surrounds the Sun, the Oort Cloud of comets. We have divided objects orbiting the Sun into five families. That's our pedagogical paradigm. What matters is asking what properties objects have in common. A third grader can learn that the gas giants are big and low in density—it's an excuse to learn the word "density." These are big and gaseous, like beach balls. Saturn has lower density than water. If you took a piece of Saturn and put it in your bathtub, it would float.

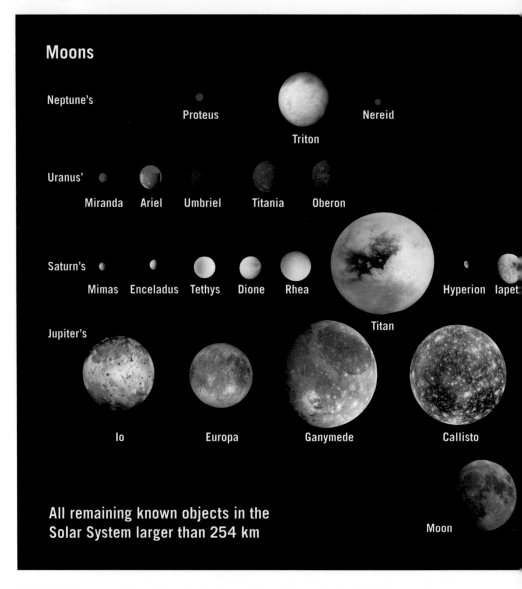

FIGURE 9.4. Solar system objects (other than the Sun and planets) larger than 254 km in diameter shown to scale (with Earth for comparison). *Photo credit:* Adapted from J. Richard Gott, Robert J. Vanderbei (*Sizing Up the Universe*, National Geographic, 2011)

I always wanted a Saturn toy rather than a rubber ducky when I was growing up—I thought that would be so cool.

Pluto, I think, is happier now, in the Kuiper belt where it belongs. To consider it a red-blooded planet overlooks its fundamental properties. If you were to move Pluto to where Earth is now, it would grow a tail just like a comet, and that is certainly no kind of behavior for a planet.

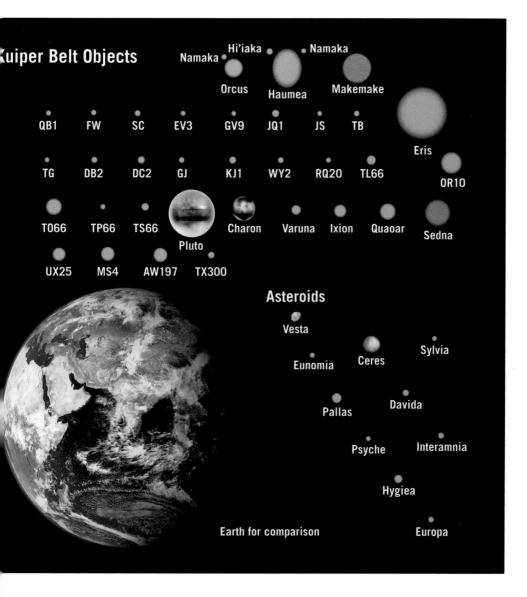

Before we make Pluto feel bad for its diminutive size, I offer you a humbling thought: Jupiter is bigger compared to Earth than Earth is compared to Pluto. (Compare figure 9.3 and figure 9.4.) Which means that if you polled people (or whatever kind of creatures) living on Jupiter—you went to some Jovians and asked, "How many planets are there in the solar system?"—what answer would they give you? Four. You'd say: "Oh, what about all those other planets? Earth, ..." The Jovians would say: "Those chunks of rock? That debris? Those vagabonds of the solar system?" So, my argument, our argument, for removing Pluto is not specifically a size-based argument. It's more an argument of physical and orbital properties.

In 2005. Mike Brown and his team at Caltech (the California Institute of Technology) discovered a Kuiper Belt object, named Eris, which was nearly identical in diameter to Pluto and 27% more massive (see Fig. 9.4). Eris has a small moon, Dysnomia, whose orbital parameters allow us to accurately estimate Eris's mass, and it is clearly more massive than Pluto. This brought the argument to a head. If Pluto was a planet, then surely Eris must be also. Either you had to demote Pluto or promote Eris. The International Astronomical Union (IAU), the official body for deciding such definitions, held a special session in a 2006 meeting to vote on Pluto's planetary status, as well as that of Eris and the other Kuiper Belt objects. The result? Pluto was demoted to dwarf planet from its previous planetary status. The story made news around the world. Textbook writers took note. To be a planet an object had to (1) orbit the Sun, (2) be massive enough for gravity to have made it assume a hydrostatic equilibrium (nearly round) shape, and (3) have cleared the neighborhood around its orbit of debris. Pluto failed the third criterion, as did Ceres—they are accompanied by other objects whose total mass is comparable with their own. Most astronomers, including Mike Brown, interpret "cleared the neighborhood" to mean that the planet must now dominate the mass in the neighborhood of its orbit. Jupiter, after all, is accompanied by more than 5,000 Trojan asteroids, clustered around stable *Lagrange points* either 60° ahead or 60° behind Jupiter in its orbit, but these asteroids in total are miniscule in mass in comparison with Jupiter itself. The IAU was not demoting Jupiter. The IAU affirmed that there are eight planets in the solar system: Mercury, Venus, Earth, Mars, Jupiter, Saturn, Uranus, and Neptune. My Very Excellent Mother Just Served Us Nachos. Pluto, Eris, and Ceres, meeting the first two criteria, were minor planets that could use the name "dwarf planet." The Rose Center was thus 6 years ahead of its time in demoting Pluto. I wrote a book, *The Pluto Files: The Rise and Fall of America's Favorite Planet* (2009), chronicling my experiences. Mike Brown has written a charming book on his discovery of Eris, titled *How I Killed Pluto and Why It Had It Coming* (2010). We have now discovered four smaller moons orbiting Pluto, in addition to Charon. Eris has one moon, and the Kuiper Belt object Haumea (now also designated by the IAU as a dwarf planet) has two. In 2006, NASA launched the New Horizons spacecraft toward Pluto. Some of Clyde Tombaugh's ashes are aboard. The spacecraft flew by Pluto and Charon in 2015, snapping the beautiful picture of them both shown in figure 9.5. A heart-shaped icy region is visible on Pluto, which has been tentatively named "Tombaugh Regio," and the pole of Charon has a dark area which has been unofficially named "Mordor" after the shadow lands in the *Lord of the Rings*. So all is well with Pluto.

FIGURE 9.5. Pluto and Charon taken by the *New Horizons* spacecraft during its 2015 flyby. *Photo credit:* NASA

10

THE SEARCH FOR LIFE
IN THE GALAXY

NEIL DeGRASSE TYSON

Because we are alive, we harbor a special interest in life in the universe. If we are to look around the universe and wonder whether some particular star has planets orbiting it, and whether those planets could have life, it's sensible to shape questions based on life as we know it—life on Earth. Living things all seem to have a set of properties in common. First, life as we know it requires liquid water. Second, life consumes energy. In chemical terms, we have a metabolism. And the fun part, number three, life has a way to reproduce itself. I'll focus on the first one, because that one has a chance of getting addressed by the methods and tools of astrophysics. All we need to do is to explore the universe in search of liquid water.

Ever since the story of Goldilocks, we've known (and agreed) that things can be too hot, too cold, or just right, where life is concerned. Take the Sun. We know it has a certain luminosity; the closer to the Sun you are, the hotter things get, and the farther away you are, the cooler things get. If life requires liquid water, and you take water and move too close to the Sun, the water evaporates. Too far away? It freezes. This leads us to conclude that an orbital swath exists—a zone exists—where a planet can sustain liquid water. Closer to the Sun we have steam, farther away we have ice, and in between we can have liquid water. People have a name for this, the *habitable zone*. This concept has pretty much dominated our paradigm for more than a half century, beginning in the 1960s, when it originated. Different stars, depending on their luminosities, have habitable zones of different sizes, which gave us something to think about. The astrophysicist Frank Drake took this concept a little further and constructed what we now know as the *Drake equation*. It is not so much an equation the way Newton's laws give us equations, but rather a way to organize our ignorance about the prevalence of intelligent life in the universe.

Before I show you the Drake equation, let's just say that, based on everything we know about life, we think life needs a planet. It's going to need a planet orbiting a star. You have to make the star, and then the planet, and then, remembering that life on Earth evolves slowly, you need billions of years of evolution to produce intelligent life. Therefore, the star has to be long lived. Not all stars live a long time. Some stars don't live as long as a billion years, or even as long as a hundred million years. Your most massive stars are dead after 10 million years or less—not much hope for intelligent life on a planet around those stars, if what happened on Earth is any indication. We need a star that is long lived and a planet, but not just any planet—a planet in the star's habitable zone.

So far, we know to look for a long-lived star, a planet in its habitable zone, and one having life, but not just any life—intelligent life. For most of Earth's history, potent microbes called "cyanobacteria," wreaked havoc on Earth's atmosphere. People today complain that humans are polluting the environment, creating the ozone holes, and adding greenhouse gases like CO_2. But our influence pales compared with the effect of cyanobacteria on Earth's atmosphere 3 billion years ago. Back then, Earth had a lot of carbon dioxide in its atmosphere, and it was happy. Then the cyanobacteria came along, ate the CO_2, and churned out oxygen, completely switching the chemical composition and balance of the atmosphere. Earth was left with an oxygen-rich atmosphere, and very little CO_2. Oxygen was actually toxic to many of the anaerobic organisms at the time. Carbon dioxide is a greenhouse gas, so with less of it, the greenhouse effect decreased, and Earth started to cool dramatically. If there had been an environmental movement back then, it might have protested, "Stop the oxygen production! This is bad for Earth"—because it was a change. Earth got colder and froze over completely several times. Meanwhile, the Sun was slowly but steadily getting more luminous as it evolved over billions of years, and the snowball Earth episodes ended. Ultimately, the oxygen in the atmosphere allowed diverse animal life to emerge, including humans. Not all changes are bad for all organisms.

We worry that the next asteroid is going to take us out. I'm telling you. It's going to happen. I don't know when, but it's going to happen, and it will be a bad day on Earth. Consider the last time when Earth got hit in a big way, 65 million years ago, when an asteroid took out the dinosaurs. There were our rodent-sized mammal ancestors, scurrying in the underbrush, just barely surviving, basically serving as hors d'oeuvres for *T. rex* and other scary predators. *T. rex* gets knocked out in the aftermath of the asteroid strike, enabling mammals to evolve into something more ambitious. These events set in motion a series of events that ultimately fostered the culture and society we now have, giving us life while simultaneously taking it away from the ferocious dinosaurs. So, I tend to take a more holistic view of change on Earth.

The implication of this story is that if you want converse with beings on a planet that might have life, it is not good enough for it to just have life. That life has got to be intelligent. Actually, you need more than that. Isaac Newton was intelligent, but you couldn't have a conversation with him across the galaxy. What he lacked in his day was some kind of technology enabling him to send signals across vast distances of space. The intelligent life we are looking for has to be technologically proficient at the epoch when we observe it. In other words, if it is 1,000 light-years away, it must have been transmitting signals across space exactly 1,000 years ago, for that signal to reach us just now. Now imagine that technology contains the seeds for its own undoing. Suppose technology in the hands of ignorant and irresponsible people enables you to destroy yourself more efficiently than any natural catastrophe. How long is that period before your blundering power causes your own extinction? That might be only a hundred years. Then, if you are looking around the galaxy, you must be lucky enough to see another planet during a hundred-year slice out of the 5-billion-year history of that planet in orbit around its star, which makes the probability of finding a cosmic pen-pal look really slender.

Frank Drake took all these arguments and wrote them into the Drake equation. This formed a starting point for the Search for Extraterrestrial Intelligence, known as SETI. He wanted to estimate the number of communicating civilizations in the galaxy we could be hearing from now: N_c. To get there, he introduced a series of fractions into one equation, with each term representing a discrete estimate based on modern astrophysics:

$$N_c = N_s \times f_{HP} \times f_L \times f_i \times f_c \times (L_c / \text{age of the galaxy}),$$

where

> N_c = number of communicating civilizations we could observe in the galaxy today;
>
> N_s = number of stars in the galaxy, ~300 billion;
>
> f_{HP} = fraction of stars suitable with a planet in the habitable zone, ~0.006;
>
> f_L = fraction of these planets where life develops, unknown, but perhaps near 1;
>
> f_i = fraction with life that develop intelligent life, unknown, but probably small;
>
> f_c = fraction with intelligent life that develop technology to communicate over interstellar distances, unknown, but perhaps near 1;
>
> L_c = average lifetime of communicating civilizations, unknown, but maybe small compared to the age of the galaxy; and
>
> age of the galaxy, ~10 billion years.

Let's start out with the number of stars in the Milky Way, about 300 billion. Because not every star in the galaxy would be suitable, you have to multiply by the fraction of stars that are long lived (long enough to develop intelligent life) and that also have a planet in the habitable zone (f_{HP}). That reduces the total number available to us, where we might look for intelligent life. As of the date of this publication, after heroic efforts surveying more than 150,000 stars, we have confirmed the existence of over 3,000 exoplanets. This has been quite a revolution.

Stars with planets turn out to be common, and many stars have multiple planets. Among stars with planets, we want to discover those with a happy planet in the habitable zone. We can find exoplanets by the gravitational tug they exert on their stars, which causes a wobble in the radial velocity of the star that we can observe. Planets closer in exert more of a tug, causing a larger radial velocity wobble of the star, which is easier to detect. It is thus relatively easy to find planets close to their stars, but such planets will be too hot to have liquid water—not ones we want for the Drake equation. The largest survey of exoplanets has been conducted by NASA's Kepler satellite (named, of course, after Johannes Kepler) which finds planets by measuring the small drop in the light of the star that occurs when the planet passes directly in front of the star in your line of sight. More generally we call this a *transit*. Jupiter has a radius about 10% that of the Sun. Its cross-sectional area (πr^2) is 1% that of the Sun.

So when a Jupiter-sized planet transits in front of its solar type star, it causes a temporary 1% drop in the light of the star. An Earth-sized planet, whose radius is 1% that of the Sun, will cause only a 0.01% drop in the light of a solar-type star. The Kepler satellite was designed to be sensitive enough in principle to detect such a diminution in the light of a star, since its main mission was to search for Earthlike planets, but that is close to its limits. Many Kepler detections are Jupiter or Neptune-sized planets (not suitable for life as we know it) but we also find many smaller planets down to sizes comparable with Earth. Figure 10.1 shows confirmed Kepler planets as dots: the vertical coordinate of a dot is the radius of the planet in terms of Earth's radius, and the horizontal

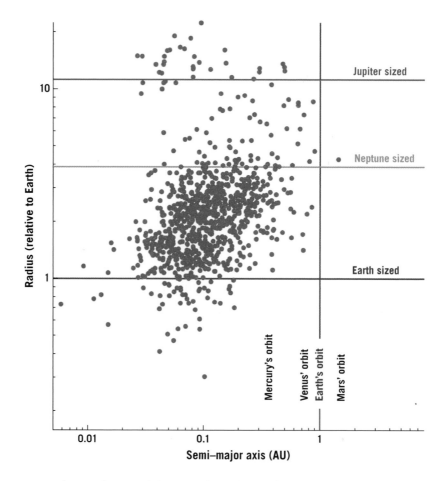

FIGURE 10.1. Exoplanets with measured planetary radii and distances from their star found by the Kepler satellite, as of February, 2016. More than 1,100 confirmed exoplanets are shown as dots, whose vertical position indicates their radius (in Earth radii), and whose horizontal position indicates the distance from their star (in Astronomical Units, AU). These exoplanets are discovered when they transit in front of their star, slightly diminishing its light. The blue crosshairs show the position Earth would occupy on this graph. *Credit:* Michael A. Strauss, NASA

coordinate is the radius of the planet's orbit in AU. Most of the Kepler planets orbit solar-type stars. The blue cross-hairs show where Earth would appear on this diagram. We are looking for exoplanets near this location in the diagram.

Transits are more likely to occur when the planet is close to its star. Thus, most of the Kepler planets discovered so far are too hot to support life. If the planet is far enough away to have a habitable temperature, then its orbit has to line up just right for us to see it transit, and because its orbital period is longer, it makes fewer transits, which lowers our chance of finding it. So far, the Kepler satellite has found only about ten confirmed exoplanets between 1 and 2 Earth diameters that are illuminated by an amount of radiation from its star within a factor of 4 of what Earth receives from the Sun. This number is low simply because these planets are harder to find using the transit technique.

One such promising candidate is the planet Kepler 62e (see figure 10.2 for an artist's conception). It is one of five planets orbiting a K star (named Kepler 62), located about 1,200 light-years from us. The star's surface temperature is 4,900 K. Planet Kepler 62e has a radius 1.61 times as large as Earth and receives only 20% more radiation per square meter from its star than Earth does from the Sun. It should be in the habitable zone. It may be either a rocky planet, or an icy planet with an ocean covering its surface. This multiplanet system is roughly 2.5 billion years older than our solar system.

What fraction of stars (f_{HP}) have a suitable planet in the habitable zone? G stars like the Sun make up nearly 8% of the stars in the Milky Way. We know they're okay for life, because the Sun is one of them. Stars much more luminous than the Sun exhaust their fuel too quickly to give their planets the time needed to evolve complex, intelligent life, something that required billions of years on Earth. Dimmer K stars and M stars are even longer lived than the Sun, so they fulfill this requirement nicely.

But main sequence M stars have such low luminosity that, to be in the habitable zone, the planet would have to huddle so close to the M star to keep warm that it would be tidally locked, with one face always pointing toward the star. Tidal forces are stronger close in. These tides force the planet into a slightly ellipsoidal shape, and its rotation is slowed till the ellipsoidal shape is locked, pointing in the direction of the parent star. (Our Moon is tidally locked in this way, with one face always pointing toward Earth, because of just this effect.) The planet probably doesn't care about this, but any life on its surface would:

FIGURE 10.2. Kepler 62e compared with Earth. Kepler 62e is on the right, and Earth is on the left. Kepler 62e is an artist's conception, but its relative size is correct. Its orbit appears to place it in the habitable zone, and thus it could have water oceans. *Photo credit:* PHL@UPRArecibo

the side of the planet constantly facing the M star would be too hot, while the other side would be too cold. An Earthlike atmosphere would freeze out on the cold side. Meanwhile the atmosphere from the hot side would expand to the cold side and freeze out too, in a runaway process. Eventually all of the atmosphere would end up frozen out on the cold side, ending chances for life. The only hope for life on the planet is to have a very thick atmosphere that circulates the air, reducing the extreme temperature variation from one side to the other. Such an atmosphere would have a very high pressure at its surface. Also, M stars have many more giant flares than do stars like the Sun, which could prove fatal to life. These things may not make life impossible, but they do make it more difficult for life to evolve.

For these reasons, G and K stars are the best candidates, and they make up a respectable 20% of all the stars in the Milky Way.

Given such a star, what is the chance of finding a planet in its habitable zone?

I'm now going to show you one of the most beautiful calculations in the cosmos, but perhaps you should judge that for yourself. I just want to show you how empowered you are with all the tools necessary for this calculation.

The Sun has a luminosity. Earth has a luminosity too; we have a temperature, and due to that temperature, Earth emits radiation, primarily in the infrared part of the spectrum—what is typically called *thermal radiation*. Since Earth has a temperature, it will radiate across the spectrum with a Planck curve corresponding to that temperature. The total luminosity of Earth is going to be the energy emitted per unit area times Earth's surface area. Let's take Earth's surface area first, $4\pi r_E^2$, and multiply it by the energy emitted per unit area for Earth, which is σT_E^4 (according to the Stefan–Boltzmann law for thermal radiation). Earth's luminosity is therefore $L_E = 4\pi r_E^2 \sigma T_E^4$. We can do the same for the Sun: $L_S = 4\pi r_S^2 \sigma T_S^4$. We now ask how much of that luminosity from the Sun is actually reaching Earth. Although Earth's temperature varies, it varies around a very steady average. In equilibrium, the energy that Earth receives from the Sun should be in balance with the energy emitted from Earth's surface. This must be true, otherwise Earth would rapidly get either hotter or colder with time, instead of maintaining the average that we observe. We've seen these equations before, but now I've got a new goal for them—to calculate the equilibrium temperature for Earth.

The solar luminosity, L_S, is *not* all hitting Earth. Rather than the total amount of energy emanating from the Sun in every direction, we care only about the energy that is going to hit Earth. All this energy from the Sun ultimately crosses through a spherical surface equal in radius to Earth's orbit (1 AU). We need to find out what fraction of this entire spherical surface Earth blocks. The part that matters for Earth—the part intercepted by Earth—equals Earth's cross-section.

The fraction of the Sun's radiation hitting Earth is, therefore, the area of that circular cross-section of Earth, πr_E^2, divided by the area of the big sphere of radius 1 AU through which all the Sun's radiation passes: $4\pi(1\,\text{AU})^2$. That fraction is $\pi r_E^2 / 4\pi(1\,\text{AU})^2$. The total solar luminosity hitting Earth is therefore $L_S \pi r_E^2 / 4\pi(1\,\text{AU})^2$ or, after substituting, using our formula for the luminosity of the Sun, $4\pi r_S^2 \sigma T_S^4 \,\pi r_E^2 / 4\pi(1\,\text{AU})^2$. If we are in equilibrium, I am allowed to set this equal to the luminosity Earth is giving off, $4\pi r_E^2 \sigma T_E^4$. Let's set them equal to one another: $4\pi r_S^2 \sigma T_S^4 \,\pi r_E^2 / 4\pi(1\,\text{AU})^2 = 4\pi r_E^2 \sigma T_E^4$. On the left side, there is a $4\pi/4\pi$, which cancels out. The πr_E^2 appearing on both sides of the equation also cancels out, and finally, the σ on both sides of the equation cancels out, which reduces to: $r_S^2 T_S^4 / (1\,\text{AU})^2 = 4 T_E^4$.

We can now calculate Earth's equilibrium temperature, T_E. First, I'll write the equation as: $T_E^4 = r_S^2 T_S^4 / 4(1\,\text{AU})^2$. To make this look prettier, I'm going to take the fourth root of both sides of the equation, which gives us

$$T_{\text{Earth}} = T_{\text{Sun}} \sqrt{[r_{\text{Sun}} / (2\,\text{AU})]}.$$

That's the simplest form this equation takes. But it's just what we need, an equation for the temperature of Earth. Let's substitute in the equation: the radius of the Sun is 696,000 km, and 2 AU = 300,000,000 km. Take the radius of the Sun, 696,000 km, and divide it by 300,000,000. What's the answer? 0.00232. What's the square root of that? 0.048. What's the surface temperature of the Sun? 5,778 K. Multiply by 0.048, and what do you get? The equilibrium temperature of Earth is 278 K. We know 273 K is 0° C, the freezing point of water. Therefore, our estimate for the temperature of Earth is 5° C, or 41° F. The average temperature of Earth is actually near that. But wait a minute, there's something I didn't include. I've been treating Earth as if it were a blackbody, but Earth doesn't absorb all the energy it receives—it has white clouds, it also has reflective snowy ice caps. In fact, Earth reflects back into space 40% of the Sun's energy that strikes it. That much never even gets absorbed, and never feeds Earth's temperature. If you put that factor into this equation, Earth's equilibrium temperature drops. And if you do the math, Earth's equilibrium temperature drops to below freezing. Yes, you read that correctly—Earth's natural equilibrium temperature in space at our distance from the Sun is below the freezing point of water. According to our earlier argument, there should be no life and no liquid water on Earth. But of course we have liquid water. And we're teeming with life. So something else is raising the temperature. You guessed it: the greenhouse effect. The infrared radiation radiated by Earth's surface does not escape directly into space, it is absorbed by the atmosphere, which heats the atmosphere, as we discussed in chapter 2. The trapping of the infrared radiation then raises the surface temperature of Earth. The greenhouse effect caused by Earth's atmosphere thus raises Earth's surface temperature. Turns out that the greenhouse effect on Earth approximately compensates for the reflectivity of Earth, so our calculation is fine after all.

From our wonderful equation, $T_E = T_S \sqrt{[r_S/(2\,\text{AU})]}$, it is clear that for a given star, the temperature of a given planet (with a particular reflectivity and greenhouse effect) would be proportional to one over the square root of its distance from its star. This equation allows us to calculate the inner and outer edges of the habitable zone for that particular planet; call these limits r_{min} and r_{max}. At the inner edge of the planet's habitable zone, at a distance r_{min} from the star, water is just about to boil on its surface. If it has an atmospheric pressure like Earth, water boils at 100° C or 373 K. At r_{min}, the inner edge of the habitable zone, the planet has a surface temperature of 373 K. Water freezes at 0° C or 273 K: that occurs at the outer edge of its habitable zone. Thus a planet at the inner edge of its habitable zone is hotter than the same planet at the outer edge of its habitable zone by a factor of 373/273. The ratio of r_{max}/r_{min} will be the square of (373/273), or 1.87. So the outer edge of the habitable zone for this particular planet is only 87% larger than the inner edge. That's a narrow range.

After adjustment for observational selection effects, the Kepler data tell us that around 10% of solar-type (G and K) stars have an Earth-sized planet (with a radius between 1 and 2 times that of Earth) with a stellar radiation flux between 1/4 and 4 times what Earth receives. Thus, around 10% of stars similar to the Sun would have Earth-sized planets within 0.5 AU to 2 AU of the star. That's because solar radiation falls off like the square of the distance. A planet at 2 AU receives 1/4 of the solar illumination we get, and a planet at 0.5 AU receives 4 times the solar illumination we receive. The Kepler data suggest that the separations of Earth-sized planets from their host stars are distributed uniformly with the logarithm of the separation. What does this mean? Of the planets between 0.5 AU and 2 AU, we expect half to be between 0.5 AU and 1 AU, and the other half between 1 AU and 2 AU. The span from 0.5 AU to 1 AU is a factor of 2. The factor span from 1 AU to 2 AU is also a factor of 2. Equal numbers of planets fall in intervals with equal factors. Planets at a distance 0.5 AU from a solar-type star would possibly be habitable if they had high reflectivity and a low greenhouse effect. But if you put Earth there, its oceans would boil. Similarly if you put Earth at 2 AU, it would freeze over. However, if you put a planet with a low reflectivity and a high greenhouse effect there, it could stay warm enough to support life. The r_{max}/r_{min} limit for a given planet, given its particular reflectivity and greenhouse effect, is narrow: 1.87. Now $1.87^{2.2} \approx 4$. Thus, approximately 2.2 factors of 1.87 multiply to give a factor of 4, the entire range from 0.5 AU to 2 AU for a solar-type star. If equal numbers of planets fall in each of those 1.87 factors, that means that if you are a random Earth-sized planet somewhere between 0.5 AU and 2 AU, there is about one chance in 2.2 (or about 45%) that you will, by chance, find yourself in the range $r_{max}/r_{min} = 1.87$ that you need to be habitable, given your particular reflectivity and greenhouse effect.

If 20% of the stars in the galaxy are suitable—G and K stars—and if around 10% of such solar-type stars have Earth-sized planets with a solar illumination between 1/4 and 4 times what we receive on Earth, *and* if about 45% of those will find themselves in the radius range they need to be habitable (having liquid water on their surface, given their reflectivity and greenhouse effect), then the fraction $f_{HP} = 0.2 \times 0.1 \times 0.45 = 0.009$.

That exercise was simultaneously exhausting and yet illuminating. We are using math and astrophysics to filter locations around stars where we might find life as we know it.

But for a planet to be a candidate, other criteria must be satisfied as well. It must have a reasonable atmosphere. If the planet is small like the Moon, its gravity will be so weak that molecules from its atmosphere at a temperature of 278 K will escape into space, and the planet will lose its atmosphere, which is why the Moon has hardly any atmosphere at all. But we are already talking about planets with radii between 1 and 2 times that of Earth, so they should retain their atmospheres. The planet's orbit can't be too eccentric. If its orbit is a Kepler ellipse with an eccentricity e, the ratio of its maximum distance from its star r_{max} to its minimum distance from the star r_{min} is $r_{max}/r_{min} = (1 + e)/(1 - e)$. Equivalently, one can say $e = ([r_{max}/r_{min}] - 1)/(r_{max}/r_{min}] + 1)$. The way this plays out is that if the planet's orbit is a perfect circle, $e = 0$. If it is very elongated, e approaches 1. (That's true for many comets.) You might see where this is going: the planet's orbit can't have a value giving $r_{max}/r_{min} > 1.87$, or else the oceans would alternately boil and freeze over. That means that the planet's orbit must have an eccentricity $e < 0.30$, so that it never wanders out of its habitable zone, lest it freezes or boils its precious liquid water. If you meet an extraterrestrial, you can say, "I bet your home planet has an eccentricity that is less than 0.30." He or she, or more likely, it, will be duly impressed.

The eccentricity of Earth's orbit is only $e = 0.017$. That's no accident—it gives us a good climate without wild fluctuations. Or more accurately, it's no accident that we evolved on a planet with a low-eccentricity orbit. Fortunately for the search for life, most Earth-like planets discovered by the Kepler satellite have low eccentricities. They often appear in multiple-planet systems, where the orbital interactions between and among the planets tend to circularize their orbits over time. The planets settle down into orbits that stay away from one another. In multiple-planet systems, the Kepler satellite has found that successive planets often have orbital periods that on average are larger than the preceding one by a factor of at least 2. Using Kepler's third law ($P^2 = a^3$), that means that successive planets have orbits that are each larger than the preceding one by a factor on average of at least $2^{2/3}$, or 1.6. That's a factor about the 1.87, r_{max}/r_{min} width of the habitable zone for a particular planet. You can get lucky and have two planets in closer orbits than that or have a high-reflectivity, low-greenhouse-effect planet close in and a low-reflectivity, high-greenhouse-effect planet farther out, but on average, we're looking to find at most one habitable planet per star system.

People once thought that binary star systems would have no planets. Since more than half the stars in the galaxy are in binaries, this would cut our fraction of candidates by a factor of 2. But the Kepler satellite has found planets in binary star systems. This is okay for your habitability if you have two solar-type stars orbiting each other with a separation of 0.1 AU and you are $\sqrt{2}$ AU = 1.41 AU away. You will then get the same illumination as we do on Earth. You will just have two stars in your sky (like the planet Tatooine does in *Star Wars IV*). The two stars are such a tight pair that they will not disturb your dynamics. But if you have two solar-type stars with a separation of 1 AU, it will be hard to find a habitable place that sustains a stable planetary orbit, since your gravitational allegiance will continually be swapped from one star to the other. If, however, the two solar-type stars orbit with a separation greater than 10 AU, that is fine again too, because you can just orbit one star at 1 AU and see the other one in the distance. Being so far away, the other star will not make your orbit unstable, and it won't make you too hot. Of course, you don't want to be in a star system with even one massive star, because it will become a red giant and die before you have had time to evolve intelligence.

These three additional factors—atmosphere, eccentricity, and binary troubles—each lower the probability of a star having a planet in a habitable zone, but combined, they probably don't lower f_{HP} by a factor of 2. Thus, I will just lower f_{HP} from 0.009 by a little to $f_{HP} \sim 0.006$.

When Frank Drake first wrote his equation in the 1960s, we had not yet discovered any planets orbiting other stars. So f_{HP} was just anyone's guess. But now we have the data to refine our estimate. This is how the equation is supposed to work. It encourages us to get the data and find the factors.

The result that $f_{HP} \sim 0.006$ is empowering. Let's see what we can do with this. The nearest star is 4 light-years away. Go 10 times as far away, out to 40 light-years. That sphere of radius 40 light-years will have 1,000 times as much volume as a sphere of radius 4 light-years, and within that sphere you will find of order 1,000 stars. With $f_{HP} \sim 0.006$ you expect, on average, to find at least six habitable planets within this radius. Yes, within 40 light-years of the Sun, we expect to find habitable planets orbiting other stars! That means TV episodes of *Star Trek* in its first season, wafting outward at the speed of light as TV signals, have probably already washed over another habitable planet with liquid water on its surface.

In the 1970s, the British Interplanetary Society conducted a study, called Project Daedalus, of the possibility of an interstellar spacecraft. It envisioned a 190-meter-tall, two-stage spacecraft, powered by nuclear fusion, using 50,000 tons of deuterium and helium-3. That is about twice as tall and 16 times as massive as the *Saturn V* rocket we used to send astronauts to the Moon. This enormous fusion-powered rocket could achieve a velocity of 12% of the speed of light. It would have a 500-ton scientific payload, including two 5-meter optical telescopes as well as two 20-meter radio telescopes. It would take this ship 333 years to go 40 light-years. Given what we now know, this ship could reach a habitable planet within 333 years. Telemetry from the flyby would reach Earth after another 40 years, for a total of 373 years to hear back from it.

Or better yet, take the same size rocket and use matter and antimatter fuel instead. This would be quite an engineering challenge—keeping the matter and antimatter safely apart until combining them in the engines—but it would convert 100% of the mass of the fuel into energy via Einstein's equation $E = mc^2$. This is much more efficient than deuterium–helium-3 fusion, which produces helium-4 and hydrogen by converting only 0.5% of the fuel's mass into energy. With matter-antimatter fuel, the same-sized rocket could take ten astronauts and land them on a habitable planet, 40 light-years away. They would start by accelerating at 1 g (9.8 meters per second per second, the gravitational acceleration we experience on Earth's surface) for a period of 4.93 years using the matter-antimatter fuel. This would be comfortable for the astronauts, who could walk about the cabin just as they do on Earth. The craft would achieve a velocity of 98% of the speed of light. Then it would coast at a speed of 98% of the speed of light for 32.65 years, and finally turn the rocket around and decelerate at 1 g for 4.93 years. The astronauts would slow to a stop and arrive at the star 42.5 years after launch. Due to relativity effects discovered by Einstein (about which Rich will have much more to say in chapters 17 and 18), by traveling so close to the speed of light, the astronauts would age only 11.1 years during the trip, while Earth would have passed 42.5 years into the future. Even if it took an extra two centuries (after the nuclear rocket launched) to develop the matter-antimatter technology, the matter-antimatter–fueled ship would still beat the nuclear rocket to the goal.

For all of this calculation to matter, you must first find a habitable planet. Forty light-years is 12 parsecs. A planet at 1 AU from its star, 40 light-years away, will be 1/12 of a second of arc away from the star in the sky. The Hubble Space Telescope, 2.4 meters in diameter, already has a resolution of 0.1 second of arc. A 12-meter-diameter space telescope would have a resolution of 1/50 of a second of arc. With the bright image of the star blocked out by a special occulting disk, designed to minimize the flooding effects of scattered star-light, it could in principle pick out a planet only 1/12 of a second of arc away from its star. The James Webb Space Telescope, under construction now and scheduled for launch in 2018, has a segmented mirror 6.5 meters in diameter. The next generation space telescope after that may be able to find and take pictures of Earthlike planets in the habitable zone out to a distance of 40 light-years. It may be green—it may have vegetation. It may be blue—it may have oceans. We could take its spectrum and determine whether it has oxygen in its atmosphere—a kind of biomarker, a by-product of photosynthesis and other chemical reactions that can betray the existence of life.

If you multiply the number of stars in the galaxy (300 billion) by f_{HP} (0.006) and stop your calculation here, you get a nice looking number, the number of planets in the habitable zone: 1.8 billion. That's huge. But not all of them count. Of those in the habitable zone, we are looking for the fraction f_L of those that have any life at all. But, not just life, intelligent life. What fraction f_i of those planets with life go on to have intelligent life? I will return to these terms of the equation shortly.

What are we up to now? So far we have the fraction of long-lived stars with planets orbiting in their habitable zone that harbor intelligent life, *and* the fraction f_c of those that develop a technology capable of communicating across interstellar distances.

The last fraction in the Drake equation is the fraction of these civilizations that are communicating at the epoch we are observing them now. That's the fraction of the time during the age of the galaxy that they are "on." If we look randomly throughout the Milky Way, we will randomly hit some planets that were just born, some that are middle-aged, and some that are old. The chance of catching a planet during its communicating phase at some random time during the life of the galaxy is equal to the average longevity of radio-transmitting civilizations divided by the age of the galaxy. That's a fraction too. Our last fraction. Multiplying all these fractions together, times our original number of stars, we arrive at N_c, the number of civilizations in the galaxy from which we can receive communications—now.

And therein are the seeds and the essence of the Drake equation. Some of these fractions we know well. For example, we know what fraction of stars are long lived, from our understanding of the main sequence on the HR diagram. And we've looked around, and we now have now discovered many planets. All good so far. What fraction of these planets are Earth-sized and live in a habitable zone? We have now just estimated that one using statistics from the Kepler satellite. Things are coming along quite nicely.

We've also discovered a loophole in the habitable zone argument. Europa, a moon of Jupiter, has an 80-kilometer-deep ocean of liquid water covered by a 10-kilometer-thick ice sheet. As already noted, Europa's moon-wide ocean contains more water than all the oceans of Earth. Yet Europa is far outside the Sun's habitable zone. How did it get warm? It orbits Jupiter with three other large moons. The other moons perturb its orbit according to Newton's laws, driving it sometimes a bit closer to Jupiter and sometimes pulling it farther away. When Europa is closer to Jupiter, the tidal gravitational forces from Jupiter squeeze the poor moon into a more oblong shape. When Europa is farther away, it relaxes into a more spherical shape. This steady kneading of Europa heats it, melting its ice and sustaining the liquid ocean. Somebody needs to spend the money and send a probe to Europa to drill down through the ice layer to the ocean below and try their hand at ice fishing. (This could be done using a small probe heated by plutonium, which could melt its way down through the ice.) See if they catch anything. If we found life forms there, we would have to call them "Europeans!" Saturn's moon Enceladus also has an ocean beneath an ice layer. So if we estimate the fraction f_{HP} by just counting planets properly heated by their stars, we have to up the estimate in some sensible way to account for tidally heated moons like Europa living far outside the habitable zone yet also sustaining liquid water. We have to broaden our concept of what a habitable zone means.

What fraction of those habitable places have life? What is f_L? Our only measure of this—our only data—comes from Earth. Biologists boast of life's diversity on Earth. But I suspect that if we find an alien, that alien will differ more from life on Earth than any two species on Earth differ from each other.

Just how diverse are we here on Earth? Line us up—it's quite a zoo. Tiny bacteria here, even tinier viruses over there, a jellyfish (now called "sea jellies," I'm told), a lobster, a polar bear. Here's another example. Suppose you've never been to Earth and someone comes to you, and after visiting, says frantically, "I just saw an exotic life form. It senses its prey by detecting infrared rays. It doesn't have any arms or legs, yet it's a deadly predator that stalks its prey. You know what else? It can eat creatures five times bigger than its head." You promptly say, "Quit lying." But what have I just described? A snake. A snake has no arms, no legs, and gets along in life just fine—for a snake—stretching its jaws open, eating stuff bigger than its head.

What else? Oak trees, and people. My point is that all this diversity shares the same planet. And we all have common DNA, like it or not. All life on Earth shares some percentage of its DNA with other life forms. We are all connected, chemically and biologically.

Earth is now about 4.6 billion years old. In the early solar system, debris left over from the epoch of formation wreaked havoc on planetary surfaces, because large rocks and iceballs were still raining down, depositing enormous amounts of energy. Kinetic energy was getting converted into heat, liquefying the surfaces of the rocky planets, and thereby sterilizing them. That went on for about 600 million years. When you want to start your life clock for Earth, it's not fair to start it at 4.6 billion years ago, because Earth's surface was supremely hostile to life. If you want see how quickly life formed, don't start there; instead, start about 4 billion years ago, when Earth's surface became cool enough to sustain liquid water and enable complex molecules to form. That's when you should start your stopwatch.

In the old days, stopwatches for timing athletic events had a button on the top that you'd push, and it would start, and these things called hands would spin around until you pushed the button on top again and the watch would stop—a *stop*watch, get it? If you watch the time-honored CBS news program *60 Minutes* on Sunday nights, they still use this mechanical museum piece, which begins and ends their show. It is the only TV program with an opening theme that uses something other than music. Just the ticks of the stopwatch.

Start your stopwatch at 4 billion years ago: 200 million years later, you'll see the first evidence of life on Earth. We have evidence of cyanobacteria 3.8 billion years ago. The fraction of planets in the habitable zone around long-lived stars that might have life is looking pretty good, because, given the chance, our planet took only a very small percentage of the total available time to make life in the first place. We still don't know exactly how this process happened—it remains a biological research frontier—but I assure you, top people are working on it. We do know it took only about 200 million out of 4 billion years to accomplish. If making life were long and hard for nature to accomplish, maybe then life would have taken a billion years, or several billion years to form on Earth. But no. It took just a couple of hundred million years, which gives us confidence that this fraction f_L in the Drake equation might be quite high, perhaps near 1.

Of course, we're limiting ourselves to life as we know it. In some circles, that reference is known by its acronym, "LAWKI." We just don't know how else to think about the problem with confidence. We could write on life as we don't know it. But that might take many volumes to cover. Maybe they've got seven legs, three eyes, two mouths, and are made of plutonium. Maybe life out there is as we don't know it, but we can't figure out how to pose the right questions. It's a practical matter, not a philosophical one. We have an example of life as we know it, that's us—it's an example of one, but it constitutes an existence proof. You're trying to prove that something exists, and you have one example of that thing staring at you on your selfie screen. The proof is already there. So, let's start with that and work our way from there. We also know that we are made out of atoms that are pretty common in the universe.

In one episode of *Star Trek,* the original TV series, the *Enterprise* crew encounters a life form based on silicon rather than on carbon. We're carbon-based life, but silicon is also pretty common in the universe. In the *Star Trek* episode, the silicon creature was basically a short pile of rocks that was alive and sort of waddled when it moved. A creative story-telling leap this was. The *Star Trek* producers were trying to broaden the paradigm of what kind of life the crew would find in the galaxy. Turns out that silicon is right below carbon on the periodic table. You might also remember from chemistry that elements in the same column all have similar outer orbital structures of their electrons. And if they have similar orbital structures, they can bond similarly with other elements. If you already know that carbon-based life exists, why not imagine silicon-based life? Nothing stopping you in principle. But in practice, carbon is about ten times more abundant in the universe than silicon. Also, silicon molecules tend to stay tightly bound, making them unwilling players in the world of experimental chemistry that is life. Carbon dioxide is a gas, whereas silicon dioxide is a solid (sand). We have even discovered complex, long-chain carbon molecules in interstellar space, such as H-C≡C-C≡C-C≡C-C≡N (with its alternating single and triple bonds). We've got acetone, $(CH_3)_2CO$; benzene, C_6H_6; acetic acid, CH_3COOH; and many other carbon molecules out there just floating in interstellar space. Gas clouds forged these molecules all by themselves. The comet Lovejoy was even discovered to be outgassing alcohol. Silicon does not form such complex molecules, and thus has far less interesting chemistry than does carbon. So, if you want to base life on a certain kind of chemistry, carbon is your element. There's no doubt about it. Whatever life forms populate the galaxy, even if we don't look alike, it's a good bet our chemistry will be similar, just because of the abundance of carbon across the cosmos and its bonding properties.

Earth is our one example of life having formed in the solar system, so I'm comfortable with the estimated number: $(f_L) \sim 0.5$. It's midway between 0 and 1, not a sure thing—a 50-50 chance. What's next? The fraction of planets in orbit around long-lived stars in the habitable zone that have life *but also have intelligence*. This doesn't look too good.

By whatever scheme you devise to measure intelligence on Earth, humans tend to sit at the top. Big brains seem to matter, and we have big brains, but elephants and whales have even bigger brains; so maybe it's not just big brains. Maybe it's a ratio. The ratio of your brain's mass to your body mass. Perhaps that's what really determines intelligence. Humans have the biggest brains compared to our bodies of any animal in the animal kingdom. We define it, so we get to put ourselves at the top. But perhaps our hubris prevents us from thinking about it any other way. Let's assert that we are intelligent, and let's define intelligence as, for example, the capacity of a species to do algebra. If intelligence, which we claim we have, is defined this way, then we're the only intelligent species on Earth. Porpoises are not doing algebra underwater. No matter how complex and thoughtful their behavior seems to be, they're not doing algebra. No other species in the history of the world, besides us, has ever done algebra, so we're intelligent. For the sake of this conversation, let's define it that way. Suppose we are looking for life that we can have a conversation with. We would not use English, but some language that we presume is cosmic: that would be the language of science, the language of mathematics.

If intelligence is important for species' survival, don't you think that feature would have shown up more often in the fossil record? It hasn't. Just because we have it doesn't make it something really important for survival. You know, after the next global catastrophe, the roaches will likely still be here, right alongside the rats, and we will be extinct. A lot of good our brains will have done us then.

Now, maybe our intelligence gives us the chance to alter this fate, as we might have altered the fate of the dinosaurs. There's a *New Yorker* cartoon by Frank Cotham showing two lumbering dinosaurs hanging out together, and one of them says to the other, "All I'm saying is *now* is the time to develop the technology to deflect an asteroid." Meanwhile, as we know, an asteroid is headed toward them to take them out—permanently. Perhaps we can use our intelligence to prolong the natural life expectancy of our species, by going out to space and batting asteroids out of the way before they destroy us—if we are willing to give NASA the money to do it. But that's not the only threat. There's also the threat of unforeseen emergent diseases. Look at what happened to the elm tree in America. Most elm trees in New England were killed off by a fungus carried by the elm bark beetle. Imagine if something like that were to attack us. A novel virulent flu virus might be all it would take to do us all in.

Intelligence is no guarantee of survival. Sight, however, seems to be pretty important. Organs of sight have evolved by natural selection in many different species of animals. The human eye has nothing in common structurally with the fly's eye, which has nothing in common structurally with the eye of a sea scallop. Although there seems to be just one primordial gene for making eyes, these different kinds of eyes arose along different evolutionary paths. Sight must be pretty important for survival. What about locomotion, having some way to get around? Maple trees don't have legs to run with, but they bear seeds with little wings to help the wind spread them far and wide. Locomotion seems to be important, because we see all kinds of ways of making it happen: snakes slither, lobsters walk, jellyfish use jet propulsion, bacteria use flagella. Many insects and most birds fly. People walk, run, swim, take cars, trains, boats, airplanes, and rocket ships, so we really get around. But we are still the only ones alive on Earth doing algebra, which doesn't give me much confidence that intelligence is an inevitable consequence of the tree of life. Evolutionary biologist Stephen Jay Gould has expressed similar views. This all suggests the fraction f_i might be small. To indicate that, set $f_i < 0.1$, realizing it could be much smaller. That's a different opinion compared with some of my colleagues, some of whom work at the SETI institute. They pretty much require that this fraction f_i be high—otherwise, what are they looking for? They know they won't be talking to bacteria.

Once you've evolved intelligence, maybe technology becomes inevitable. I might even assume that: $f_c \sim 1$. You can do algebra, you have a curious brain, you want to make life easier, you want to have vacations, you want to watch HBO, and so forth; with such motivations, the fraction of intelligent beings who create technology might be high. After all, the only species we know that is able to do algebra did go on to develop technology for communicating across interstellar distances. But if technology contains the seeds of its own undoing (e.g., by the invention of ever-cleverer ways of killing one another and destroying our planet), then I'm sorry, the duration of your technological, communicating culture may be a small fraction of the age of the galaxy. Rich has an argument based on the Copernican Principle (i.e., your location among citizens of radio-transmitting civilizations is not likely to be special), which he discusses in the last chapter of this book, suggesting that the mean longevity of radio-transmitting civilizations is likely to be less than 12,000 years. If you divide by the age of the galaxy, that is a tiny fraction.

The point is, you put your best numbers into the Drake equation and at the end find what your estimated number of communicating civilizations might be. Whole textbooks have been written analyzing the terms of this equation. This is how we organize our thoughts about the search for life.

The Drake equation makes a cameo in the movie *Contact*, the 1997 film based on a story by Carl Sagan and his wife Ann Druyan. (I recently hosted a new version of the TV series *Cosmos* with her and colleague Steven Soter, co-writers with Sagan of the original series from 1980.) *Contact* was clever enough to avoid actually portraying the aliens, because what would they look like? What should they look like? We don't know. In 1950s B-movies, there was always some actor in a suit playing an alien, and all the aliens from other planets always had a head, two arms, two legs, and walked bipedally. In the 1982 film *ET*, the extraterrestrial is a cute, funny-looking creature, but still has two eyes, two nostrils, teeth, arms, neck, legs, knees, feet, and fingers. Compared to a jellyfish, ET is identical to a human being. That's poor imagination from Hollywood. As already noted, if you're going to come up with a new life form, it had better be more different from anything on Earth than any two life forms on Earth are from one another. Even the 1979 space thriller *Alien*, which had a creature that was a bit different, which showed some creative investment, still had a head and teeth.

Back to *Contact*. My first attendance at a world premiere of a film ever was for *Contact*. A personal invitation was extended to me, because I was friendly with Carl Sagan and Ann Druyan from years past. There were two embarrassing moments for me, which arose simply because I don't usually hang out in Hollywood. You walk the red carpet, which is lined with photographers, and once you enter the theater, it's all done up in movie posters and other themed ornamentation. And of course, there is popcorn and soda on display. So I reach for a tub of popcorn, and ask the guy behind the counter, "how much?" and he replied, "fifty dollars," temporarily freaking me out. After briefly basking in my despair, he declared, "of course it's free." And after 5 seconds of rational analysis, I said to myself, of course it's free, it's got to be free. Why would they charge you for popcorn at a World Premiere? I begged forgiveness, quickly confessing that I'm clueless, from the East Coast. And then after the screening, there was a reception where every cocktail table had a small telescope, or other quaint astronomical instrument. I thought, this was a classy touch, and wondered where they got these table ornaments; some amateur astronomy group must have loaned them these telescopes. I had to know, because it must have been a pretty active astronomy group to own that much hardware. So I went up to the event organizer, and asked, "Where did you get these telescopes?" He looked at me in a way such that the first part of his next sentence surely began silently with the phrase "You idiot," followed by an audible, "we got it from a prop house." My second stupid East Coast question of the night. Okay, prop houses have everything, including telescopes, apparently.

In the movie, Jodie Foster, who stars in the film, has a scene in which she and Matthew McConaughey, her co-star, are sitting there with the stars above them, while she points out stars and planets to him. Then they slide a little closer, and she starts to recite an abbreviated version of the Drake equation. She starts with 400 billion stars in the Milky Way. Close enough. I've given you 300 billion, but that is in the noise of what we're doing, not important. She goes on to say—and by the way, her character is a scientist who's searching for intelligent life in the universe—"There are 400 billion stars out there, just in our galaxy alone; if only one out of a million of those have planets, and just one in a million of those had life, and only one in a million of those had intelligent life, there would be literally millions of civilizations out there."

The first one out of a million cuts 400 billion down to 400 thousand. The second one in a million does what? It cuts it down again, to 0.4. The third one out of a million? So, Jodie, I'm sorry, that leaves you with 0.0000004 civilizations in the galaxy, not millions. This was the world premiere, and guess who is sitting right over there, one row ahead of me in the theater—Frank Drake, himself. I'm apoplectic about this arithmetic error, and it turned out Frank was completely unfazed by it. Maybe he was into the romance. Right after Jodie delivers these lines, she and McConaughey kiss, and in the next scene they're in bed. So uttering the Drake equation in that moment was indeed a bit of intense geek romance. No denying that. But the difference between my reaction and that of Frank Drake alerted me that perhaps I occasionally overreact to things like this.

Apparently, Jodie Foster had been told about the error, much too late for anything to be done about it. She was flustered, because she had studied that line so hard, and worked on how to deliver it, keeping both its rhythm and the romance in motion. But who's to blame here? It turns out Jodie Foster read the script correctly. Do you go back and blame the screen writer? Maybe. The script supervisor? Possibly. Do you blame Carl Sagan, when he had been dead for a year? Of course not. Someone made a mistake.[1]

Overall, I thought it was a brilliant film; riding the religion–science edge intelligently (McConaughey's character is a religious philosopher), recognizing that there are many people who feel all kinds of ways about these things. It also accurately captured the extent to which pop culture, crackpots included, would react to the discovery of alien intelligence. Generally crackpots react even when we don't discover stuff. I have boxes of mail from people who send me their latest theories of the universe. I have one postcard that says, "when I gaze upon the Moon at night, it makes my beer taste better than it ought to. What should I do?"

Just for fun, and realizing their uncertainty, let's put the numbers we have discussed into the Drake equation, and consummate the calculation:

$$N_c = N_s \times f_{HP} \times f_L \times f_i \times f_c \times (L_c \,/\,\text{Age of the Galaxy}).$$
$$N_c = 300 \text{ billion} \times (0.006) \times (0.5) \times (<0.1) \times 1 \times (<12,000 \text{ years}/10 \text{ billion years}).$$
$$N_c < 108.$$

According to our latest estimates for each term in the equation, we might expect to find *up to* a hundred civilizations in the galaxy communicating with radio waves now. Our biggest radio telescopes can detect versions of themselves—their extraterrestrial counterparts—all the way across the galaxy. So we have a chance. We have only begun to search.

Besides, there are about 50 million other galaxies like ours within 2.5 billion light-years of us; that multiplies the number by 50 million, giving us possibly *up to* 5 billion extragalactic, radio-broadcasting civilizations. All the galaxies in this bunch are billions of years old at the epoch we are seeing them—plenty of time for intelligent life to have developed within them, if it were to develop at all. The most distant of these extragalactic civilizations (2.5 billion light-years away) would be roughly 40,000 times as distant as the most distant ones we might find in our own galaxy (62,500 light-years away). The inverse-square law would tell us that a typical extragalactic civilization would have a radio brightness only 1/1,600,000,000 as large as those in our galaxy. That's why people usually consider looking for extraterrestrial civilizations only in our own galaxy.

The hunt for extragalactic civilizations is not as hopeless as it might at first appear. Intelligent civilizations could beam their signals all over the sky or take the same amount of energy to beam a focused, more intense signal to a tiny region of sky. A civilization could make itself appear 10 times as luminous by beaming all its energy toward 1/10 of the sky. A civilization could make itself appear 50 million times as luminous, by beaming to only 1/50-millionth of the sky. Most observers would miss its signal, but for those few within its beam, it would be visible out to great distances. In fact, Frank Drake himself used this strategy in 1974, when he co-opted the 1,000-foot-diameter Arecibo radio telescope to send a narrow-beam radio signal to the globular cluster M13. (Turns out they did not send the signal to where M 13 would be when the signal arrives. The cluster's motion in orbit around the Milky Way will have taken it out of the beam by the time the beam arrives. So the signal will miss the globular cluster entirely, but that detail is not important here.) If civilizations adopt various beaming patterns, some emitting in all directions and some in narrow beams, this naturally leads to a very broad distribution in apparent luminosity called *Zipf's law*, where the signal with the highest apparent luminosity exceeds the Nth highest one by about a factor of N. This means that with 50 million galaxies, the civilization with the highest apparent luminosity will appear about 50 million times as luminous as the highest apparent luminosity radio transmitter in our own galaxy. With 50 million times as many chances, we just might get lucky and fall in someone's really bright, narrow beam. Thus, the brightest extragalactic civilization might have an apparent brightness 1/32 (= 50,000,000/1,600,000,000) times as bright as the brightest one we see in our galaxy. On this reasoning, searches for extragalactic civilizations should be undertaken as well.

Finally, some caveats about the Drake equation. The habitable zone may be even narrower than we have figured. If Earth were farther out than it is, our planet would be colder and form more polar ice; the reflectivity of Earth's surface would go up, reducing the absorption of solar flux, and Earth would get colder still. You could trigger a runaway ice age. If you were to place Earth closer to the Sun, the ice would melt, the reflectivity would go down, and Earth would get hotter still. Methane trapped in peat would be released, further adding to the greenhouse effect.

The Sun is getting hotter as it evolves over billions of years. To compensate, the greenhouse effects will have to lessen or the reflectivity will have to increase, to keep the range of temperatures on which all of civilization is based. If a star evolves in luminosity over billions of years, the habitable zone will move outward, and a planet would need to remain in the habitable zone long enough for intelligent life to develop. As mentioned before, we think that the planet has to be continuously habitable for billions of years for life to have sufficient time to evolve intelligence.

Interestingly, life itself can affect the balance too. If the star is a main sequence M star, and evolves not much at all in 10 billion years, the planet might be habitable for simple life at the beginning, but when that life turns its carbon dioxide (CO_2) atmosphere into an oxygen-rich atmosphere, the greenhouse effect will decrease, perhaps sending it into a permanent ice age. This is another reason M stars may not be ideal for forming intelligent life.

Life can affect the habitable zone in other ways. Carbon dioxide from the atmosphere can be captured in the form of calcium carbonate in the shells of sea animals and deposited in sedimentary rock (limestone) when they die, thereby lessening the greenhouse effect. Vulcanism (volcanic activity) can pump CO_2 into the atmosphere, increasing the greenhouse effect. And, of course, life forms like humans can dig up long-buried fossil fuels like oil and coal from ancient organisms, and burn them, pumping more CO_2 into the atmosphere. Estimates of the habitable zone for a given planet therefore depend intimately on its geology, meteorology, and even its biology.

GALAXIES

11

THE INTERSTELLAR MEDIUM

MICHAEL A. STRAUSS

We now move from the study of individual stars and planets to a broader view of how stars fit into our own Milky Way galaxy and the interaction between stars and what we call the *interstellar medium*. So far we have talked about the space between the stars as if it were essentially empty, but I want to convince you in this chapter that the huge volume of space between the stars actually contains a large amount of material—it's just thinly spread out. *Interstellar*, of course, means between the stars, while *medium* means "stuff." So the interstellar medium is the "stuff between the stars."

Let's take a look at the interstellar medium, which has produced many of the most beautiful images in astronomy.

Figure 11.1 is a composite picture compiled from a variety of images of the Milky Way. It illustrates the full sphere of the sky, projected in a clever way onto a flat plane. The band of light called the Milky Way that we sometimes see arching across the night sky actually stretches in a full circle all the way around the celestial sphere, tracing out what we call the *galactic equator*. Our Milky Way galaxy is a disk of stars, and because we are located in this disk, when we look out into it, we see a band of light circling the sky. The brightest part of the Milky Way (toward the galactic center, in the center of this picture) is not clearly visible from mid-latitudes in the Northern Hemisphere. If you ever get to the Southern Hemisphere, on a clear moonless night away from city lights, look up! Especially from March through July, the southern view of the Milky Way is tremendously dramatic, much brighter than what those of us who live in the North can see.

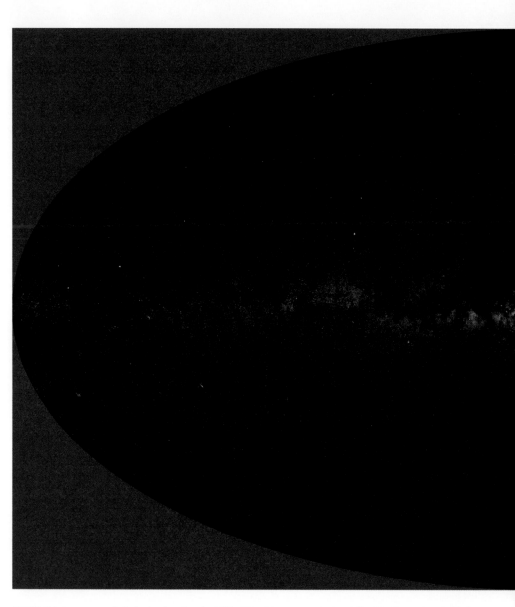

FIGURE 11.1. Panorama of whole sky showing the Milky Way. Distant stars in the Milky Way form a band of light that circles the sky along the galactic equator, mapped as a straight horizontal line across the center of the map. The Milky Way galaxy's center is in the center of this figure. Note the dark lanes and patches along the Milky Way where background stars are obscured by dust. *Photo credit:* Adapted from J. Richard Gott, Robert J. Vanderbei (*Sizing Up the Universe*, National Geographic, 2011) Based on data from Main Sequence Software.

Since we are halfway out from the center in the disk, the picture gives the appearance that we are looking at the Milky Way edge-on from outside. One of the things you will notice right away is that the Milky Way is not smooth, but seems to have black blotches or patches in it. If you look at it with a telescope,

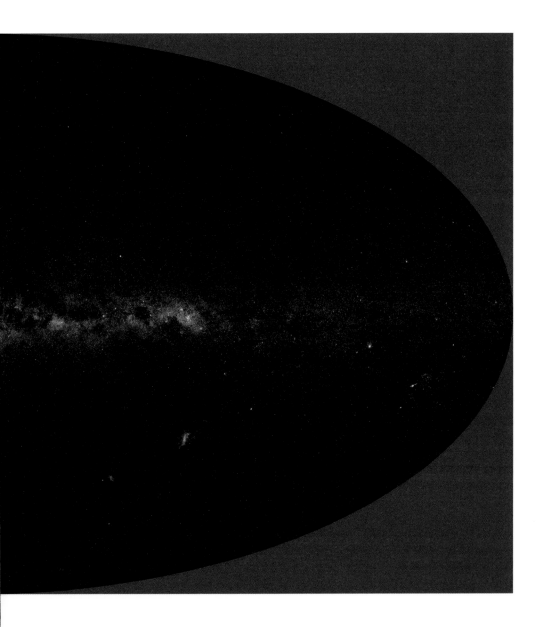

you'll see (as Galileo famously did) that the diffuse light of the Milky Way actually is due to combined light of myriad stars, but there are regions (dark lanes) in which no stars are apparent. One hundred years ago, astronomers argued about what could account for these lanes. One possibility they considered was that the distribution of stars is intrinsically patchy, and the dark regions are simply those places in which there happen to be very few stars. Alternatively (and this is the right idea), there is a smooth distribution of stars, but something is blocking our view of them. That something is indeed the interstellar medium.

One way the interstellar medium manifests itself is by being quite opaque. It is thin stuff, but the volume of space it occupies is large. In Earth's atmosphere, even a very thin haze or a small amount of smoke can obscure distant objects. The interstellar medium has tiny dust particles just like those in smoke. Indeed, "dust" is the technical term astronomers use to refer to these particles, but perhaps "smoke" would be a better word. This material is highly rarified, but over enormous distances, its effects mount up and can absorb the light of background stars. In some directions, the cumulative effect of the dust is so pronounced as to completely obscure the light of background stars. Because of dust, for example, the very center of the Milky Way is completely obscured to visible light.

It turns out that the dust obscures light of short wavelengths more than it does long wavelengths. At longer infrared wavelengths, the dust causes far less absorption than it does in visible light, and one can get a largely unobscured view of the Milky Way. Figure 11.2 is a close-up picture of the center of our Galaxy taken with the Two Micron All-Sky Survey (or 2MASS, an appropriate acronym, since this survey was led by astronomers at U Mass, the University of Massachusetts). As the name implies, the 2MASS uses infrared wavelengths of about 2 microns (2×10^{-6} meters), considerably longer than visible light (0.4–0.7 microns). You can see that the light in the picture is coming from individual stars. The effects of dust are still apparent but are not nearly as extreme as in visible light. Preferentially suppressing blue light from an object makes it appear red. Therefore, when we see stars through dust they appear "reddened" relative to their normal color. The brightest small red clump peeking out from behind the dust at the upper left is the galactic center itself, a tight group of stars which harbors a 4-million-solar-mass black hole.

Figure 11.3 shows a dark region called the Coalsack Nebula, a big dust cloud that completely obscures the stars behind it, leaving a blank spot in the sky that is quite apparent to the naked eye. Australian Aboriginal astronomers have known of the Coalsack Nebula for almost 40,000 years. It forms the head of the Emu, a dark pattern in the Milky Way famous in Aboriginal lore.

Thus, the interstellar medium is far from smooth, containing many clumps or clouds that are particularly dense. In addition to dust, it contains gas composed of hydrogen, oxygen, and other elements. We refer to the various fuzzy, or cloudlike, objects we see in the sky (as opposed to the pointlike stars) as *nebulae*, from the Middle English *nebule* for cloud or mist, which in turn comes

FIGURE 11.2. The Milky Way center. The dust of the Milky Way obscures short-wavelength light more than longer wavelengths, giving the stars behind that dust a distinctly reddish tinge. There are about 10 million stars in this image, which measures about 4,000 light-years across. The exact center of the Milky Way is the densest red spot in the upper left. *Photo credit:* Atlas image obtained as part of the Two Micron All Sky Survey, a joint project of the University of Massachusetts and the Infrared Processing and Analysis Center/California Institute of Technology, funded by NASA and the NSF

FIGURE 11.3. The Coalsack Nebula. This is a region of the Milky Way completely obscured by a dense foreground cloud of dust. *Photo credit:* Vic Winter and Jen Winter

FIGURE 11.4. Orion Nebula. The bright colors in this star-forming region are caused by fluorescing gas illuminated by the young bright embedded stars. Filaments of dust are also visible. *Photo credit:* NASA, ESA, T. Megeath (University of Toronto), and M. Robberto (STScI)

from the Latin *nebula* for mist or fog. These gas clouds do more than just obscure our view of the stars. Figure 11.4 shows the Orion Nebula, which can be seen with the naked eye. It is at the bottom of Orion's sword hanging down from his belt. Even through binoculars, it appears markedly fuzzy, not sharp like a star. Ultraviolet light from hot stars can excite the gas in the interstellar medium. The photons from the hot luminous young stars in the nebula excite atoms in the gas to high energy levels. As the electrons drop back to lower levels, the atoms emit photons at specific wavelengths, as we saw in chapter 4, giving rise to the colorful nebulosity we see. This fluorescence is the same process that happens inside a neon bulb, and in fact, neon is one of the elements present in the interstellar medium.

The Orion Nebula is an example of an *emission nebula*, which is to say that its spectrum is dominated by emission lines corresponding to various electronic transitions in the atoms. We can identify the specific elements in the nebula by the wavelengths of their emission lines. The reddish color in the image is due to emission of photons when electrons drop from the $n = 3$ to the $n = 2$ energy level in hydrogen (Hα, one of the Balmer lines, described in chapter 6). The hint of green color is due to oxygen, and other elements produce the remaining light. The dark regions are caused by patchy dust, mixed in with the gas.

FIGURE 11.5. Trifid Nebula. The red light is fluorescing gas shining in hydrogen alpha (Hα) emission , while the blue light is mostly starlight reflected from the abundant dust. *Photo credit:* Adam Block, Mt. Lemmon SkyCenter, University of Arizona

The object in figure 11.5 is called the Trifid Nebula, because it is divided by lanes of dust into three parts. These dust lanes hide the emission that would otherwise make the nebula appear smooth. As before, embedded hot stars are causing the gas to glow, and the red emission is due to Hα. The extended blue emission to the right is light from blue stars reflected off the dust, which acts as a sort of mirror. We call this part a *reflection nebula*. Remember that blue light going through dust is absorbed, which is why stars seen through a dust cloud appear reddish. That blue light has to go somewhere; it tends to be either absorbed or reflected in a different direction. So reflection nebulae tend to be blue.

The Pleiades is a young cluster of stars which is easily visible to the naked eye. Pictures taken with a large telescope (see figure 7.2) show that its stars are illuminating dust, resulting in a blue reflection nebula. Each blue star is surrounded by a blue fuzzy patch.

The interstellar medium is the raw material from which stars are made, as we touched on in chapter 8. The interstellar medium is quite diffuse over most of the Milky Way galaxy, but in certain regions, such as in emission nebulae and dark clouds, it is relatively dense—these are the regions that are ripe for star formation. Gravity pulls a small knot in the cloud of dust and gas together. As it collapses, it heats up, converting its gravitational potential energy to kinetic energy as it falls inward, eventually becoming hot and dense enough for thermonuclear reactions to take place, and a star is born. The core of the Trifid Nebula is filled with massive hot blue stars. These stars live fast and die young. So these stars must have been born recently.

The scale in which this is happening is enormous. In the Orion Nebula, we have observed about 700 stars in the process of forming, many of which have disks of gas and dust surrounding them that may eventually form into planets. As in the Orion and Trifid Nebulae, stars tend to form in large groups rather than in isolation. With time, the radiation and stellar winds from the young stars evaporate and blow away the dust surrounding them, gradually unveiling the stars. Young stars also often emit winds, consisting of hot gases emitted from their surface, analogous to—but stronger than—the solar wind that our own Sun emits. These winds sculpt the gas and dust around them, which can give some nebulae a windswept look.

The full details of how stars form are poorly understood; it is one of the most important unsolved problems in astronomy. Not all dense regions of the interstellar medium are collapsing to form stars; we don't have a complete picture of why star formation proceeds in some regions of the Milky Way but not in others. We do know that the winds from the first stars forming in a region tend to blow away the gas and dust that surrounds them, which keeps additional stars from forming from this material. A star like the Sun has a random motion relative to its neighbors of about 20 km/sec. In the 4.6 billion years since the Sun formed, it has wandered far from its *stellar nursery*, where it was born. (Yes, that is the actual term astronomers use!) It is thus not possible to determine which stars are its siblings—those born along with it. Over hundreds of millions of years, groups of stars dissipate and spread around the Milky Way; most older stars in the Milky Way disk are either single (like the Sun), in pairs, or in groups of just a few stars.

We have now painted the broad outlines of the birth and life cycles of stars. Stars are formed out of the interstellar medium. The lowest-mass stars are all still burning their original store of hydrogen; they are frugal enough to continue doing so for more than a trillion years. Stars with masses similar to the Sun or somewhat higher will become red giants and eventually return some of their material to the interstellar medium in the form of planetary nebulae. Stars with cores more than twice the mass of the Sun (total main-sequence mass more than 8 times that of the Sun) will explode much more dramatically as supernovae, sending the heavier elements they have created into the interstellar medium. These heavier elements can then be incorporated into the next generation of stars. By this process, the interstellar medium becomes more and more enriched by elements heavier than hydrogen and helium. These heavier elements make up most of the world around us. Earth, for example, is made up mostly of iron, oxygen, silicon, and magnesium. Our own bodies are mostly hydrogen, carbon, oxygen, and nitrogen, together with smaller amounts of other heavy elements. Heavy elements up to iron are made by fusion in the cores of dying stars. The rest of the naturally occurring elements all the way to uranium form as heavy nuclei merge with neutrons in the cores of red supergiant stars, or in the envelopes of stars about to explode as supernovae, or in the collision of two neutron stars in a tight binary. The details of these processes are still not well understood and are a current area of research.

The Milky Way is like a living ecosystem, with stars living and dying. Each generation of stars contributes material to the interstellar medium, which gets incorporated into the next generation. The heavy elements are the raw material out of which planets—locations where life can exist—form. It's both humbling and awesome to realize that most of the material in our bodies and everything that surrounds us has been produced through thermonuclear processing in stars.

I mentioned that one way to create elements heavier than iron is in the collision of two neutron stars in a tight orbit. We know that such tight neutron-star binaries exist. Russell Hulse and Joe Taylor discovered two neutron stars, each with a mass of 1.4 solar masses, orbiting each other once every 7.75 hours. The diameter of the orbit is about 3 light-seconds, somewhat smaller than the diameter of the Sun. The two neutron stars are slowly inspiraling due to emission of gravitational waves, an effect predicted by Einstein's theory of general relativity. Indeed, their measurements are in beautiful agreement with general relativity's predictions, and Taylor and Hulse were awarded the Physics Nobel Prize for their discovery in 1993. The two neutron stars will continue their slow death spiral toward each other until they eventually collide and merge, about 300 million years from now. Enrico Ramirez-Ruiz of the University of California, Santa Cruz, estimates that one such collision could eject a Jupiter-mass of gold. Think of it: the atoms of gold in my wedding ring could have been born in a collision of two neutron stars billions of years ago!

12

OUR MILKY WAY

MICHAEL A. STRAUSS

Most of the stars that you can see with your naked eye are tens, hundreds, or thousands of light-years away. Until we were able to see and understand the nature of more distant objects through telescopes, that was the full extent of the known universe. The history of astronomy has been a progression of ever-greater understanding of just how large the universe is.

Back in the time of Copernicus, our universe consisted of the solar system, surrounded by distant stars, about which we knew very little. Galileo Galilei, who was the first to point a telescope to the heavens, saw that the light from the Milky Way comprises myriad (indeed, billions of) individual stars. Astronomers quickly understood that our concept of the universe needed to be much broader than previously realized.

In 1785, William Herschel (who also discovered Uranus) counted the number of stars visible through his telescope in different directions to make a map of the Milky Way galaxy. He reasoned that the number of stars seen in any direction was a reflection of the extent of the Milky Way in that direction. From his observations, he concluded that the Milky Way had a flattened lens shape, and we were located near the center. In 1922, the Dutch astronomer Jacobus Kapteyn completed a more comprehensive survey of the Milky Way. It is amazing that the Netherlands, which is famous for its cloudy weather, has produced so many distinguished astronomers! Like Herschel, Kapteyn made accurate counts of the stars in different directions, now using much more sensitive astronomical photographs taken in different directions of the sky.

This is a tricky business, of course. Remember the inverse-square relationship $B = L/(4\pi d^2)$ between the brightness B, distance d, and luminosity L of a star. If we see a bright star, we don't know a priori whether it is a very luminous star at a great distance, or a less luminous star that is closer. Kapteyn did the bulk of his work before Hertzsprung and Russell demonstrated that the color of a main-sequence star enables one to infer its luminosity (see chapter 7). Kapteyn did the best he could, and after many years of careful measurement, came up with a model for the known universe similar to that of Herschel: it was shaped like a lens 40,000 light-years in diameter, with the Sun just 2,000 light-years from the center.

Previous to Copernicus, people thought Earth was the center of the universe. After Copernicus, the Sun became the new center of the known universe. In the centuries that followed, astronomers began to understand that the Sun was a star, like all those seen at night, but Kapteyn still located the Sun more or less at the center of their distribution. But about the time Kapteyn was doing his work, scientists started to understand the effects of dust in the interstellar medium on the apparent brightnesses of stars (see chapter 11). The obscuring effects of that dust, if not properly taken into account, could distort one's understanding of the distribution of stars. Where dust dims the stars in a region of sky, you will see fewer stars. If the dust is so thick that the stars become completely invisible, you may be fooled into thinking that there is a hole in the distribution of stars. As astronomers began to understand how extensively dust is distributed in the Milky Way, they realized that Kapteyn's picture of the universe was amiss.

Harvard professor Harlow Shapley took a different approach. Sprinkled around the Milky Way are about 150 globular clusters, agglomerations of up to a million stars each. Globular clusters are beautiful objects, as illustrated by the picture of M13 in figure 7.3. In 1918, Shapley was able to estimate distances to the globular clusters and thus map their distribution in three dimensions. Given that these clusters are one of the components of the Milky Way, you might guess that they would be distributed more or less centered on the distribution of stars that Kapteyn was trying to map—that is, more or less symmetrically around the Sun. Instead, what Shapley found changed our conception of the universe: the center of the distribution of globular clusters was (to use the modern value) about 25,000 light-years from the Sun. The Sun was definitely off center. Shapley's globular clusters showed that the Sun was not in the middle of the known universe (which was the Milky Way, according to Shapley), but rather on its outskirts, and the full extent of the Milky Way was several times larger than Kapteyn had realized. Kapteyn had been horribly misled by all that dust. It turns out that the dust in the Milky Way is mostly concentrated to its central disk, or *galactic plane*, whereas the globular clusters mostly lie above or below this disk. Since the globular clusters are out of the galactic plane, the dust affected Shapley's analysis much less than it did Kapteyn's. Shapley was in effect the new Copernicus, showing that the Sun was not at the center of the Milky Way, that is, not at the center of the universe we could observe.

This was the extent of the known universe as it was understood by Shapley roughly 100 years ago: a flattened structure (the Milky Way), perhaps 100,000 light-years across, whose center lay 25,000 light-years from the Sun. These scales are enormous: one light-year is 10 trillion kilometers, so 100,000 light-years seems incomprehensibly large. But major discoveries in the 1920s, as discussed in chapter 13, made it clear that the visible universe is many orders of magnitude larger than even our enormous Milky Way galaxy.

Let's try to visualize just how large the Milky Way is. The nearest stars are about 4 light-years away, or 4×10^{13} kilometers. Divide this by the diameter of the Sun, 1.4 million kilometers. This will tell us how many Suns we would have to lay side by side to reach to the nearest star: 30 million. Placing 30 million Suns next to each other suggests an enormous distance indeed. The Sun itself is about 100 times the diameter of Earth. In other words, the distance to the nearest stars is 3 billion times the diameter of Earth.

The stars are tiny specks compared to the enormous distances between them. In *Star Trek*, every time they turn around, the *Enterprise* and its crew just happen to be passing by a "class M planet"; the writers of that show seem to have forgotten the huge separations between the stars. Perhaps this is why they must rely so much on their warp drive! (And we won't even talk about the fact that the aliens always speak perfect American English, even in the Delta Quadrant!)

It turns out that the distance of 4 light-years is a typical distance between stars in our galaxy. We now know that our Milky Way is a very flattened structure, a circular disk, roughly 100,000 light-years in diameter, but only 1,000 or so light-years thick. A thousand light years is a huge distance by human standards, and yet, relative to the full extent of the Milky Way, it is actually quite tiny. Most of the dust and the interstellar medium in the Milky Way are found in the disk. The extent of the Milky Way is about 25,000 times larger than the typical distances between stars, or 75 trillion times Earth's diameter.

The constellation of Sagittarius lies in the direction of the galactic center. With the dust of the interstellar medium concentrated in the disk of the Milky Way, the center of the Milky Way is heavily shrouded by dust, obscuring our view of it. In photographs of the Milky Way, we find regions in the disk of the Galaxy that show few stars, indicating particularly dense patches of dust that hide the stars behind them. The Sun lies in this disk, but if we look in directions away from the disk of the Milky Way, there is little dust obscuration, and we get a clear view of the universe beyond our galaxy.

Earth and our Sun lie close to the midplane of the Milky Way. Because the stars in the Milky Way are also largely concentrated in the flattened disk, we see the highest concentration of stars in a band that stretches in a full circle all around the celestial sphere. We can only see part of that full circle above the horizon at a given time; the remainder is beneath our feet, our view of it blocked by Earth itself. In the Northern Hemisphere, we get the best view of that part of the Milky Way lying in the direction away from the center of our galaxy. Because Earth and the Sun lie far from the center, relatively few of the Milky Way's stars lie in that direction, and we get a relatively sparse view. From the Southern Hemisphere, however, one can look directly toward the heart of the Milky Way, and the view is much more dramatic, in spite of the obscuring effects of dust. On a clear moonless night in May, away from city lights, in Chile, the view is breathtaking. Among my fondest memories are those times I spent looking up at the sky at Cerro Tololo Observatory in Chile next to the woman whom I would later marry, with the Milky Way dramatically splayed out across the sky over our heads.

An even better view is available if we look at the Milky Way in infrared light. We've seen already that dust obscures red light less than blue light, and infrared light is even less affected (see chapter 11). Figure 12.2 shows an infrared map of the entire sky made with the 2MASS telescopes (the same survey that made the stunning image of the galactic center in figure 11.2). The thin disk of the Milky Way dominates the image, and a central bulge is now apparent in the middle.

This map of the infrared sky is analogous to that in figure 11.1, taken in visible light. The horizontal "equator" in the middle of this projection is the *galactic plane*; the disk of the Milky Way, which is a full circle on the sphere of the sky, appears as a horizontal straight line in the figure. Although figure 12.2 is based

FIGURE 12.1. The Milky Way over Cerro Tololo. The night sky as seen from the Cerro Tololo Inter-American Observatory in the Chilean Andes. The large dome in the center of the picture houses the 4-meter-diameter Victor Blanco Telescope. The center of the Milky Way appears near the right edge of the picture. The Large and Small Magellanic Clouds, companion galaxies to the Milky Way roughly 150,000 light-years away, are apparent on the left. *Photo credit:* Roger Smith, AURA, NOAO, NSF

on data in the infrared, dust in the Milky Way still has some obscuring effects, and the patchiness one can see along the disk is due to dust. Finally, note the bulge in the center of the Milky Way; its slightly lumpy appearance is a clue to the fact that it is potato-shaped, rather than spherical as was originally believed. The Large and Small Magellanic Clouds, satellite galaxies of the Milky Way, can be seen below the plane of the galaxy and to the right.

Harlow Shapley realized he needed to look away from the plane (where the obscuring effects of dust are overwhelming) to understand the three-dimensional structure of the Milky Way. The globular clusters in the Milky Way are not concentrated in this plane, and thus are visible all over the sky. Shapley

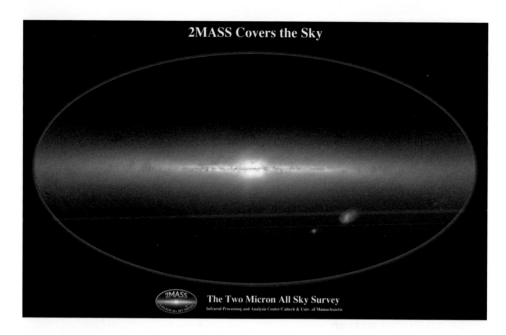

FIGURE 12.2. The Milky Way in the infrared. Shown is the distribution of stars over the entire sky, as measured by the Two-Micron wavelength All-Sky Survey (2MASS), a wavelength at which the obscuration due to dust is modest. The plane of the Milky Way galaxy stretches horizontally across the center of the map along the galactic equator. The Large and Small Magellanic clouds are below it. *Photo credit:* Atlas Image mosaic obtained as part of the Two Micron All Sky Survey (2MASS), a joint project of the University of Massachusetts and the Infrared Processing and Analysis Center/California Institute of Technology, funded by NASA and the NSF

wanted to make a three-dimensional map of their distribution, and therefore he needed to measure their distances. The way to do this was straightforward in principle, using the inverse-square law that relates brightness and luminosity: $B = L/(4\pi d^2)$. Thus if we measure the brightness of any star in a globular cluster (which is straightforward) and if we know the star's intrinsic luminosity (that's the hard part), we can determine its distance d. The correction for the effects of dust will be relatively small, because we're looking at a globular cluster away from the plane of the Milky Way.

How are we going to determine the luminosity of any given star? The main sequence shows a relationship between the color of a star and its luminosity (see figure 7.1). Assuming our observations are sensitive enough to identify main-sequence stars in the globular cluster, the colors of the main-sequence stars allow us to infer their luminosities; combining these with measurements of their brightness via the inverse-square law gives us the distance to the globular cluster.

Ah, if life were only that simple. The easiest stars to measure in a globular cluster are of course the brightest ones. All stars in the cluster are roughly the same distance from us, so the brightest stars we see are also the intrinsically most luminous stars in the cluster. But these are *not* main-sequence stars, but rather red giants, which display a large range of luminosities at a given color (because they vary greatly in size at a given color). With modern telescopes, we now have the sensitivity to observe the significantly fainter main-sequence stars in globular clusters, but in 1918, when Shapley was working, this was beyond the capabilities of the telescopes and instruments available to him. Instead, he used a type of star called an *RR Lyrae variable star*, a star about 50 times as luminous as the Sun, which varies periodically in brightness.

Variable stars are those whose luminosities (and therefore observed brightnesses) are not constant. RR Lyrae variable stars change their brightness by a factor of 2 on timescales of less than a day. They are pulsating, with their radius regularly increasing and decreasing. These are the typical variables found in globular clusters.

We know that stars are in equilibrium between the gravity holding them together and the pressure from their interior heat pushing outward. However, after becoming red giants, some stars become bluer and move quickly across the HR diagram. During this time, they undergo a stage where helium burns in the core, hydrogen burns in a shell, and the star's equilibrium is affected by the way the energy generated in the interior finds its way out of the star. This causes the internal pressure to oscillate, leading to corresponding changes in the size, and thus the luminosity (and brightness), of the star.

Although astronomers tend to give simple names to the objects they study ("red giant," "white dwarf," etc.), variable stars are an exception to this. When astronomers first started cataloging variable stars in the early 1800s, they gave them Latinized names for the constellation in which they resided. The first variable star discovered in the constellation Lyra, the Harp, was called R Lyrae; the letters A through Q had already been taken for other sorts of stars. When a second one was discovered in Lyra, they called it, naturally, S Lyrae, then T Lyrae, and so on; by then they realized that they were running out of letters, so after Z Lyrae, the next one was called RR Lyrae (the eponym for the entire class of variable stars like itself), then RS Lyrae, and so on, all the way to ZZ Lyrae. Even those names weren't enough, so they cycled back to AA Lyrae, AB Lyrae, and so on, running out at QZ Lyrae (and for some reason skipping the letter J). That gets you 334 combinations; but variable stars are even more common than that! The next variable star discovered in the constellation Lyra was called V335 Lyrae. Astronomers are up to V826 Lyrae as of this writing. There are many types of variable stars known, and the terminology for them can get complicated indeed: AM Canum Venaticorum stars, FU Orionis stars, BL Lacertae stars (which actually turn out to be a bizarre type of galaxy with a variable galactic nucleus), ZZ Ceti stars, and so on, each class of stars named after the first example discovered. Cepheid variables, which will be a key part of our study of distant galaxies in chapter 13, are named after their prototype Delta Cephei, discovered in the late 1700s.

Shapley used RR Lyrae variable stars as *standard candles* to measure the distances to globular clusters, using the fact that the luminosities of all RR Lyraes (after averaging over their variability) are all roughly the same. Measuring the (average) brightness of an RR Lyrae in a globular cluster, and knowing the luminosity, allowed him to infer the distance to the star, and thus to the globular cluster in which it was embedded. With the resulting three-dimensional map of the globular clusters, he could determine the position of the center of their distribution, finding that the Sun lies far from the center of the Milky Way.

Applying the standard candle approach to map the distribution of stars in the plane of the Milky Way (which is where most of the stars are found) is significantly more difficult because of the effects of dust. With much work over many decades, we now have a moderately complete picture of the overall structure of the Milky Way. Most of the stars are in a very flattened disk, with a diameter of about 100,000 light-years. It has no well-defined outer edge, but rather the density of stars just drops off steadily as one goes farther out. In the center of this disk, we find a thicker, roughly potato-shaped distribution of stars, about 20,000 light-years long, which we call the *bulge* of the Milky Way. The stars in the disk are arrayed along a series of spiral arms that radiate out from the bulge. Most of the stars that you can see with the naked eye are within a few thousand light-years of the Sun, in the same spiral arm in which the Sun is embedded.

Although the Milky Way is a spiral galaxy, we can't see this pinwheel structure in the sky, because we're embedded in the disk ourselves, and the pinwheel only becomes apparent when the distances to individual stars are measured, allowing a three-dimensional view of the Galaxy's structure. If we could some-how view the Milky Way from a vantage point a couple of hundred thousand light-years from here, where we could see it face-on, it would look like the artist's conception shown in figure 12.3. The Sun is in a spiral arm about half-way out, directly below the center (at 6 o'clock in the diagram). Our galaxy is a *barred spiral*, because its bulge has a bar shape. The spiral arms start from the ends of the bar.

Soon after we were married, my wife insisted that I no longer wear any of my nerdy T-shirts from my college days. The one I miss the most is the one with a picture of a galaxy, spiral arms and all, and an arrow pointing to a spot halfway out from the center, labeled "You Are Here."

Not all stars in the Milky Way are located in the spiral arms and the bulge. We have already seen that the globular clusters are distributed in a more-or-less

FIGURE 12.3. Simulated view of the Milky Way from above.
Photo credit: NASA Chandra Satellite

spherical distribution extending above and below the plane of the disk. In addition, a sprinkling of stars, far more sparse than those in the disk and also spherically distributed, extends about 50,000 light-years from the center of the Milky Way. We refer to this as the *halo* of our galaxy. We used to think that the stars in this halo were pretty smoothly distributed, with a concentration falling off gradually from the center of the Milky Way, but as astronomers have made ever-more accurate maps of the distribution of faint stars, they have found that the halo is anything but smooth. It has lumps and streams in it, believed to be the remnants of smaller companion galaxies that have fallen into the Milky Way and been torn apart by its tidal gravitational forces.

The stars in the bulge, and especially in the halo, tend to be old stars that formed billions of years ago. Thus the hottest main-sequence O and B stars, with their lifetimes measured in mere millions of years, are simply not found there; no star formation at all has been occurring in the halo of our galaxy for billions of years. The young hot stars are found almost exclusively in the spiral arms in the disk, where star formation is occurring now.

The spiral, or pinwheel, structure of the disk suggests that the whole structure is rotating. Indeed, this is exactly what is happening. The entire disk is rotating around its central axis, and the Sun in particular is moving in a roughly circular orbit with a speed of 220 km/sec. Just as the gravity of the Sun holds Earth in its yearly orbit, so also does the gravity of the Milky Way (at least that part within the radius of the Sun's orbit) pull the Sun and the planets around the galactic center in an orbit. Given a speed of 220 km/sec, and the radius of the orbit—a whopping 25,000 light-years, it is straightforward to calculate that the Sun goes around the Milky Way once every 250 million years. Thus the Sun has completed something like 18 orbits of the galaxy in the roughly 4.6 billion (Earth) years since it formed.

To calculate the gravitational force of the Milky Way on the Sun, we can treat the Milky Way's mass as concentrated at its center, 25,000 light-years away, just as the gravity of Earth acts as if all its mass were concentrated at its center, 6,400 kilometers beneath our feet. The mass that counts is the mass of the galaxy out to the radius of the Sun's orbit. The gravitational attraction due to the material outside that radius—each part pulling in different directions— roughly cancels out.

This suggests a calculation. Given Newton's laws of motion and law of gravity, we found in chapter 3 a relationship between the mass of the Sun M_{Sun}, the orbital velocity v_E of Earth around the Sun, and the radius r_E of Earth's orbit around the Sun:

$$GM_{Sun}/r_E^2 = v_E^2/r_E,$$

where G is Newton's gravitational constant. Multiply both sides of the equation by r_E^2 and we get:

$$GM_{Sun} = v_E^2 r_E.$$

We can write down a similar equation relating the mass of the Milky Way, M_{MW}, the velocity v_S of the Sun, and the radius R_S of the Sun's orbit around the center of the Milky Way:

$$GM_{MW} = v_S^2 R_S.$$

Divide the second equation by the first, and the quantity G drops out:

$$M_{MW}/M_{Sun} = (v_S/v_E)^2 \, (R_S/r_E).$$

The ratio of velocities is $v_S/v_E = (220 \text{ km/sec})/(30 \text{ km/sec})$, or about 7. The ratio of distances is: $R_S/r_E = 25{,}000 \text{ light-years}/1 \text{ AU}$. There are about 60,000 AU in a light-year, so this ratio is $25{,}000 \times 60{,}000 = 1.5 \times 10^9$. Thus

$$M_{MW}/M_{Sun} = 7^2 \times 1.5 \times 10^9 \sim 10^{11}.$$

Thus the mass of the Milky Way (within the radius of the Sun's orbit) is roughly 100 billion times the mass of the Sun.

The Milky Way is made of stars, and so we can say that the Milky Way contains roughly 100 billion stars, under the crude approximation that all stars have the same mass as the Sun. Indeed, the typical star in the Milky Way has a mass somewhat less than that of the Sun, and we haven't accounted for those stars farther out from the center of the Milky Way than the Sun, so a better estimate is that there are roughly 300 billion stars in the Milky Way. Carl Sagan in his classic TV series *Cosmos* often referred to "billions and billions" of stars, in his distinctive voice. Sagan wasn't exaggerating; the Milky Way indeed has billions and billions—roughly 300 billion—stars in it. We've used that number in Drake's equation.

The stars in the disk are all on approximately circular orbits. Stars are like cars on a circular racetrack. The ones on the inner lanes are passing the ones on the outside lanes. The spiral pattern we see is due to traffic jams in the stars as they circle. If you are on an expressway and approach a traffic jam where the cars are going more slowly than average, you will slow down too. Eventually you pass through the traffic jam, and then you can speed up as the cars speed up around you. The traffic jam represents a *density wave* in the pattern of cars. The cars are most densely packed in the traffic jam—although individual cars are continually moving through the traffic jam and passing out of it. In the same way, a spiral density wave in the galaxy represents a gravitational traffic jam of stars, whose gravity pulls even more stars toward it. Furthermore, as the stars crowd together, the interstellar gas is pulled together by the extra gravitational force there, causing clouds of gas to collapse gravitationally and form new stars. So the spiral arms are regions in which stars are actively forming. Among the newly formed stars are massive luminous blue stars, whose lifetimes are shorter than the time it takes for them to drift out of the traffic jam of the spiral arm. Thus, the spiral arms in galaxies are brightly lit up by newly born, massive, blue stars. Stars do not travel on spiral paths—instead the spiral arms shine brightly due to star formation caused by these traffic jams of stars circling the galactic center.

The mass of 100 billion Suns that we've just estimated represents that part of the Milky Way within the Sun's orbit. The gravitational forces from different parts of the Milky Way *beyond* the Sun's orbit pull us in opposite directions: material just outside the Sun's orbit on our side of the galaxy pulls us outward, whereas stuff outside the circle of the Sun's orbit but on the other side of the galactic center pulls us inward. These opposing forces effectively cancel each other out and have no net effect on the orbit of the Sun. The stuff inside the Sun's orbit, like the mass of Earth, acts as if it were located at the center. So if we can measure the orbital speed of stars at *different* distances from the center of the Milky Way, we can map out a mass profile of the Milky Way as a function of distance from the center of the Milky Way.

What do we expect to find? The Sun is roughly halfway out to the edge of the Milky Way, and the density of stars drops off considerably the farther out you go beyond the Sun. Star counts suggest the majority of the mass of the Milky Way is contained within the Sun's orbit. So we can apply the equation we've just used:

$$GM(<R) = v^2R,$$

where $M(<R)$ is the mass interior to radius R. If there is not much mass beyond the radius of the Sun's orbit, $M(<R)$ becomes constant, and beyond the Sun's

orbit we expect v^2R to be approximately constant, and v^2 to be proportional to $1/R$. Thus the orbital velocities v outside the solar orbit should scale proportional to $1/\sqrt{R}$. This behavior is seen in the solar system; the outer planets feel a weaker gravitational pull from the Sun and thus are moving more slowly in their orbits than the inner planets do. We expected to find stellar orbital velocities falling off with radius beyond the Sun's orbit.

Making these measurements in the Milky Way is difficult, and it wasn't until the mid-1980s that astronomers had determined the orbital speeds of stars and gas over a range of distances from the center of the Milky Way. To their very great surprise, they found that the orbital speeds did *not* decrease in the outer Milky Way, but rather stayed just about constant, as far out as the measurements went.

So what is wrong with our reasoning? We see little starlight when we look farther out from the center of the Milky Way than the Sun, and we inferred therefore that there was little contribution to the mass at those distances. We need to question that inference. We have used the orbit of the Sun to infer the mass of the Milky Way within its orbit; similarly, we can use speeds of stars orbiting even farther out in the Milky Way to measure the mass enclosed in those larger orbits. Using our equation $GM(<R) = v^2R$, we can see that if the velocity v stays constant, the mass interior to radius R goes up linearly with R. The farther out you go, the more mass you find. There is a significant component of the mass of the Milky Way outside the orbit of the Sun that is simply not visible in the form of stars. We call it *dark matter*. We have inferred its presence solely through its gravitational effect on stellar orbits.

How much dark matter does the Milky Way contain? The answer depends on how far out we think the Milky Way extends. The stars mostly peter out 40,000 light-years or so from the center, but the orbital speeds of the rare stars or clouds of gas even farther out are about the same as that of the Sun, 220 km/sec. Our best modern estimates tell us that the stars and interstellar medium in the Milky Way represent only a small fraction, perhaps 10%, of the entire mass of the galaxy. The vast majority of the Milky Way's mass, roughly a trillion times the mass of the Sun, is in the form of dark matter, extending perhaps 250,000 light-years from the center. We infer the same mass by calculating the mutual orbit of the Milky Way and its companion galaxy, the Andromeda Galaxy, once again using Newton's law of gravity. The two were once moving apart from one another as part of the general expansion of the Universe and are now falling together at a speed of about 100 km/sec, set to collide about 4 billion years from now.

The Caltech astronomer Fritz Zwicky was the first to discover dark matter, in 1933, when he measured the total mass of the Coma Cluster of galaxies with a sophisticated version of the $GM = v^2R$ formula, using the radius of the cluster and the velocities of the individual galaxies moving in the gravitational field of the cluster as a whole. He concluded that the cluster was significantly more massive than the total of the stars and gas making up the individual galaxies we could see. He dubbed the rest *dunkle Materie* in his native German, that is, "dark matter." As we will describe in chapter 15, this dark matter is almost certainly not composed of ordinary atoms but rather of elementary particles we have yet to identify.

Another very interesting form of nonluminous matter in the Milky Way occurs right at its center. Infrared observations of the center of the Milky Way can penetrate the obscuring dust. The stars in the very center of the galaxy are seen to be moving on Keplerian elliptical orbits, with semi-major axes as small as 1,000 AU (1/60 of a light-year) and periods of 20 years or so. The object about which they are all orbiting is invisible, but again Newton's Laws allow us to determine its mass: a whopping 4 million times the mass of the Sun. It is very small (certainly smaller than the orbits of the stars around it) and thus extraordinarily dense, and invisible. It turns out to be a black hole, one of the universe's most fascinating objects, which we will consider in detail in chapters 16 and 20. Thus, our study of the Milky Way has led us to the frontiers of physics, from new elementary particles populating the outskirts of our galaxy to a massive black hole lurking in its center.

13

THE UNIVERSE OF GALAXIES

MICHAEL A. STRAUSS

A century ago, when Harlow Shapley was determining the dimensions of the Milky Way and our place in it, astronomers' consensus understanding of the extent of the universe was the Milky Way itself. Indeed, when Shapley demonstrated that the Milky Way had an extent of tens of thousands of light-years, he was convinced that this enormous number proved that he had, in effect, mapped the entire universe. However, astronomers had long been intrigued by the nebulae they saw in their telescopes; while a star appears as a point of light through a telescope, nebulae are extended and often fuzzy looking. We have already encountered a variety of nebulae in this book, including planetary nebulae, which result when red giants throw off their outer layers; the Orion Nebula, a region of intense star formation in which the surrounding gas fluoresces due to light from hot young stars; and even dark nebulae, the dust clouds that block the light from background stars. However, there is another class, called *spiral nebulae* because of their shape, whose members closely resemble the Milky Way as we now understand it. The Milky Way is fuzzy looking itself. However, the spiral structure of the Milky Way disk was certainly not known 100 years ago, because, living in the disk itself, we didn't have a good understanding of its three-dimensional structure, making it harder to detect its resemblance to that larger class of objects. Remember that we have no depth perception in an astronomical image; we cannot tell a priori by looking at a particular nebula whether it is an intrinsically small object at a distance, of say, a few hundred light-years, or a truly enormous structure millions of light-years away.

Figure 13.1 shows a typical spiral nebula, M101, seen face-on. You can clearly see its spiral arms—like a pinwheel—so astronomers just call it the Pinwheel galaxy.

The physical nature, distance, and size of the spiral nebulae were among the most important questions facing astronomers in the first decade of the twentieth century. The German philosopher Immanuel Kant had speculated as early as 1755 that the spiral nebulae were other "island universes," that is, objects as large as the entire known universe, the Milky Way. Given Shapley's determination of the extent of the Milky Way and the small apparent angular size of spiral nebulae, if that was true, this meant that they must be astonishingly distant, millions or tens of millions of light-years away.

Shapley himself found this notion to be completely implausible, and in 1920, took part in a public debate with astronomer Heber Curtis of Lick Observatory in California on the nature of the spiral nebulae. Curtis was convinced that the hypothesis that the spiral nebula were galaxies like the Milky Way was correct, while Shapley said that the implied distances of the spiral nebulae were far too large to be believable. As is often true in science, controversies such as these only get settled with the advent of new and better data, and the debate itself was inconclusive. The astronomer who made the observations that settled the question once and for all was Edwin Hubble, of Mount Wilson Observatories in California. He used variable stars (a technique discussed in chapter 12) to determine the distance to the Andromeda nebula, the brightest spiral nebula in the sky (figure 13.2).

The Andromeda nebula is visible to the naked eye under ideal conditions (a clear, moonless night away from city lights), and indeed it was known to the ancients.

Mount Wilson Observatory, which sits in the San Gabriel Mountains overlooking the Los Angeles basin, had the largest telescope in the world at the time; its primary mirror is 100 inches (2.5 meters) in diameter. When Hubble took pictures of the Andromeda nebula with this telescope, he found that its diffuse light resolved into individual stars, just as Galileo had found when he pointed his primitive telescope at the Milky Way 300 years previously. That observation already told Hubble that Andromeda must be quite distant, but to get real numbers, he had more work to do. Based on repeated observations of the Andromeda nebula, Hubble was able to identify several stars that periodically brightened and dimmed, which he understood to be Cepheid variable stars. These are variable stars that are more luminous than RR Lyrae stars and

FIGURE 13.1. M101, the Pinwheel galaxy. *Photo credit:* NASA/HST

FIGURE 13.2. Andromeda galaxy from the Sloan Digital Sky Survey. The Andromeda galaxy is a spiral seen almost edge on, accompanied by two small elliptical satellite galaxies (M32 below, NGC205 above).

Photo credit: Sloan Digital Sky Survey and Doug Finkbeiner

FIGURE 13.3 Henrietta Leavitt, who discovered the relationship between the period and luminosity of Cepheid variable stars, key for measuring the distances to nearby galaxies. *Photo credit:* American Institute of Physics, Emilio Segrè Visual Archives

have pulsation periods from days to months. In 1912, Henrietta Leavitt, who worked at Harvard (see chapter 7), found a relation between a Cepheid's period of variability and its luminosity (figure 13.3). Hubble was able to measure their periods, use Leavitt's relation to infer their luminosities, and, by measuring their brightness, find their distances. The conclusion was stunning: the Andromeda nebula lay at the then inconceivably large distance of almost a million light-years, putting it well beyond the known extent of the Milky Way.

Images of the Andromeda nebula, out to its outer edges, showed an angular diameter of 2° on the sky. The circumference of a circle is 2π (a bit more than 6) times its radius. So a giant circle with a radius of a bit less than 1 million light-years will have a circumference of about 6 million light-years. Two degrees covers 1/180 of that complete 360° circle, from which Hubble could deduce that the diameter of the Andromeda galaxy must be about 6 million light-years/180, or about 30,000 light-years across. Hubble was therefore able to infer two compelling facts: (1) the Andromeda nebula is almost as large as the Milky Way itself, and (2) Andromeda lies well beyond the boundaries of the Milky Way.

Moreover, the sky was filled with other spiral nebulae, which appeared much smaller in angular size and fainter than Andromeda. If they were similar to the Andromeda nebula, they must be even farther away. This was a pivotal moment in the history of our understanding of the cosmos. Hubble had shown that the Andromeda nebula, and by extension, the other spiral nebulae, were roughly the same size as the entire Milky Way galaxy, and that they were situated at inconceivably large distances from us. Kant's hypothesis that the spiral nebulae were "island universes," as large as the Milky Way itself, was proven correct. The boundaries of the known universe had just taken a dramatic leap outward.

Two decades later, astronomers realized that there was more than one kind of Cepheid variable in the sky. When everything got straightened out, it turned out that Hubble had actually significantly *underestimated* the distance to the Andromeda nebula. Our modern estimate of its distance is 2.5 million light-years. Furthermore, modern photographs using digital cameras on telescopes (instead of film) show Andromeda's outer fainter regions extending to a diameter of about 3° in the sky. With these larger values, we infer that the diameter of the Andromeda galaxy (and indeed, we now call it a galaxy, not a nebula) is about 130,000 light-years, somewhat larger than the Milky Way. Still, Hubble's estimate was in the right ballpark, and his conclusion that Andromeda was another galaxy like the Milky Way was correct. Even a rough estimate was good enough to answer the big question posed by the Shapley–Curtis debate. Shapley was wrong and Curtis was right.

The Andromeda galaxy is only the nearest large galaxy. The images Hubble took with the telescopes at Mount Wilson Observatory showed that the sky was filled with galaxies. The Andromeda galaxy is indeed a spiral, but the spiral arms are indistinct and difficult to follow, partly because we are seeing the disk nearly edge on. But other galaxies have much more dramatic and coherent spiral arms.

Consider the Pinwheel galaxy shown earlier (see figure 13.1). We are seeing this galaxy nearly face-on, making the spiral arms clearly visible. It shows the same basic features as the Milky Way, including a central bulge (somewhat smaller than the Milky Way's bulge) and three spiral arms radiating out from the center. The Pinwheel's spiral arms are quite blue, a sign they include a significant number of hot, and therefore young, massive stars. This tells us that star formation is ongoing in the spiral arms, just as it is in the Milky Way. You can also see some thin dark "veins" along the spiral arms; these are dust clouds, which are confined to the disk and arms of the galaxy, as they are in the Milky Way. The central bulge is yellowish, indicating that the stars there are lower in temperature on average than those in the arms. The hot young stars seen in the arms are simply not present in the bulge. This is a general trend seen in most spiral galaxies, including the Milky Way and Andromeda: younger stars and active star formation are seen in the disk and arms; older stars are in the bulge.

Peppering the entire picture of the Pinwheel galaxy are many points of light. These stars are not part of the Pinwheel galaxy; at its distance (20 million light-years), individual stars would be much fainter than this. Rather, these are stars in our own Milky Way, at distances of perhaps a few thousand light-years, which show up along the line of sight. They are like raindrops on the windshield of your car. This reminds us again that when we look at the sky, we are seeing it as if projected onto two dimensions; with no depth perception, we do not know which objects are close by and which are far away. Indeed, some of the faint objects in the periphery of this figure are not stars but are themselves other galaxies in the background, at distances not of millions, but of billions, of light-years. The angular diameter of the Pinwheel galaxy is about half a degree on the sky; at a distance of 20 million light-years, its diameter is about 170,000 light-years, about twice the size of the Milky Way.

The galaxy in figure 13.4, called the Sombrero galaxy, has a huge bulge (much larger than that of the Milky Way), which completely dominates it and suggests the crown of a wide-brimmed hat. The galaxy is oriented such that the disk is seen almost edge-on, clearly showing how thin it is but obscuring the spiral structure. The edge-on view of this galaxy allows us to see the effects of dust, which lies in the plane of the disk and gives rise to the beautiful dark lanes in the disk (the "fringe" of the hat brim), exactly as we saw in our own Milky Way.

Not all galaxies have a disk—some are pure bulge, dominated by old stars, with very little gas or dust. Hubble called these *elliptical galaxies*.

Figure 13.5 shows the Perseus cluster of galaxies, in which hundreds of elliptical galaxies congregate in a region of space about a million light-years across. Indeed, almost every galaxy in this image is an elliptical. In addition, we see many stars in the foreground, because the Perseus Cluster sits behind a dense screen of stars in our own Milky Way.

Most luminous galaxies are either ellipticals or spirals, but some galaxies don't fit into either category, and we simply call them *irregular galaxies* because of their irregular shapes. The Large Magellanic Cloud, a small satellite galaxy (14,000 light-years across) that orbits the Milky Way at a distance of about 160,000 light years, falls into this category. It appears at the far left edge of figure 12.1, next to the observatory dome. Indeed, it is so close that it is easily visible to the naked eye.

The distance between the Milky Way and the Andromeda galaxy—2.5 million light-years—is roughly 25 times larger than the size of the two galaxies themselves. Galaxies are separated by distances that are considerably larger than their diameters, meaning that most of the volume of the universe

FIGURE 13.4. The Sombrero galaxy. The Sombrero galaxy is a spiral with a large bulge seen nearly edge-on.

Photo credit: NASA and the Hubble Heritage Team (AURA/STScI) Hubble Space Telescope, ACS STScI-03–28

FIGURE 13.5. Center of the Perseus cluster of galaxies from the Sloan Digital Sky Survey.

Photo credit: Sloan Digital Sky Survey and Robert Lupton

is *intergalactic space*—the space between the galaxies. However, we found in chapter 12 that the distance from the Sun to the nearest star is about 30 million solar diameters. Yet the distance to the next big galaxy is only 25 Milky Way diameters. Even when you get your head around the sizes of individual stars, the distances between stars are difficult to fathom. But if you can get used to the sizes of galaxies, the distances between them are not all that much larger. Given that galaxies are fairly close to one another, relative to their size, you will not be surprised to learn that they often collide with one another.

The Tadpole galaxy (figure 13.6), some 400 million light-years from Earth, is the result of a collision of a large and small spiral, the smaller of which appears, greatly distorted, within the arms of the larger one at the upper left. The gravitational interaction of the two has pulled one of the spiral arms of the larger galaxy into a long tail, about 300,000 light-years long, sprinkled with hot blue stars. The center of the bigger galaxy is fairly dusty, as can be seen by the dark dust lanes. Currently, the Milky Way and the Andromeda galaxy are falling together under the influence of their mutual gravitational attraction. When they collide with each other, about four billion years from now, gravitational tidal forces are likely to pull streamers of stars out of them like those seen in the Tadpole galaxy.

Astronomers have been arguing for decades about what happens in such galaxy mergers. After they settle down in a few hundred million years, do they turn into elliptical galaxies? This indeed brings up the basic question of how galaxies formed in the first place. The stars in elliptical galaxies tend to be older than those in spirals, suggesting that elliptical galaxies formed earlier in the history of the Universe. The bulges of spirals share similar properties with elliptical galaxies, suggesting that they may have formed in similar ways. Gas that falls later onto an already-formed elliptical galaxy can cool before it has had time to form stars. The cooling causes the gas to lose energy, but not angular momentum, which can make it form a thin rotating disk. This process could make a spiral galaxy with an elliptical bulge. The details of this process are still poorly understood and hotly debated.

There is even more to this picture of the Tadpole galaxy. If you look closely, you'll see many much smaller galaxies sprinkling the frame. These are full-sized galaxies, which just happen to be much more distant (thus appearing fainter and smaller). Some are billions of light-years away. Their light has thus taken billions of years to reach us: we're not seeing these galaxies as they are today,

FIGURE 13.6. Tadpole galaxy from the Hubble Space Telescope. This is actually two galaxies that have merged, and are in the process throwing out a long tail. Many faint and much more distant galaxies are also visible in this image. *Photo credit:* ACS Science and Engineering Team, NASA

but as they were when the universe was much younger. Telescopes are time machines: they show us the distant past and allow us to study the processes by which galaxies evolve through cosmic time. Of course, we see any given galaxy at only one epoch of its lifetime, but by comparing the properties of distant galaxies with those we see in the nearby universe, we can ask how the galaxy population has changed over billions of years and start to address the questions of when the galaxies formed, and why some are spirals and some are ellipticals.

Very long exposures with the Hubble Space Telescope have shown thousands of faint, distant galaxies in a region of sky just a few arc-minutes across (see figure 7.7). So there are of order 100 billion galaxies in the observable universe. Each of these barely resolved dots of light is a full galaxy, as large as the Milky Way, containing more than 100 billion stars. With 10^{11} stars in each of 10^{11} galaxies, we infer that the observable universe contains something like 10^{22} stars, a mind-boggling number indeed. What do we mean by "the observable universe?" How did all these galaxies form? To answer questions like these, we need to understand how the universe itself evolves, a subject to which we now turn.

14

THE EXPANSION OF
THE UNIVERSE

MICHAEL A. STRAUSS

In astronomy, we have two basic strategies for learning about the nature of objects in the sky. One is by taking pictures of them and measuring their sizes and brightnesses. The other is by measuring their spectra. We have seen that the spectra of stars allow us to infer their surface temperatures and elemental makeup. Using this, and our understanding of the HR diagram, we have been able to determine their sizes, masses, and states of evolution.

What can the spectra of galaxies teach us about their physical nature? Astronomers started measuring the spectra of galaxies about 100 years ago, around 1915. Galaxies are faint, and the telescopes back then were smaller, and the instruments far less sensitive, than those we have today. So measuring a spectrum of a galaxy required exposures many hours long. But these first spectra showed absorption lines just like those seen in stars (in particular, G and K stars), which immediately told the astronomers that galaxies are made of stars. Edwin Hubble came to the same conclusion when he resolved individual stars in his detailed photographic images of the Andromeda nebula a decade later (as described in chapter 13). The spectra of galaxies were refreshingly familiar to those astronomers accustomed to studying the spectra of stars. However, they quickly noticed a significant difference. The absorption lines, from such elements as calcium, magnesium, and sodium, were at wavelengths somewhat different from those seen in stars. Typically, all the spectral lines from an individual galaxy were shifted systematically to the red. We call this phenomenon the *redshift*.

We can understand how the redshift works by simply standing on a busy street corner and listening to a motorcycle drive by. You will hear a high-pitched whine as it comes toward you. Then as it reaches you and begins to travel away, the pitch of the engine's roar drops noticeably as it zips past you. The whole thing sounds like "Neeeeyaoooowwww!"

The sound we hear from the motorcycle is a pressure wave in the air, which (like light) has a certain wavelength and frequency; the higher the frequency is (and the shorter the wavelength), the higher will be the pitch your ear perceives. As the approaching motorcycle emits a succession of wave crests, it is moving closer and closer, crowding the successive wave crests together and giving a higher pitch. Conversely, the waves that reach you as the motorcycle is moving away from you are stretched out by the motion and thus have a lower pitch. This effect, first described by the Austrian Christian Doppler in 1842, works for light waves as well as sound waves: the motion of a distant star or galaxy will imprint itself as a systematic shift in the wavelengths of features in its spectrum. Thus we interpret the redshifts of galaxies as resulting from the Doppler effect: the galaxies are moving away from us. The fractional change in the wavelength of a wave emitted by an object moving at some speed is equal the object's speed, divided by the speed of sound (if we're talking about sound waves) or the speed of light (if we're measuring the light from some object). The speed of sound in air here on Earth is roughly 1,200 kilometers per hour, and a fast motorcycle can easily travel a tenth that speed. The corresponding change in pitch as the motorcycle passes you (from a velocity of 10% of the speed of sound approaching you to a velocity of 10% of the speed of sound receding from you) is about 20%—quite noticeable, equivalent to a musical interval of a minor third.

The wavelength of light is related to its color, and an object going away, with its light shifted to longer wavelengths, will be redder. The effect would be perceptible (at least to your naked eye) only for speeds that are an appreciable fraction of the speed of light. The motorcycle is traveling at a tiny fraction of the speed of light, which is why you don't notice the color of the motorcycle changing from blue to red as it zips by you. We can't watch stars or galaxies whiz past us at high speed, but they have specific spectral features, absorption lines corresponding to their elemental makeup, whose wavelengths we know accurately from laboratory measurements here on Earth. We can measure the wavelengths of the same features in a given star or galaxy; the difference between the wavelengths from these elements seen on Earth and in the star or galaxy, interpreted as a Doppler shift, tells us how fast that star or galaxy is moving relative to us.

By 1915, Vesto Slipher, working at the Lowell Observatory (where Pluto was later discovered) had measured the Doppler shifts of 15 galaxies. Andromeda and two other galaxies were blue shifted, showing these galaxies were moving toward us, but all the rest were redshifted and thus moving away from us. We define the redshift z to be the quantity $(\lambda_{observed} - \lambda_{lab})/\lambda_{lab}$, where λ_{lab} is the wavelength of the emission or absorption line of an element in the lab on Earth, and $\lambda_{observed}$ is the wavelength observed for that element's line in the galaxy spectrum. The redshift z of a nearby galaxy is related to its recessional velocity v by the formula: $z \approx v/c$. Thus, a galaxy with a recessional velocity of 1% of the speed of light will have a redshift of $z = 0.01$ and will have the wavelengths of all its spectral lines shifted to longer wavelengths by 1%. The astronomical community has now measured the spectra of more than two million galaxies; with only a handful of exceptions like Andromeda, all of them show a redshift. We thus conclude that essentially all galaxies in the universe are moving away from the Milky Way. I once saw a silly cartoon showing a mad scientist at the telescope, waving his arms in the air, saying, "The galaxies are fleeing because they hate us!" That's not the correct explanation, but it is remarkable that we seem to be in a special position, at the center of the motion of all the galaxies. What's really going on? It was Hubble who again made the critical measurements, in the late 1920s and early 1930s, that led to our modern understanding of these redshifts.

After measuring the distance to the Andromeda nebula using Cepheid variable stars, he continued this effort with other galaxies, using a variety of estimates to determine their distances. This gets increasingly difficult for more distant galaxies; it is harder and harder to distinguish individual stars the more distant a galaxy is. His measurements were crude by modern standards, but by the late 1920s, he had rough measurements of the distances of a number of galaxies, for which the spectra—and thus the redshift and inferred speed—had also been measured. He then made a simple plot, comparing the distances of galaxies to their speeds. What he saw was a trend: the more distant the galaxy, the higher the speed. Indeed, despite the substantial measurement errors, he was able to conclude that speed v and distance d appeared to be proportional to one another:

$$v = H_0 d.$$

This proportionality between speed and distance is now known as *Hubble's law*, and the constant of proportionality H_0 ("H naught") is now called the *Hubble constant* in his honor. The Hubble constant is indeed constant throughout the universe at any given point in time, but as we'll see later, it does change with cosmic epoch. The quantity H_0 refers to the value of the Hubble constant at present.

In retrospect, it is remarkable that Hubble was able to infer the proportionality between redshift and distance, given the rather poor quality of his data (remember that his measurement of the distance to the Andromeda galaxy was too small by a factor of 2.5). Telescopes and techniques have gotten much better since 1929. Indeed, one of the key projects the Hubble Space Telescope was designed to do was to make accurate measurements of the distances of galaxies, using, among other techniques, the measurement of Cepheid variable stars, just as Hubble did. These measurements have demonstrated that Hubble was right, and galaxy redshifts and distances are indeed accurately proportional. It is often true that groundbreaking discoveries are made from poor data at the leading edge of what is possible with the technology of the time. Hubble's first plot only included galaxies out to a velocity v of about 1,000 km/sec, corresponding to a modern distance of about 50 million light-years. By 1931, Hubble and his colleague Milton Humason had extended the plot to include galaxies receding at 20,000 km/sec. That really cinched the case.

Is it really true that the Milky Way galaxy occupies a special position in the universe, a point away from which all other galaxies are moving? Such a notion would go against a recurring theme we have encountered, sometimes termed the *Copernican Principle*: that Earth is not in a special place in the universe. Ptolemy and the ancients put Earth at the center of the universe, but Copernicus demonstrated that Earth orbits the Sun. We then learned that the Sun is an ordinary main-sequence star, and although Kapteyn first thought that the Sun lay at a special place near the center of the Milky Way, Shapley's more accurate work demonstrated that the Sun lies about halfway out from the center. The measurements of redshifts, at first glance, seem to put the Milky Way at a special place relative to the other galaxies—at the center of the expansion. But that is not the case.

Consider four galaxies equally spaced along a line: Galaxy 1 is on the left, next comes the Milky Way at a distance of 100 million light-years, with Galaxy 3 at a

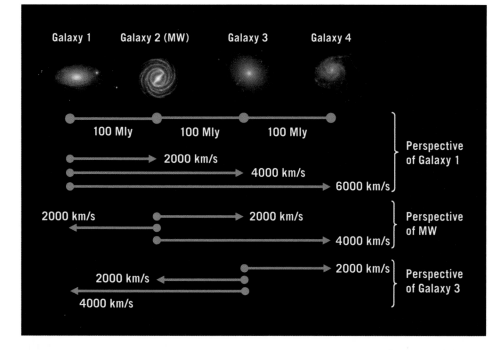

FIGURE 14.1. Galaxies in an expanding line, illustrating that no galaxy is at the center of an expanding universe. Four galaxies are shown across the top; the second galaxy is a representation of the Milky Way. They are separated by 100 million light-years (Mly) each. Because of Hubble's Law, they are moving apart from one another as the line expands; the first set of three arrows shows the relative speeds, as seen from the perspective of Galaxy 1. Because motions are relative, an astronomer in the Milky Way thinks she is at rest and the other three galaxies are moving away from her at speeds proportional to their distances (next set of arrows). The same is true from the perspective of Galaxy 3; all observers separately conclude that they are at rest, and all galaxies are moving away from them at speeds following Hubble's Law. *Photo credit:* Michael Strauss, Milky Way (schematic artist's conception from NASA); other galaxy images (courtesy Sloan Digital Sky Survey and Robert Lupton)

distance of 100 million light-years farther on, and Galaxy 4 another 100 million light-years farther still (i.e., 300 million light-years away from Galaxy 1). Hubble's law states that, as seen by Galaxy 1, the Milky Way is receding at a velocity of about 2,000 km/sec (see the first set of arrows in figure 14.1). Galaxy 3 (twice as far away from Galaxy 1 as the Milky Way) is receding from Galaxy 1 at a velocity of 4,000 km/sec—twice as fast, and Galaxy 4, three times as far away, is receding from Galaxy 1 at 6,000 km/sec. How does it look to us on the Milky Way? That is shown in the second set of arrows. We are separating from Galaxy 1 at 2,000 km/sec, but we measure motions relative to our frame of reference, and so we see Galaxy 1 moving away from us, to the left, at 2,000 km/sec. We see Galaxy 3 moving away from us at 2,000 km/sec in the opposite direction, to the right. The two galaxies are equidistant from us and are moving away from us at equal speeds. Galaxy 4 is moving away from us at a relative velocity of

4,000 km/sec. It's twice as far from us, and it is moving away from us at twice the speed. We see all the galaxies fleeing from us, and the farther away they are, the faster they are fleeing—our observations also fit Hubble's law.

Now take the perspective of an alien on a planet in Galaxy 3. All that counts in the Doppler shift is the relative speed of galaxies. From the alien's perspective, she sees the Milky Way, at a distance of 100 million light-years, moving away (toward the left) at a speed of 2,000 km/sec. Galaxy 4, at a distance of 100 million light-years in the other direction, is moving away from her (in the other direction) at a relative speed of 2,000 km/sec. Finally, Galaxy 1 is moving away from her at a relative speed of 4,000 km/sec. That alien sees all the galaxies moving away from her, and concludes that *she* sits at the center of the motion. The alien thinks she is at rest, and all the galaxies are fleeing from her, just as we concluded here on the Milky Way that we were at rest and all the galaxies were fleeing from us. We and the alien both conclude that speed is proportional to distance, and neither the Milky Way nor Galaxy 3 is in a special location.

Hubble's Law really tells us two things. First, the distance between any two galaxies is increasing: all galaxies are moving apart from one another. Hubble discovered that the universe is expanding! Second, there is no single galaxy that is uniquely at the center of the expansion. Sitting on any single galaxy, we come to the conclusion that all the other galaxies are moving away from us. It is like the galaxies are beads attached to a stretching rubber band, with all the beads moving away from each other. To fully conclude that there is no center to the expansion, we really need one more ingredient: the confidence that there is no edge to the distribution of galaxies. Rich will return to this topic in all its nuances in chapter 22, when he discusses Einstein's theory of general relativity as applied to cosmology.

The Milky Way is about 100,000 light-years across but is only one of 100 billion (10^{11}) galaxies in the observable universe, each with of order 100 billion stars. Andromeda, the nearest big galaxy, is at a distance of 2.5 million light-years from the Milky Way; most galaxies are much farther away still, with distances measured in billions of light-years.

Edwin Hubble discovered that galaxies are moving apart from one another with speeds proportional to the distances separating them; this speed can be an appreciable fraction of the speed of light for a distant galaxy. From this, we are able to conclude that the universe as a whole is expanding. This was truly one of the great scientific discoveries of the twentieth century, on a par with the discovery of the structure of DNA and its role in transmitting the genetic code or Einstein's development of his theories of relativity.

Hubble's law gives us an easy method for measuring the distances to galaxies. Given the proportionality between the redshift and distance of a galaxy, a measurement of a galaxy's redshift (easy to obtain if you can measure its spectrum) directly leads to an estimate of its distance (which is otherwise hard to measure). This works fine, as long as you know the proportionality constant H_0 relating the two. To determine its value, you must first accurately measure the distances to a sample of galaxies in an independent way.

As we have seen before, measuring the distance to an astronomical object is an essential step in understanding it. Knowing its distance, we can determine many key quantities about an object, including its luminosity and size. Much of the story of astronomy is therefore centered on the sundry clever means scientists have developed for measuring distance. The measurement of the AU (the distance between Earth and the Sun) in physical units (i.e., meters) was one of the preeminent scientific problems in the eighteenth and nineteenth centuries, finally properly solved with observations of Venus transiting across the face of the Sun, and Mars passing near distant stars, as seen from different locations on Earth (see chapter 2). This parallax effect allowed the distances to Venus and Mars, and thus the AU, to be determined by triangulation. The AU sets the distance scale for the entire solar system and also allows us to use the parallax effect due to Earth's orbit around the Sun to determine distances to nearby stars. For stars that are too distant to show a measurable parallax—beyond a few hundred light-years[1]—we use the inverse-square law relating a star's intrinsic luminosity to its observed brightness in the sky. The fainter an object of known luminosity appears in the sky, the farther away it is.

The tricky part of this is knowing the luminosity of your object. We have talked about Cepheid variable stars, which are an example of a *standard candle*, an object whose true luminosity can be determined, allowing the inverse-square law to be applied to infer distances. A good standard candle must

1. be luminous enough to be seen at large distances;
2. be readily identifiable and distinguishable from other objects; and
3. have comparable examples close by, so that its absolute luminosity can be calibrated (e.g., by the parallax effect or other methods).

Cepheid variable stars satisfy the first two of these requirements; they are very luminous, and their variability allows them to be identified in a dense star field. However, few Cepheid variables are close enough to have accurately measured parallaxes, which has led to controversies about their true luminosities. Indeed, it was a miscalibration of the distance to Henrietta Leavitt's Cepheid variables, resulting from others misidentifying nearby counterparts, that caused Hubble to underestimate the distance to the Andromeda galaxy. The closest Cepheid variable star in the sky is Polaris, the North Star, at a distance of about 400 light-years.

We have seen that stars on the main sequence show a direct relationship between temperature and luminosity. Thus if we can measure the temperature of a star (e.g., from its spectrum), we can make a good estimate of its luminosity; then using its apparent brightness, we can measure its distance. This standard candle has been fairly well calibrated from measurements of nearby stars by parallax, and it can be used on more distant stars—stars so distant that their parallax is too small to measure. Only the most luminous stars can be seen at large distances, but these very luminous stars are also rare, and therefore, few of them are close enough to allow their parallax to be measured.

This basic approach of using the main sequence to define standard candles can be done for a whole group of stars instead of one star at a time. For example, all the stars in a globular cluster are effectively at the same distance. Therefore if we compare the main sequence of the stars in a globular cluster today with the (calibrated) main sequence of nearby stars, we can determine the distance to the cluster directly. In doing so, we can find the distance to the relatively rare stars in the cluster for which there are no near-enough examples available to measure parallax.

Like stars, galaxies come in a large range of luminosities. Something roughly analogous to a main sequence seems to exist for spiral galaxies, in which there is a correlation between the speed at which the galaxy rotates (measurable from its spectrum by the Doppler effect) and its luminosity. We can calibrate this rotation–luminosity relation for nearby spirals. Then we can use measurements of the rotation of more distant spirals to estimate their intrinsic luminosities, and thus (given additional measurement of their brightnesses), we can determine their distances.

This series of steps—whereby measurements of the distance to one type of object is used to infer the distance to another, more luminous but rarer type of object, which can in turn be used to measure objects at greater distances still—is called the *cosmic distance ladder*. If this ladder is starting to sound a little rickety, it is indeed, and the uncertainties multiply as we move out to larger distances. Thus the determination of the Hubble constant H_0, relating the redshift and distance of galaxies, has been quite controversial.

Hubble's law, $v = H_0 d$, implies that the Hubble constant H_0 has units of a velocity v moving away from us (usually measured in kilometers per second) divided by a distance d, measured in megaparsecs (Mpcs; i.e., millions of parsecs). Hubble's estimate of the Hubble constant was about 500 (km/sec)/Mpc (too big, as we've seen, because of his underestimate of the distance to the Andromeda galaxy, due to the miscalibration of Cepheids by others). Hubble passed away in 1953, soon after the great 200-inch (5-meter) diameter telescope at Palomar Mountain near San Diego was completed. His former assistant Allan Sandage took over his program of determining the distances to galaxies.

Over the following decades, Sandage and his collaborators used the Palomar 200-inch telescope and other telescopes around the world to make tremendous advances in our understanding of galaxies. By the early 1970s, Sandage really only had one important rival in the determination of the distances to galaxies and, therefore, the Hubble constant: a University of Texas astronomer named Gérard de Vaucouleurs. In the 1970s, the groups headed by Sandage and de Vaucouleurs each wrote a monumental series of papers outlining their "steps to the Hubble Constant." Sandage's answer was about 50 (km/sec)/Mpc (a full factor of 10 smaller than Hubble's original estimate), whereas de Vaucouleurs' estimate was about 100 (km/sec)/Mpc. They differed on every detail and step of the cosmic distance ladder. Everyone in the astronomical community cared deeply about the result, because the value of the Hubble constant sets the scale for our universe. The redshift of a galaxy is straightforward to measure from its spectrum; if we also know the Hubble Constant, we can translate that redshift into a distance.

Finally, in the early 1980s, various younger astronomers dared to enter the fray, with the introduction of new kinds of standard candles and improved observational techniques. The Hubble Space Telescope was designed in part to address this question: free from atmospheric interference, its superior resolving power allows it to identify and accurately measure the properties of Cepheid variable stars in galaxies 30 to 40 million light-years away. A team led by Wendy Freedman (who was for many years director of the Carnegie

Observatories in Pasadena, where Sandage worked) carried out an extensive observing campaign with the Hubble Space Telescope. They published their results in 2001, finding $H_0 = 72 \pm 8$ (km/sec)/Mpc, a value almost exactly halfway between the Sandage and de Vaucouleurs results. Interestingly, Rich Gott and his colleagues in 2001 made an estimate of the Hubble constant by combining all published measurements of its value in papers up to that time (which had used a wide variety of methods) and taking the median, or middle, value: 67 (km/sec)/Mpc. The median is often a surprisingly good indicator, being less influenced by errant values than a straight average. The best estimate today, more than a decade later—using measurements of the cosmic microwave background (CMB) by the Planck satellite—is 67 ± 1 (km/sec)/Mpc. As we'll discuss in chapter 23, this value has been confirmed by an measurement of 67.3 ± 1.1 (km/sec)/Mpc by the Sloan Digital Sky Survey team combining results from supernovae, galaxy clustering, and the CMB.

Allan Sandage passed away in 2010 at the age of 84, one of the giants of our field. In his last paper on the subject in 2007, he said that the Hubble constant probably lay in the range from 53 to 70 (km/sec)/Mpc; thus, he was willing to countenance a value as high as we measure today.

With the value of the Hubble constant now nailed down, we can return to exploring the ramifications of Hubble's law and the expansion of the universe.

You can picture the universe as an enormous loaf of raisin bread rising in the oven. The galaxies are the raisins, and the dough is the space between them. As the bread rises (as the dough expands), each raisin moves farther apart from every other raisin, and so from the standpoint of each raisin, all the other raisins are receding from it. Therefore each raisin (galaxy) could (erroneously) conclude that it is at the center of the raisin bread (the universe). Furthermore, a raisin that was twice as far away from the first raisin would be receding twice as fast, because there would be twice as much expanding dough in between. The raisin bread universe follows Hubble's law.

This analogy is not perfect. Whereas our raisin bread has a well-defined center, which we can locate because it has a crust, the real universe appears (as far as we can measure) to be infinite in extent, with no edge that would allow us to define a center. We'll return to the question of the shape or geometry of the universe in chapter 22.

Hubble's law tells us that galaxies in general are moving apart from one another, leading to the conclusion that the universe is expanding. Does this mean that the individual galaxies are expanding, with the stars moving apart from one another? Is the solar system expanding? The Sun? Our very bodies? Those of us trying to lose weight may say yes to the last of these, but in fact, the Hubble expansion of the universe holds only on the scale of the distances between galaxies. Galaxies, like the raisins, don't themselves expand, but rather it's the space between the raisins that's expanding. Objects held together by gravity or other forces, including individual galaxies, individual stars and planets, and even ourselves, are not expanding. In fact, even the Milky Way and the Andromeda galaxies are gravitationally bound to each other, and therefore falling together, not moving apart. The Andromeda galaxy is thus one of the handful of galaxies that exhibit a blueshift.

We've already mentioned that the Milky Way and Andromeda galaxies will collide in roughly 4 billion years (before our own Sun exhausts the hydrogen in its core and becomes a red giant). However, the distance between individual stars in each galaxy is so vast compared to the stars' sizes, that the two galaxies will pass through each other with essentially no collisions among stars. Thus Hollywood is unlikely to make a blockbuster disaster flick called *Galaxies in Collision*—well, actually, they might; it would not be the first time that a film played havoc with scientific fact for dramatic effect!

If the universe is expanding now, and the space between the galaxies is growing with time, then the galaxies were closer together in the past. Consider a galaxy that lies at a distance d away from us. It is moving away from us with a speed, given by Hubble's law, of $H_0 d$. Crudely assuming that this speed remains constant with time, we can ask, how long does it take this galaxy to travel the distance d? Equivalently, how long ago was that galaxy right here on top of us? If a city is 500 miles away, and someone comes to visit me from that city driving at 50 miles per hour, the time it will take him to traverse that distance is just the distance divided by the velocity: 500 miles/50 mph = 10 hours. In our case, we want to know how long ago the galaxy was on top of us. That length of time t in the past is equal to the distance the galaxy traveled d divided by its velocity v (which is equal to $H_0 d$ by Hubble's law):

$$t = d/v = d/(H_0 d) = 1/H_0.$$

This seems like a simple result, and indeed it is. But it has quite a bit to tell us. Notice that the time t does not depend on the distance d to that galaxy. Thus we would find the same time t in the past at which any galaxy would have been on top of us. It seems as if the galaxies were all together at some single time in the past. Before exploring that thought any further, let us keep in mind that this *still* doesn't mean that we are at the center of the expansion; we could have gone through the same argument but centered our calculation on any other galaxy and obtained the same result. We are led to conclude that there was a time in the universe when all matter in the universe was compressed together. All the "raisins" were compacted together. And we know when that time was! It is a time $1/H_0$ ago. This is another reason people care so deeply about the value of the Hubble constant. It tells us the age of the universe.

Let's do the calculation. The best current estimate of the Hubble constant from the Planck satellite team is 67 (km/sec)/Mpc, so its inverse $1/H_0$ is $(1/67)$ sec $\cdot$ Mpc/km. A megaparsec is equal to 3.086×10^{19} kilometers, so substituting that number for Mpc/km and dividing by 67, we find that $1/H_0$ is 4.6×10^{17} seconds. Converting from seconds to years, that time when all the galaxies were on top of each other was roughly 14.6 billion years ago.

We refer to this time as the *Big Bang*, a term coined by Fred Hoyle in the late 1940s. Even though Hoyle was a lifelong opponent of the Big Bang model and went to his grave convinced that the idea was wrong, the term has stuck ever since. In 1994, Carl Sagan, science journalist Timothy Ferris, and television broadcaster Hugh Downs felt, like Calvin (of comic strip fame; figure 14.2), that such an important concept, core to our modern understanding of cosmology, deserved a more evocative name than "Big Bang." They held an international competition, asking people to suggest alternative names. They received more than 13,000 suggestions, sifted through them all, and then completely backed down, deciding, after considering the alternatives, that "Big Bang" was good enough after all.

Hubble's law has led us to the conclusion that at a specific time, about 14.6 billion years ago, all the universe was crushed together; it has been expanding ever since. Our calculation of the time since the Big Bang was crude, as we assumed that each galaxy moves at a constant velocity, but the modern value based on a more sophisticated calculation is close, about 13.8 billion years. Does this estimate for the age of the universe (for this is what we're really talking about) make sense? We know the age of the solar system, mostly from measurements of radioactivity in rocks from the Moon and meteorites; it is about 4.6 billion years. This is in the same ballpark as the expansion age of the universe, although comfortably smaller. The Sun and the solar system

FIGURE 14.2. The "Horrendous Space Kablooie," *Calvin and Hobbes* cartoon.

have been enriched in heavy elements formed in earlier supernovae, so we would not expect the Sun to be among the earliest stars formed. We have also described how we can use the position of the turnoff of the main sequence in the HR diagram of globular clusters to determine their age: the oldest globular clusters are between 12 and 13 billion years old.

It is absolutely amazing that these three different, and completely independent, ways of estimating the age of the universe (and the oldest objects in it) are consistent with one another! We should marvel that these estimates agree with one another to within a factor of three; that is a great triumph for our fundamental ideas of how the universe is put together. That they are all in the same ballpark (and that the oldest objects we know about in the universe have an age less than the time since the Big Bang) gives us real confidence that our basic physical ideas are correct.

Let us now imagine what the universe was like in the past. Because the universe is expanding, its density is decreasing with time, a given amount of mass occupying a larger volume at later times. So at earlier times it was denser. Just as for stars, the denser you make things, the hotter they tend to be, so in the past, the universe was much hotter than it is now (in chapter 15 we'll talk about what we mean by the temperature of the universe; it turns out that this concept is quite well defined). Indeed, using the simplest extrapolations, there appears to be a moment, about 13.8 billion years ago, when the entire observable universe would have been infinitely hot and infinitely dense, and from that time to the present, it has been expanding and cooling. We cannot trace back any further than 13.8 billion years ago—this is our definition of the birth of the universe. The expansion started with a Big Bang and is still observed in the universe today in the form of Hubble's law.

At the time of the Big Bang, then, the universe appears to have been infinitely dense and infinitely hot. Was it also infinitely small? This is where things get tricky. The answer is: not really, in the sense that we would usually use the word "small" in English. Let us assume that the universe today is infinitely large. "Wait a minute!" you might protest. "You've been telling us throughout this book that the observable Universe is finite, with a radius of a few tens of billions of light-years!" Indeed. We'll make a distinction between the universe as a whole, and the *observable* universe, the part we can see today; it is the latter that has a finite size. The universe is expanding, and therefore decreasing in density, but if it is infinite in size today, shrinking it in the past still leaves it infinite in size, and that was true all the way back to the Big Bang. In the beginning, that makes the universe infinite in extent, infinitely dense, and infinitely hot. It had no center, and certainly no edge outside of which one could look at the universe as a whole.

This may all sound like semantics, but it is the simplest way to think about our modern understanding of the early universe. What we are doing here is putting into words the results one finds when solving the appropriate equations of Einstein's theory of general relativity, as we'll explore in later chapters. The Big Bang was not an explosion, as it is sometimes erroneously depicted, of something very small and dense expanding into empty space. It is not like a bomb. Because the universe has no edge, there is no empty space "out there" for it to expand into. It is the *space itself* that is expanding.

If there is no such thing as an outer edge to the universe, can we ask what existed before the Big Bang? Unfortunately, our equations do not let us do so. Yes, it is a reasonable question, but no, general relativity doesn't have an answer for you. The equations of general relativity predict an infinite density at the moment of the Big Bang. In science, when your equations yield a result of infinity, you know that your theory is incomplete; there is more physics going on than the equations describe.

Thus the equations of general relativity break down at the moment of the Big Bang, which is why we can't extend them to a time before the Big Bang. "What happened before the Big Bang?" is a question that cosmologists are asked all the time, and unfortunately, they often respond by saying that this is a meaningless question, implying that the questioner is silly for asking. They are not silly for asking: that the equations break down at the Big Bang is a sign of a problem with the theory, not the question! We return to these questions in chapters 22 and 23, when we ask about the overall geometry of the universe and what might have started the Big Bang itself.

In any case, because of this ignorance, cosmologists consider time to start at the Big Bang. It is our own creation myth, but as we've seen, it is drawn from direct observations of the universe and our understanding of physics. The universe appears infinite in extent but has a finite age. This finite age, because of the finite velocity of light, means that there is only a finite part of the universe that we can observe. Consider, for example, our situation today, just 13.8 billion years after the Big Bang, sitting in the Milky Way galaxy. The universe is infinite around us, but we can't see it all, because light travels at a finite speed. Light from the most distant material we can see now has been traveling toward us for 13.8 billion years, and has traversed only 13.8 billion light-years in distance through the ever-expanding space between that material and us. But we are seeing that stuff as it was in the past—where it used to be. Where is it now? The expansion of the universe in the meantime has carried that material (now formed into galaxies) out to a distance of 45 billion light-years from us by now. This represents the boundary of the present-day observable universe. Beyond those galaxies lie other, more distant, galaxies from which we have never received photons. As we'll see in future chapters, the space between them and us is simply expanding so fast that light from them has not had time to traverse it. So beyond the edge of the observable universe, there is much more universe out there, indeed, an infinite amount, if we are to believe our current measurements of the geometry of the observable universe and our cosmological models. You might consider this the greatest extrapolation in science: we carry out observations within our finite observable universe with a current radius of "only" 45 billion light-years, and extrapolate to an infinite universe!

15

THE EARLY UNIVERSE

MICHAEL A. STRAUSS

The early universe just after the Big Bang was very hot and dense, but it was expanding and cooling. Our equations allow us to do detailed calculations of the expected state of matter in the early universe; it is a fertile area for physicists, because it involves the calculation of the properties of matter at extremely high temperatures and densities. Moreover, nuclear reactions in the early universe leave telltale traces in the chemical abundances of the elements we see in the universe today. We'll see that these predictions of light-element abundances from Big Bang physics accord beautifully with observations, giving us confidence that we actually understand what happened in the first moments after the Big Bang. Let's start the story about 1 second after the Big Bang. The universe was tremendously hot, about 10^{10} K (10 billion kelvins!) and terribly dense by human standards, about 450,000 times the density of water. Galaxies, stars, and planets didn't exist yet. Indeed, it was much too hot for atoms or molecules, or even atomic nuclei to form. The ordinary material of the universe at this point consisted of electrons, positrons, protons, neutrons, neutrinos, and, of course, lots of blackbody radiation (i.e., photons). And if, as is currently thought, dark matter consists of as-yet undiscovered elementary particles, we would expect those particles to also exist in large numbers in the universe at this time.

But two and a half minutes later, the universe has cooled down to a temperature of "only" a billion kelvins; at this time, the photons have a blackbody spectrum peaking in gamma rays. A billion kelvins is cool enough to permit nuclear fusion reactions in which neutrons and protons are able to stick together. In the Sun, we found that under high temperatures and densities, protons fuse together to make helium nuclei (see chapter 7). In the center of stars like the Sun, it takes billions of years to turn 10% of the hydrogen into helium. The nuclear reactions taking place in the early universe go much faster, because free neutrons as well as protons are present. Proton–proton collisions require high energy, because both protons are positively charged and they repel each other, making actual collisions infrequent. Neutrons are electrically neutral (and thus are not repelled by protons), so neutron–proton collisions occur more often. Fusion can occur by adding neutrons to protons, on the way to producing helium. This allows the slow first steps of the solar fusion process (proton–proton collisions) to be skipped.

Protons and neutrons can transmute into each other. A neutron plus a positron can combine to give a proton plus an anti-electron neutrino, and vice versa. A neutron plus an electron neutrino can combine to give a proton plus an electron, and vice versa. And a neutron can decay into a proton by emitting an electron and an anti-electron neutrino. At 10 billion kelvins (the temperature when the Universe is 1 second old), these processes are in balance. Neutrons are slightly more massive than protons, meaning they take slightly more energy to make and so there are slightly fewer neutrons than protons 1 second after the Big Bang. But by the time the universe cools to a billion kelvins as it continues to expand, this balance changes so that more neutrons are converted into lighter protons, yielding seven protons for every neutron. At a temperature of a mere billion kelvins, less thermal energy is available to make up the ($E = mc^2$) mass difference between protons and neutrons; neutrons, therefore, become rarer relative to protons. At this point, the universe has cooled enough for a neutron and a proton to collide and stick together to form a *deuteron* (the nucleus of heavy hydrogen—deuterium) without the deuteron immediately coming apart when it collides with the next particle. A deuteron can then participate in additional nuclear reactions to add an additional neutron and an additional proton to form a helium nucleus (two neutrons and two protons). After just a few minutes of nuclear burning, essentially every neutron is incorporated into a helium nucleus, and by that time, the universe has cooled and thinned enough that these nuclear reactions stop.

Let's calculate how many helium nuclei result. There are two neutrons in each helium nucleus. With a ratio of one neutron for every seven protons, those two neutrons are paired with 14 protons. Two of those protons are also included in the helium nucleus, with 12 protons left over. This predicts that one helium nucleus forms for every 12 protons (these, of course, are hydrogen nuclei). After these first few minutes, the universe becomes too cool and thin for further nuclear reactions to take place. Thus, a significant number of helium nuclei are made in the Big Bang, along with trace amounts of leftover deuterons (the nuclei of deuterium), lithium and beryllium nuclei (which decay into lithium), and no heavier elements.

This basic calculation was first done by George Gamow and his student Ralph Alpher in the 1940s. They couldn't resist the temptation to add Hans Bethe's name as a co-author in their famous Alpher-Bethe-Gamow ("α-β-γ") paper describing some of their results. One helium nucleus for every 12 hydrogen nuclei is excellent agreement with the results, dating back to the work of Cecilia Payne-Gaposchkin, that the stars are composed of about 90% hydrogen and 8% helium (see chapter 6). Thus, our predictions for the conditions in the universe just a few minutes after the Big Bang have given us a basic explanation of why hydrogen and helium are the two most abundant elements in the universe, and why they are found in the proportions we see! This is an astonishing success of the Big Bang model and gives us a strong justification for extrapolating the expansion of the universe to a time just a few minutes after the Big Bang, when temperatures were above a billion kelvins.

Gamow and Alpher hoped originally to explain the origin of all the elements from the Big Bang, but their calculations showed that the nuclear reactions proceeded only through the lightest elements. All the heavy elements (including the carbon, nitrogen, and oxygen in our bodies, and the nickel, iron, and silicon, which contribute to the makeup of Earth) were created later by nuclear processes taking place in the cores of stars, a process we have described in chapters 7 and 8. Fred Hoyle, a rival of Gamow's, hoped to demonstrate just the opposite: that both the heavy and light elements could be created from hydrogen by nuclear cooking in the cores of stars without invoking an early hot dense phase in the universe's history, and he spent much of his career trying to do so. He developed much of our modern understanding of the formation of the heavy elements in stars. But the quantity of helium that gets made in stars is not nearly enough to explain the amount that we observe.

The fact that we see some deuterium in the universe today points to a Big Bang origin. Deuterium (having one proton and one neutron) is fragile and is destroyed by being fused into helium in the cores of stars, rather than being manufactured there. Stars can't make it. The only way we know how to make it is in the Big Bang, and calculation of the amount of deuterium created in the first few minutes after the Big Bang (one deuterium for every 40,000 ordinary hydrogen nuclei) is in excellent accord with the observed value. The nuclear burning after the Big Bang suddenly stops when the universe has thinned out sufficiently, leaving a small residual amount of deuterium that has not "finished" fusing into helium. The nonequilibrium nature of the burning, because things are changing so rapidly in the early universe, is the key to leaving a small residual amount of deuterium today. Gamow realized this. To Gamow, the observed cosmic abundance of deuterium was a smoking gun pointing to the Big Bang.

As the universe expands, space stretches, and the wavelengths of photons traveling through the cosmos stretch as well; this is just the redshift phenomenon we've already discussed. If space is expanding and we observe a distant galaxy, we will see photons from it redshifted, because it is moving away from us, and we can interpret this effect as a Doppler shift. But we could equally well interpret this as simply the stretching of space itself, stretching the distance between us and the distant galaxy, and stretching the wavelength of the photon traveling from the galaxy to us. Draw a wave on a thick rubber band and stretch the rubber band; the wavelength of the wave you have drawn will increase. Both interpretations of the redshift are equivalent: we can view the redshift as a Doppler shift from a distant object that is moving away because of the expansion of space, or we can interpret the redshift as the lengthening of the wavelength due to the stretching of space itself. Photons from the early universe retain their blackbody (Planck) spectrum, but as their wavelengths lengthen due to the expansion of space, the temperature of the photons drops. Gamow and his students, Alpher and Herman, conceived of the universe beginning with a hot Big Bang and then cooling off with time as it continued to expand.

Einstein, in thinking about the universe overall around 1917, hypothesized what we call the *cosmological principle*: on large scales, and at any given time, the universe looks more or less the same from any vantage point. If we step back far enough and look at large enough scales, the material in the universe should be smoothly distributed. We have seen one aspect of Einstein's hypothesis already—the expansion of the universe looks the same from the perspective of any given galaxy—from which we inferred that the universe has no center. In the same way, an infinite plane has no point one can label the "center," and the curved surface of a sphere has no point on its surface that can be labeled its "center," as all points on the surface of a sphere are equivalent.

Of course, we look around the universe today, and it looks anything but smooth! The mass of our solar system is concentrated into planets and the Sun. Stars are separated by distances vast relative to their sizes. Stars are gathered into galaxies, which are separated from one another by distances of millions of light-years, and galaxies group together in clusters. Einstein's cosmological principle suggests that we should step back even further, looking on scales of hundreds of thousands of galaxies, and we will see the universe as approximately uniform. Hubble's observations showed that counts of faint galaxies in different directions were the same; the universe indeed looked uniform in space on the largest scales.

Fred Hoyle took this one step further: not only is the universe homogeneous in space, more or less the same wherever we look, he claimed, but also homogeneous in time. If you go back to the past, it should look the same as it does today, Hoyle figured. The laws of physics don't change with time, so why should the universe? If you take this notion literally, then there can be no beginning or Big Bang to the universe; the universe has existed forever. Hoyle called this idea the *perfect cosmological principle*. Given that the distance between galaxies is growing with time due to the expansion of the universe, Hoyle had to hypothesize that new matter was being created in the space between galaxies, which would eventually form into new galaxies—a crazy idea, perhaps, but one that he thought was less crazy than forming the entire universe from a moment of infinite density and temperature, marking the beginning of time.

Which one of these pictures is right? As we continue to explore the predictions of the Big Bang model and compare them with what we observe, we will see that the empirical evidence for the Big Bang theory, in the form of agreement between its predictions and our observational data, is very strong indeed.

The first prediction the Big Bang model makes is that the universe should be expanding, as, of course, we observe. The model also predicts the age of the universe—13.8 billion years—in accord with the slightly smaller ages found for the oldest stars in the universe. This is an unambiguous success of the Big Bang model: if we had found stars that were a trillion years old, then we would have been forced to conclude that the Big Bang model couldn't be right. Indeed, we've been through just such a crisis in the past: Hubble's first estimate of his constant was $H_0 = 500$ (km/sec)/Mpc, corresponding to a time since the Big Bang $(1/H_0)$ of only 2 billion years. It was clear by the 1930s from radioactive dating of rocks that Earth is older than that. This age was inconsistent with the Big Bang model: Earth can't be older than the universe itself! This inconsistency was an argument in favor of Hoyle's model, because in his model, the universe was infinitely old and ever-expanding, with new galaxies being formed in intergalactic space all the time. The discrepancy was resolved in the 1950s and 1960s with much-improved measurements of the distances to galaxies, greatly reducing the value of the Hubble constant and making $(1/H_0)$ consistent with the ages of the oldest stars.

We also saw that the Big Bang predicts that there should be 12 hydrogen nuclei for every helium nucleus in the universe, and 40,000 hydrogen nuclei for every deuterium nucleus, exactly as is observed. It didn't have to be this way; indeed, before the science of spectroscopy was fully mature, and Cecilia Payne-Gaposchkin and others had determined that the Sun is mostly hydrogen, people had little idea of the relative abundances of elements in the universe.

Let us take stock of the elements a few minutes after the Big Bang. Essentially all free neutrons have been incorporated into helium nuclei. Nuclear burning ceases, as the universe is too cool and too low density at this stage for any additional reactions to take place. In addition to these helium nuclei, and trace quantities of deuterium and lithium nuclei, we also have protons, electrons, neutrinos, and photons—the positrons present earlier have annihilated with electrons to produce additional photons, leaving behind just enough electrons to balance the charge of all the protons. It is very hot and, as we know, hot things emit photons, so there are plenty of photons around too. As the universe continues to cool and drop in density, its makeup does not change for about 380,000 years.

Up to this point, the material of the universe is a *plasma* (as in the interiors of stars): the atomic nuclei and the electrons are not bound together but move independently of one another. If an electron briefly is captured by a proton, forming an atom of neutral hydrogen, it will quickly be hit by one of the many high-energy photons present, kicking the electron free of the proton. Moreover, because photons interact so strongly with free electrons (i.e., those not bound up in an atom), a photon can't travel very far before it collides with another electron and bounces (*scatters* is the technical term) off in a different direction. That is to say, the universe at that time was opaque; it was a bit like a thick fog in which you can't see very far in front of you. This is analogous to what we found in the interior of stars: the interior of the star is opaque, and energy in the form of photons generated in the core takes a very long time, of order a couple of hundred thousand years, to diffuse outward to the surface.

The story changes drastically when the temperature has dropped to 3,000 K, at a time about 380,000 years after the Big Bang. At this point, the photons no longer have enough energy to ionize hydrogen, and the electrons and protons pair up to make neutral atoms. Neutral hydrogen does not scatter photons nearly as much as individual free electrons do, and the universe suddenly becomes transparent: the fog has lifted. The photons can now travel on straight trajectories.

This suggests that we, in the present-day universe, should be able to see those photons, which have been streaming freely toward us ever since that time when the universe became transparent, 380,000 years after the Big Bang. If the universe has no edge, we should expect to be receiving these photons from every direction in the sky. That is, in every direction we look, there is material at the appropriate distance such that the photons it emitted 380,000 years after the Big Bang are just reaching us today.

These photons are emitted by gas at a temperature of 3,000 K, and thus should have a blackbody spectrum appropriate to that temperature. Such a blackbody peaks at a wavelength of about 1 micron (10^{-6} meter). However, we have to take one other important aspect of the story into account: the universe is expanding! This 3,000 K blackbody radiation is thus redshifted. The universe has expanded by a factor of about 1,000 from when it was 380,000 years old until today, 13.8 billion years later. The wavelength of the radiation is stretched by this same factor as the space expands. Thus the peak wavelength of the thermal radiation now is 1 millimeter rather than 1 micron. If the peak wavelength has increased by a factor of 1,000, the temperature has decreased by that same factor. That means today we should see this thermal radiation with a temperature of about 3 K coming to us from all directions in the sky. This radiation comes from a time when the universe was a mere 380,000 years old, 0.003% of its present age.

In 1948, Alpher and Robert Herman, another of Gamow's students, predicted that the universe today should still be filled with this thermal radiation left over from the Big Bang, and calculated that by today its temperature should have dropped to about 5 K—close to the correct value.

But by the 1960s, the Herman and Alpher prediction was largely forgotten, and Bob Dicke, Jim Peebles, Dave Wilkinson, and Peter Roll of the Princeton

University physics department went through a similar line of reasoning and came up with the same prediction. They took this one step further, realizing that blackbody radiation peaking at 1 millimeter could actually be detected with radio telescopes and sensors that Dicke had developed. (That meant they would be looking for *microwaves*, short-wavelength radio waves like those produced in a microwave oven.) They started building a microwave telescope on the rooftop of a building on the Princeton campus to see whether they could detect the blackbody radiation from the early universe that they had theorized must be there, if the Big Bang idea was correct.

In the end, they got scooped. This was 1964, very early in the Space Age, and Bell Laboratories was starting to think about the possibility of using satellites for long-distance communication. Two Bell Labs scientists, Arno Penzias and Robert Wilson, were investigating whether microwaves could be used to communicate using satellites and were trying to characterize emission from the sky at such wavelengths. They used a large radio telescope at the Bell Lab campus in Holmdel, New Jersey. To their surprise, they found microwave radiation coming from every direction in the sky toward which they pointed the telescope. Once the Princeton folks heard about this, they realized that Penzias and Wilson had discovered the cosmic microwave background (CMB) radiation that they had predicted. Their two papers—the Princeton paper making the prediction, and the Penzias and Wilson paper, making the discovery—were published back-to-back in *The Astrophysical Journal* in May 1965.

With this result, another fundamental prediction of the Big Bang model was verified observationally. The CMB was emitted throughout the entire universe when it was 380,000 years old, and thus should be observable coming from all directions in the sky with the same intensity. This is exactly what is observed. Indeed, this observation reminds us that the Big Bang happened everywhere, with no well-defined center, and thus the leftover heat radiation from the Big Bang is coming to us equally from all directions. In 1967, Penzias and Wilson published a limit on variations in the strength of the emission over the sky of a few percent. As technology has improved, the measurements have gotten much better; as we'll see below, the emission is actually uniform to an astonishing one part in 10^5.

Alpher and Herman's original paper in 1948 predicted that the temperature of the CMB blackbody spectrum should be about 5 K. Penzias and Wilson found a temperature of 3.5 K in their original paper (later refined with more precise measurements to be 2.725 K). This was astonishingly close to the original Alpher and Herman estimate. The discovery of the CMB convinced the astronomical community that the Big Bang model was correct. The unchanging universe model championed by Fred Hoyle, for example, has no natural way to explain the CMB, whereas it is an inevitable and direct prediction of the Big Bang model. This is how science proceeds. This process of continual testing is how scientists gain confidence in their ideas. Penzias and Wilson were awarded the 1978 Nobel Prize in Physics for their discovery.

Peebles and Wilkinson were just beginning their scientific careers in 1965. With the discovery of the CMB, they decided to devote their careers to cosmology, the study of the universe as a whole. Jim Peebles became one of the most important theorists working in the field. Dave Wilkinson made ever more sophisticated measurements of the CMB, first using radio telescopes here on Earth and eventually launching satellites to take data from space. (I should mention here that Wilkinson is my scientific grandfather. My PhD thesis advisor, Marc Davis, completed his PhD thesis under the aegis of Dave Wilkinson in 1974.)

The question Wilkinson wanted to address first was this: is the spectrum of the CMB really that of a blackbody? Wilkinson was one of the scientific leaders of the NASA satellite, the Cosmic Background Explorer (COBE), which was designed to measure the CMB spectrum to high accuracy. It succeeded spectacularly; the CMB spectrum that the COBE satellite measured follows the blackbody formula perfectly within the (very small) error bars. This experiment has been called the most accurate measurement of a blackbody in nature (figure 15.1).

The next big question that Wilkinson tackled was: how uniform is the CMB—that is, does it have the same intensity (or equivalently, the same temperature) in all directions? The cosmological principle, whereby the universe is hypothesized to be smooth on very large scales, predicts that the CMB should be extremely uniform. Penzias and Wilson's initial measurements could put only crude limits (of a few percent) on just how smooth it was, but by the late 1970s, Wilkinson and others discovered that the temperature of the CMB was not exactly the same in all directions, but varied smoothly across the sky, changing by about 0.006 K from one side of the sky to another. It quickly became apparent what was causing this variation. In addition to the relative motions of galaxies due to the overall expansion of the universe, galaxies can

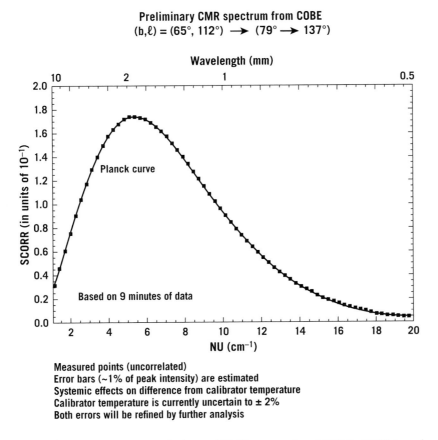

Preliminary CMR spectrum from COBE
$(b, \ell) = (65°, 112°) \longrightarrow (79° \longrightarrow 137°)$

Measured points (uncorrelated)
Error bars (~1% of peak intensity) are estimated
Systemic effects on difference from calibrator temperature
Calibrator temperature is currently uncertain to ± 2%
Both errors will be refined by further analysis

FIGURE 15.1. Preliminary cosmic microwave background (CMB) spectrum from COBE. David Wilkinson showed this spectrum of the CMB from the COBE satellite in a 1990 talk at Princeton University, and the audience burst into applause. Its match to the theoretical Planck blackbody curve for thermal radiation is spectacular. (In the diagram, the Planck blackbody curve [solid line] is plotted on linear versus linear scales, with the data showing the observational error limits as little boxes. The Planck blackbody spectra shown in chapters 4 and 5 are on logarithmic versus logarithmic scales and thus appear a bit different.) *Photo credit:* Adapted from collection of J. Richard Gott

move individually because of the mutual gravitational attraction between them. In addition, the Sun is orbiting the center of our galaxy. These motions combine to give the Sun a velocity of about 300 km/sec relative to the bulk of the matter in the universe that is giving rise to the CMB. This causes a Doppler shift of about one part in a thousand (because 300 km/sec is 1/1,000 of the speed of light) in the CMB; the CMB is slightly blueshifted in the direction of our motion, slightly redshifted in the opposite direction, and varies smoothly in between—just as we observe.

We should pause at this moment to reiterate all the ways we are in motion, despite our perception of sitting still. Earth is rotating around its axis; at North American latitudes, this corresponds to a speed of about 270 m/sec.

Earth orbits around the Sun at a speed of 30 km/sec. The Sun orbits around the center of the Milky Way at 220 km/sec, and the Milky Way and the Andromeda galaxy are falling toward each other at about 100 km/sec. Finally, the two galaxies together are moving at a speed of almost 600 km/sec with respect to the mean velocity of all the material in the observable universe. Add up all these different motions in different directions, and you get the Sun's motion of 300 km/sec relative to the CMB. It is dizzying indeed to envision all this and is an illustration of Galileo's dictum, refined in Einstein's theories of relativity, that it is *relative* motions that are important. Without sophisticated astronomical measurements, we simply perceive ourselves as sitting still.

This Doppler shift induced by our motion with respect to the CMB gives a smooth deviation from uniformity in the CMB of one part in one thousand, which has now been observed to high accuracy. So subtract that effect out. The next question Wilkinson wanted to ask was whether there are any ripples in the CMB that are intrinsic and not just a consequence of our motion. If our understanding of the Big Bang is correct, the answer must be yes. Indeed, the early universe could not have been exactly smooth, without any deviations from perfect uniformity. A perfectly uniform universe will expand uniformly, and no structure will ever form: no galaxies, no stars, no planets, no humans to look up at the sky and wonder what it all means. The fact that we live in a Universe with structure, with real deviations from uniformity—that is, a universe in which we exist—tells us that the early universe, and thus the CMB, could not have been perfectly smooth.

How did structure form in the universe? Consider a region in the early universe in which the density of matter is slightly higher than in the neighboring regions. The mass associated with that region is also slightly higher, and thus it has a slightly higher gravitational pull than the material around it. A random hydrogen atom or particle of dark matter will be attracted toward that region, thereby increasing its density at the expense of the regions around it. Material thus falls into this region, increasing its mass, and in the gravitational tug-of-war, it will be even more effective in pulling extra matter toward it. As time goes on, this process will cause subtle fluctuations in the density of matter to grow with time—enough in principle to form the structures we find around us today. Jim Peebles has a wonderfully succinct way to describe this process of *gravitational instability*: "Gravity sucks!" he likes to say.

Given the amount of structure we observe in the universe today, and given the physics of gravity, how strong should the fluctuations in the early universe (and thus the observed undulations in the CMB) be? It's a tricky calculation: the story is complicated by the fact that the universe is expanding at the same time matter is trying to clump because of gravity. You also have to understand all the components of matter, both dark matter and the ordinary stuff made of atoms. We mentioned earlier that while the universe was still completely ionized (before 380,000 years after the Big Bang), photons were continually scattering off the free electrons in the universe. The pressure from those photons kept fluctuations in the distribution of ordinary matter (electrons and protons) from growing under gravity. If this were the full story, the fluctuations could have been growing via gravity only since the time the universe became neutral, and the nonuniformities in the CMB would have to be larger than we observe.

However, as Jim Peebles realized in the 1980s, dark matter can explain the discrepancy. Dark matter is *dark*; that means it does not interact with photons, and therefore fluctuations in the dark matter can grow under gravity impervious to the pressure of photons. After the universe becomes neutral, ordinary matter can fall into the lumps of dark matter that had already been growing for some time. So, if there is dark matter, we can start off with fluctuations in the CMB that are smaller than if there were only ordinary matter present. By the 1980s, the limits on the fluctuations in the CMB were so stringent that models that didn't invoke dark matter were ruled out.

So the dark matter we infer from the rotation of galaxies is also needed to understand the CMB. What is dark matter made of? Detailed comparison of the abundance of helium and especially deuterium with the predictions of the processes occurring in the early universe tell us that the average density of ordinary matter (i.e., that made from protons, neutrons and electrons) is a mere 4×10^{-31} grams in each cubic centimeter. That's equal to one proton in every four cubic meters! We're reminded of the truly vast (and mostly empty) expanses between stars in galaxies, and from one galaxy to another. But measurements of the motions of galaxies, as well as the fluctuations in the CMB (which we're about to describe), tell us that the total density of matter in the universe is roughly six times larger. The difference is the dark matter, but we conclude that dark matter cannot be made of ordinary protons, neutrons, and electrons. We suspect that dark matter is composed of unseen elementary particles of a yet-to-be-discovered type, which were presumably formed in the extreme heat and pressure of the early universe, just as protons, neutrons, and electrons were. There are a number of speculations as to what these elementary particles might be. The theory of *supersymmetry* predicts that each particle we observe should have a massive supersymmetric partner: the *photino* for the photon, the *selectron* for the electron, the *gravitino* for the *graviton*, and so forth. The search is on at the Large Hadron Collider for such particles. If one of them is discovered, it would prove the theory of supersymmetry. In 1982, Jim Peebles proposed that dark matter is composed of weakly interacting massive particles (and yes, astronomers actually call these "WIMPs") considerably more massive than the proton. The lightest supersymmetric partner of a known particle might just fill the bill. George Blumenthal, Heinz Pagels, and Joel Primack proposed the *gravitino* as a candidate in 1982. It has to be the lightest, because the heavier ones are not stable in the theory; they decay to something lighter and so don't stick around.

Another speculation is that dark matter could be made of elementary particles called *axions*. The Large Hadron Collider, the world's most powerful particle physics experiment, which sits on the Swiss-French border, may be our best hope for finding and identifying any of these candidates. But if the mass of the Milky Way galaxy is mostly dark matter, we expect that there should be dark matter particles all around us. Dark matter particles should be passing through your body right now. But again, they are dark, which means that they don't interact much with ordinary matter (except by gravity). However, the supersymmetric or axion models for dark matter predict that on rare occasions, a particle of dark matter may interact with an atomic nucleus and cause a reaction that we could hope to observe. Experiments are underway to look for such reactions. It's a difficult game: one such experiment uses 100 kilograms of liquid xenon, looking for the flash of light expected if a dark matter particle scatters off one of the xenon nuclei. These experiments are placed in deep mines to minimize confusing interactions with normal particles. These experiments have not yet found convincing evidence of dark matter; but experimental limits on its properties are only now approaching the range the particle physics models predict. The search for the dark matter particles takes us to the forefront of particle physics.

Invoking the presence of dark matter, one predicts that the CMB should be smooth, with fluctuations at the level of one part in 100,000. The instruments on the COBE satellite had been designed with the required sensitivity. In 1992,

I remember attending a presentation Dave Wilkinson gave for the Princeton astronomical community describing the satellite's measurements. The fluctuations in the CMB that had to be there (according to our understanding of growth of structure in a hot Big Bang universe) had finally been detected by the satellite, at a level of one part in 100,000, just about the level that Peebles and others had predicted.

At that time, Wilkinson was already thinking about a next-generation satellite, with instruments capable of measuring these fluctuations (or *anisotropies*, in the jargon) with greater precision. Wilkinson put together a team, including many of the veterans from the COBE satellite, to build the Microwave Anisotropy Probe (MAP). MAP was launched in 2001, and mapped the sky for 9 years.

Sadly, Wilkinson was suffering from cancer through this period. He was able to see the early results from the satellite just before he passed away in September 2002. In February 2003, the team published the results from the first year's data. NASA decided to rename the satellite after Wilkinson; it was henceforth known as the Wilkinson Microwave Anisotropy Probe, or WMAP.

Figure 15.2 shows the map of the fluctuations in the microwave background temperature as found by the WMAP satellite after 9 years of taking data (in 2010). The elliptical shape maps the full sphere of the sky. The north galactic

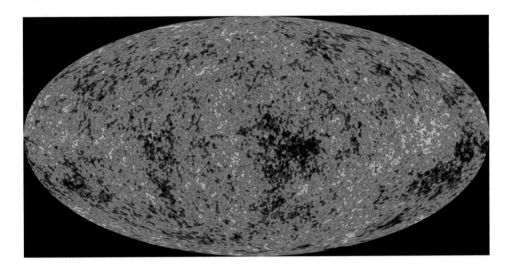

FIGURE 15.2. WMAP satellite map of the cosmic microwave background. based on 9 years of data, 2010. This is a map of the entire sky, in the same projection as figure 11.1 and figure 12.2. Microwave emission from the Milky Way itself has been subtracted off as well as the Doppler shift due to Earth's peculiar motion relative to the cosmic microwave background. Red denotes slightly above average temperature; blue, slightly below average temperature; and green, intermediate temperature. *Photo credit:* WMAP satellite, NASA

pole is at the top, the south galactic pole is at the bottom, and the galactic equator, tracing the plane of the Milky Way galaxy, is the line passing horizontally across the middle of the map. Emission from the interstellar medium in the Milky Way, as well as the one-part-in-a-thousand deviation due to our motion relative to the CMB, have been subtracted.

This is really a baby picture of the universe, our direct view of it when it was a tiny fraction of its present age. These photons have been traveling to us for all but 380,000 years of the 13.8-billion-year age of the universe. The contrast on this map has been cranked up so that the deepest reds and blues correspond to fluctuations of several times ±0.001%; the more typical values are ±0.001% (i.e., one part in 100,000).

Figure 15.3 shows the measured strength of these fluctuations as a function of the angular scale (note scale at the bottom). These measurements come from a successor satellite, called Planck, launched by the European Space Agency, as well as a variety of other telescopes on the ground.

There is a peak at an angular scale of 1°, corresponding to the typical size of the "bumps" you see in the WMAP image. The graph tells us, for example, that there are smaller variations from one 18°-wide patch to another than from one 1°-wide patch to another. Where no error bars are apparent, the observational errors are smaller than the size of the red dots.

The smooth green curve going through the points is the result of a theoretical calculation based on Big Bang theory, including the effects of dark matter, dark energy, and inflation (about which we learn much more in chapter 23). At large angular scales, the green line broadens to encompass the theoretically expected scatter in the predicted results. The agreement between the two is astonishing: the observations fall along the green theoretical curve within the observed errors. The Big Bang model has racked up another success: it predicts the detailed nature of the extremely subtle fluctuations seen in the CMB.

After recombination, material starts gathering into ever-denser lumps to make the first stars and galaxies. But given the angular size of structures we see in the CMB, we predict that there should be substantial structure in the universe on larger scales than just galaxies, which are a mere 100,000 light-years across. That is, the galaxies should not be randomly distributed in space but should be organized in larger structures. To map these structures, we return to Hubble's law. Remember that when we look at an astronomical image, we see objects as if painted on the two-dimensional dome of the sky; we have no depth

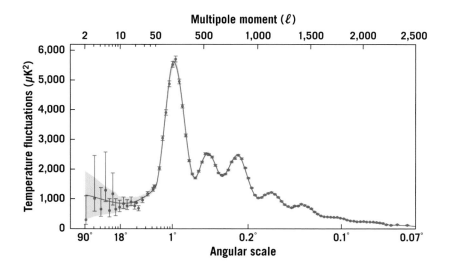

FIGURE 15.3. Strength of cosmic microwave background fluctuations as a function of angular scale (red dots) compared with theory (green curve), from the Planck Satellite Team 2013. The strength (power) in the variations of the temperature of the cosmic microwave background is plotted vertically, as a function of the scale of the fluctuations in degrees. The units on the vertical axis are micro-Kelvin squared, representing fluctuations from the uniform temperature of 2.7325 K of about one part in 100,000. The oscillations in the curve are due to sound waves traveling through the universe until the time of recombination. The solid curve going through the data points is the predicted curve given our model for the Big Bang, including the effects of dark matter, dark energy, and inflation (about which we learn much more in chapter 23); the essentially perfect agreement with the observations is stunning confirmation that the Big Bang model is correct. Data from NASA's WMAP satellite earlier resulted in much the same conclusion. *Credit:* Courtesy ESA and the Planck Collaboration

perception at all and cannot necessarily distinguish between a nearby galaxy and one lying at a much greater distance. But Hubble's law gives us a method to explore the third dimension: by measuring the redshift of each galaxy, we can determine its distance and see how the galaxies are distributed in space.

Starting in earnest in the late 1970s, astronomers began measuring the redshifts of thousands of galaxies and were able to make three-dimensional maps of their distribution. They immediately noticed that the galaxies are not at all distributed randomly in space: they found clusters of galaxies (up to 3 million light-years across) containing thousands of galaxies, and empty regions (voids 300 million light-years across) almost completely devoid of galaxies. Indeed, these early maps caused people to question the cosmological principle; there was so much structure apparent in these maps that people wondered whether there was any scale on which the universe appeared smooth, or whether ever-larger surveys of the sky would show yet larger structures. The Sloan Digital Sky Survey was designed in part to address this question. It is a telescope dedicated to mapping the sky; it has measured redshifts now for more than

2 million galaxies. Figure 15.4 is a map of a small fraction of these galaxies, those in a 4°-wide slice in Earth's equatorial plane; if we showed you all the data in one plot, the density of points in the figure would be so high that it would show solid ink, not allowing you to see any of the structure.

Each of the more than 50,000 dots in this figure represents a galaxy of 100 billion stars. It is worth taking a moment to appreciate the enormousness of these numbers.

We can see two big slices of the pie; the Milky Way galaxy sits at the center of the image. The empty regions on the left and the right have not been covered by the survey; this is the region obscured by dust from the Milky Way, making it difficult to pick out distant galaxies.

The radius of this figure is 860 Mpc, almost 3 billion light-years. Even a cluster of galaxies appears small in this picture; the majority of galaxies appear to lie along filaments, strings of galaxies hundreds of millions of light-years long. A particularly prominent filament, dubbed the Sloan Great Wall, appears somewhat above the center of the image. It has a length of 1.37 billion light-years. But no structures stretch across the entire width of either survey slice, indicating that on the very largest scales, Einstein's cosmological principle holds.

Notice in the figure that the density of galaxies does drop off dramatically near the outer edges of the map. This is not evidence that the cosmological principle is wrong: it simply reflects the fact that galaxies in these regions are the most distant from us and are therefore the faintest. Only a small fraction of the most distant galaxies are luminous enough for the Sloan Digital Sky Survey to measure their spectra, and therefore for their redshifts to be included in this map.

If we compare this to our map of the CMB from WMAP, it is not obvious, even under the process of gravitational instability, that fluctuations at the level of one part in 100,000 can evolve into the incredibly structured universe we see today in the galaxy distribution. The equations of gravitational instability (which are based on Newton's law of gravity, with the added complication of the expansion of the universe) can be solved approximately and tell us that the numbers are roughly right, but to do the calculation properly, and understand the gravitational tug of each parcel of matter in the universe on every other part, requires a large computer. One starts with a distribution of matter with subtle fluctuations at the level that we measure from the CMB map. Then one lets gravity, plus the expansion of the universe, take over and evolve the

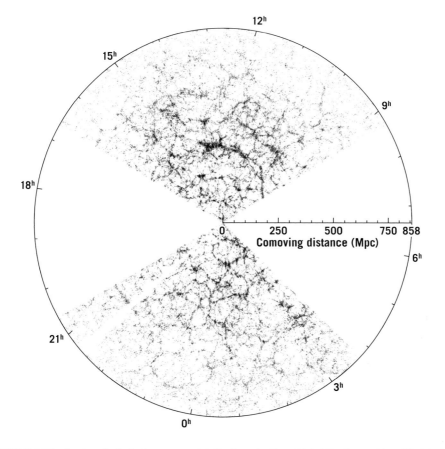

FIGURE 15.4. Distribution of galaxies in an equatorial slice from the Sloan Digital Sky Survey. The Milky Way is at the center. Each dot represents a galaxy. The two fans show galaxies in the survey region; the two blank regions are regions the survey did not cover. The radius of this diagram is about 2.8 billion light-years.

Credit: J. Richard Gott, M. Jurić, et al. 2005, *Astrophysical Journal* 624: 463–484

structure for 13.8 billion years on the computer. The resulting distribution of galaxies these computer simulations predict shows the same sort of structure we see in the galaxy maps: clusters, voids, and filaments, of just the right size and contrast to match observations.

Of course, we don't expect the computer simulations to produce the exact structures of the present-day universe, just ones with the same statistical properties. Remember that the part of the universe we're seeing in the CMB is very distant from us; we're not seeing the matter that evolves into the galaxies that are near us. But we do assume that the general properties, including the fluctuations, of the material giving rise to the CMB, are statistically similar to that matter which gives rise to the galaxies around us. Overall, large computer simulations based on the Big Bang model have been remarkably successful in reproducing the filamentary weblike structure that we see in the observations.

This, then, is the final triumph of the Big Bang model. We have explored the predictions of the model and compared them with observations in every way that we could. We inferred that the universe was born 13.8 billion years ago, in excellent accord with (i.e., slightly older than) the ages of the oldest stars. We concluded that hydrogen and helium nuclei were formed in the first few minutes after the Big Bang, in a 12:1 ratio, which is exactly what is observed, and we are able to predict the quantity of deuterium produced, also in agreement with observations. We predicted the existence of the CMB, and its various properties: its spectrum, its temperature, and its incredible smoothness; all this is exactly as observed. Perhaps most impressively, we predicted that the CMB should not be perfectly smooth but should show fluctuations at one part in 100,000, with a predicted variation depending on the angular scale that should follow a complicated curve. The WMAP and Planck satellite measurements have confirmed this prediction as well. Finally, computer models of how these fluctuations should grow under gravitational instability predict a highly structured universe today, with galaxies arrayed along filaments hundreds of millions of light-years long, just as the maps from the Sloan Digital Sky Survey reveal. The Big Bang model is far more than "just a theory": it is supported by a vast array of empirical, quantitative evidence and has passed every test we have given it with flying colors.

16

QUASARS AND SUPERMASSIVE BLACK HOLES

MICHAEL A. STRAUSS

In the 1950s, radio astronomy, the study of electromagnetic radiation that astronomical objects emit at wavelengths longer than a centimeter or so, was still in its early days. The radio telescopes of the day were making the first maps of the sky. It was a challenge to determine which astronomical objects were responsible for the radio sources seen, because the radio telescopes did not have the resolution to pinpoint accurately the position of the radio source on the sky. That is, they could specify the position of any given source only to the nearest degree or so, and it wasn't at all obvious which of the thousands of stars and galaxies lying in that region of sky were responsible for the radio emission.

The best radio maps of the sky at the time were made using a radio telescope in England; the Cambridge University astronomers who ran the survey published several catalogs of sources found in these maps. Our story starts with the 273rd entry in the third of the Cambridge catalogs, called, for short, 3C 273. The Moon's path on the sky occasionally passes over 3C 273, and by timing exactly when this radio source disappeared behind the Moon, astronomers were able to pinpoint its position to much greater accuracy. Astronomers then took images of that region of the sky in visible light, to see what was responsible for the radio emission. To their surprise, 3C 273 coincided with what appeared to be a star, one too faint to be seen by the naked eye, but certainly bright enough to be easily studied with what at the time was the largest visible-light telescope in the world, the 200-inch telescope at Palomar Observatory.

Maarten Schmidt, a young professor at Caltech in Pasadena, knew that to understand what sort of star it was, he needed to measure its spectrum. He obtained the spectrum with the 200-inch telescope in 1963, but when he first looked at the data, he couldn't make sense of what he saw.

He saw a series of very broad emission lines whose wavelengths did not correspond to any atoms he had ever seen before. His first thought was that this might be some really unusual type of white dwarf star, but then he had an "aha!" moment. He realized that the emission lines were just the familiar Balmer lines of hydrogen, which form a regular pattern well known from studies of stars. However, these lines were not at their familiar wavelengths, but were all shifted systematically to the red by an astonishing 16% (figure 16.1). That is, the wavelength of each of these features in the spectrum was 16% larger than the Balmer transitions observed here in the lab on Earth.

Could this be a redshift due to the expansion of the universe? A redshift that large corresponds (using the modern value of the Hubble constant) to a

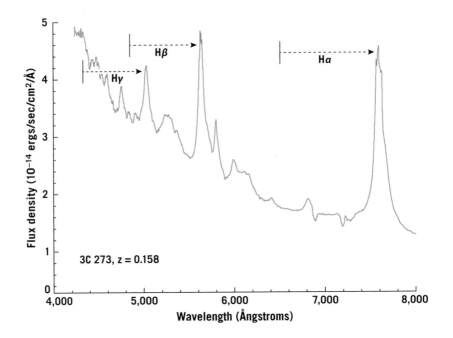

FIGURE 16.1. The spectrum of the quasar 3C 273. The strongest emission lines present are Balmer lines of hydrogen, as marked. In each case, the arrow is drawn from the rest wavelength to the observed wavelength of the line—shifted redward in each case by 15.8%. The other emission lines apparent in the spectrum are due to oxygen, helium, iron, and other elements. *Credit:* Michael A. Strauss, from data taken by the New Technology Telescope at La Silla, Chile; M. Türler et al. 2006, *Astronomy and Astrophysics* 451: L1–L4, http://isdc.unige.ch/3c273/#emmi, http://casswww.ucsd.edu/archive/public/tutorial/images/3C273z.gif

distance of about 2 billion light-years. A small number of galaxies known at the time had similarly large redshifts, but they were incredibly faint, at the limit of what telescopes were capable of measuring. Yet 3C 273 was several hundred times brighter than these faint fuzzy galaxies. Moreover, it appeared starlike, a point of light, not having an extended size like a galaxy. This left two interpretations: (1) perhaps this object was much closer than 2 billion light-years—even within our own galaxy—and the redshift had nothing to do with the expansion of the universe, or (2) this star was enormously luminous. The inverse-square law tells us that for 3C 273 to be as bright as observed, if it were really at a distance of 2 billion light-years, it would have to be hundreds of times more luminous than an entire galaxy containing 10^{11} stars!

Maarten Schmidt told his colleague Jesse Greenstein about his discovery. It turned out that Greenstein had measured the spectrum of another radio source, 3C 48; Greenstein immediately realized that this must be a similar object, at an even higher redshift of 0.37 (or 37%). Schmidt mused that many such objects must be out there to discover, and that he better get busy finding them. As he and others discovered more of these starlike radio-emitting objects with ever-larger redshifts, they needed a name for them. The first term they used was *quasi-stellar radio source*, but this was too much of a mouthful, and it was quickly shortened to *quasar*. While the first quasars were all found by their radio emissions, Allan Sandage (famous for his measurements of the Hubble constant) soon discovered similar starlike objects at high redshifts that had no associated radio emission; the majority of quasars in fact are faint in the radio part of the spectrum.

Fritz Zwicky, whom we met in chapter 12, was a colleague of Schmidt and Greenstein at Caltech. He was one of the most brilliant, and eccentric, characters in twentieth-century astronomy (figure 16.2). He made a series of discoveries so far ahead of their time that the rest of the scientific community took decades to catch up with him. We've already seen that in 1933, he was the first to infer the existence of dark matter from the motions of galaxies in clusters. The idea only took hold in the astronomical community in the 1970s, when Morton Roberts and Vera Rubin and her colleagues started measuring the rotation of the outer parts of galaxies, and Jeremiah P. Ostriker, Jim Peebles, and Amos Yahil began using stability arguments to infer the existence of large amounts of dark matter in galaxies. Zwicky and his colleague Walter Baade hypothesized (correctly!) in 1934 that neutron stars can form in supernova explosions, an idea

FIGURE 16.2. Fritz Zwicky, posing with his catalogs of galaxies. *Photo credit:* Archives Caltech

that was confirmed only three decades later with the discovery of pulsars. In fact, Zwicky and Baade coined the word *supernova*. Zwicky also predicted correctly, decades ahead of the observations that would confirm it, that Einstein's light-bending effect in general relativity could make distant galaxies act like gravitational lenses, magnifying even more distant galaxies behind them. And he claimed that he was the first to discover quasars.

Zwicky knew he was smart, and wasn't shy about expressing his views when he thought others were mistaken. Denied access to the 200-inch Palomar telescope, Zwicky did most of his work with a small, 18-inch survey telescope at Palomar, using it to discover supernovae (he found more than 100 of them in his lifetime) and make catalogs of galaxies. He noticed that some of the galaxies he tabulated were quite compact, almost appearing starlike. But because he was not allowed to observe on the 200-inch, he wasn't able to measure spectra of these galaxies and determine their physical nature. Some of the compact galaxies he had noticed turned out later to be quasars of the type that Schmidt and Sandage had subsequently discovered, and Zwicky claimed—with some justification—that he should be given credit for their discovery.

The graduate students at Caltech loved Zwicky, who shared office space with them in the sub-basement of the astronomy building on the Caltech campus. Zwicky passed away in 1974: my colleagues Jim Gunn, who was a graduate student at Caltech in the 1960s, and Rich Gott, who was a postdoc there from 1973 to 1974, remember him fondly.

Zwicky's basic insight was correct. Some compact galaxies had an incredibly luminous unresolved pointlike source of light (the quasar) coming from the center of the galaxy, which outshone the faint parts of the galaxy surrounding it, making the galaxy itself appear almost pointlike, like a star.

This phenomenon is clearly seen in images of quasars taken with the Hubble Space Telescope: its sharp images can distinguish the light from the quasar and the faint extended light around it from the galaxy in which it sits. These images were taken by my wife Sofia Kirhakos, in collaboration with her colleagues John Bahcall and Don Schneider, so I am particularly pleased to show them in this book (figure 16.3). In the center of each image is a very bright point of light; that's the quasar itself. It is surrounded by a galaxy (and in one case, a pair of galaxies that appear to be colliding): spiral arms are visible. Images such as these resolved the distance controversy: quasars really are at the distances their redshifts imply (they are not just a weird type of star in our own Milky Way galaxy), and thus they are incredibly luminous.

To understand what the quasar phenomenon is all about, let us return to the spectrum of 3C 273. The emission lines here are broad, spread over a range of wavelengths, even though we learned in chapter 6 that atomic transitions correspond to specific, precise energies and thus wavelengths. We understand this as a manifestation of the Doppler shift: within the quasar, there is gas moving at a range of speeds. The quasar overall is moving away from us at 16% of the speed of light, but relative to that overall motion, some of the gas in the quasar is moving toward us (blue shifting part of the emission line relative to the average), whereas some of the gas is moving away from us (making part of

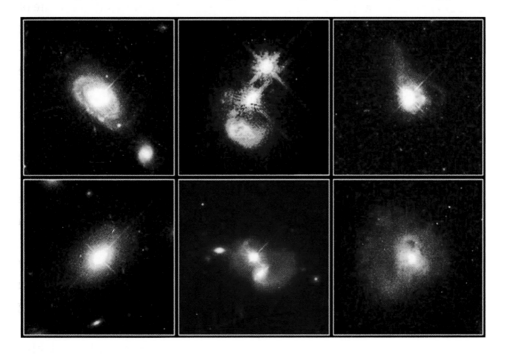

FIGURE 16.3. Quasars in their host galaxies, taken by the Hubble Space Telescope.
Photo credit: J. Bahcall and M. Disney, NASA

the emission line even more red shifted). This broadens or widens the emission line. Consider this emission to be from gas in orbit around a central mass: there is gas at every point along a circular orbit, and each of these points has a different component of motion along the line of sight, and thus a different Doppler shift. The broad emission line reflects this range of Doppler shifts.

We can take this one step further. The width of the emission line tells us how fast the gas is moving; a typical value for quasars is 6,000 km/sec. Something is causing the gas to move at this enormous speed. We will hypothesize that these motions are due to gravity—that this is gas moving in orbit around some central object, whose nature we would like to understand.

What is the radius of this orbit? If we can determine this, then we can use Newton's Laws and our knowledge of the speeds to calculate how massive that central object must be. We've already seen that quasars appear pointlike, like a star, and therefore they are smaller than what our telescopes can resolve. A clue to their true sizes became available when people discovered that quasars are *variable*; their brightness changes significantly on timescales of a month or so.

Imagine that the light from a quasar was coming from a region a light-year across. The light that reaches us from the front side of the quasar (as seen by us) would arrive a year earlier than the light from the back. Even if the whole structure were somehow to double in luminosity instantaneously, the brightness we would detect would brighten gradually over the course of a year, as first the light from the front side, then eventually the back side, reached us. Thus, the fact that quasars change their brightness on timescales of a month tells us that they can't be much bigger than a light-month in size. This size is astonishingly small: remember that stars in our Milky Way are separated from one another by several light years, and this volume a light-month across (or even smaller) is emitting as much energy as several hundred ordinary galaxies.

We now know the speed of the gas moving in the quasar, and roughly how far it is from whatever is causing it to move gravitationally. We can carry out the same calculation we did in chapter 12 when determining the mass of the Milky Way from the orbit of the Sun around it: the mass is proportional to the velocity squared times the radius. When we do this calculation for the quasar, we find a mass of an astonishing 2×10^8 times the mass of the Sun.

Let us summarize: quasars are found in the centers of galaxies, they are a light-month or smaller in diameter, they have luminosities hundreds of times larger than entire galaxies, and have masses hundreds of millions times the mass of the Sun. Huge masses in a tiny volume: could this be a black hole? And yet, black holes are supposed to be *black*—light cannot escape from them—whereas quasars are among the most luminous objects in the universe. In addition, the only way we know how to make a black hole is to collapse a massive star. The most massive stars we know of are perhaps 100 times the mass of the Sun; we can't make a 200-million-solar-mass black hole that way. What is going on?

Well, black holes can grow in mass. Consider gas falling toward a black hole. If it is going straight in, it will simply be swallowed by the black hole and disappear without a trace, adding to its mass but otherwise having no effect. However, it is more likely that the gas has a bit of sideways motion, or angular momentum, relative to the black hole. Because of this angular momentum, it will not fall straight in but will orbit around the black hole. In analogy to stars orbiting in the Milky Way, we think that the gas around a black hole lies in a flattened rotating disk. The gravity of a black hole is strong; the gas closest to the black hole is moving tremendously fast, at an appreciable fraction of the speed of light. The gas that is closer to the black hole will have a higher velocity and rub against the gas a little farther out. This friction can heat the gas up tremendously, to temperatures of hundreds of millions of degrees. And as we've seen over and over again, hot things radiate energy.

So while the black hole itself is invisible, the gas around it, before it falls all the way in, can be tremendously luminous. A quasar is a supermassive black hole, surrounded by a disk of gaseous material glowing so hot that it can outshine the entire galaxy in which it is embedded. Indeed, it is material falling in during this process that can cause a relatively small black hole, born presumably from the death of a massive star as a supernova, to grow: as material falls in, the disk material shines as a quasar and continually adds to the mass of the black hole. The quasar is powered by gravitational energy turned into kinetic energy as the gas spirals deeper and deeper into the gravitational well of the black hole. As the gas finally enters the black hole, it adds to the black hole mass. This accretion process, operating over hundreds of millions of years, can result in black holes with masses of millions, or even billions, of solar masses.

The tremendous energy associated with the disk close to the black hole causes energetic particles to be emitted. These particles are blocked by the disk itself, and thus must spurt out as a jet of material perpendicular to the disk, entrained in part by powerful magnetic fields. Such a narrow jet is seen as the faint linear feature at 5 o'clock in figure 16.4, a Hubble Space Telescope

FIGURE 16.4. Quasar 3C 273 and its jet.
Photo credit: Hubble Space Telescope, NASA

picture of 3C 273 (the sharp, straight spikes emanating from the quasar itself are artifacts of the telescope optics).

Such jets are the hallmark of black holes into which material is falling. The elliptical galaxy M87 has one of the most massive black holes in the nearby universe, 3 billion times as massive as the Sun. It is also emitting a jet, about 5,000 light-years in length.

There is a popular mental image of black holes as cosmic vacuum cleaners, slurping up everything in their vicinity. However, imagine that the Sun turned magically into a black hole (of the same mass) tomorrow. This would of course be terrible news for us, because we would no longer receive heat and light from the Sun and Earth would freeze. But Earth's orbit would remain unchanged. The angular momentum of Earth in its orbit around the Sun will keep us circling, as we have for the past 4.6 billion years. Similarly, stars in orbit around the black hole in the center of the Milky Way are not going to be swallowed up by that black hole any time soon. This black hole probably went through a quasar phase in the distant past, when it grew to its current size of 4 million solar masses. We can measure its mass today by plotting the orbits of individual stars we see orbiting around it. However, there is no material falling into it now to form a disk, so it is currently quiescent and is not shining as a quasar.

Quasars are rare in the nearby universe. Indeed, 3C 273, 2 billion light-years away, is one of the nearest luminous quasars. Quasars were much more common in the early universe; most quasars are at high redshift, and therefore at great distances. The light from these distant quasars has traveled for billions of years to reach us. We are thus seeing them at a time when the universe was significantly younger than it is today. The fact that the number of quasars in the universe has changed with time is direct evidence for an evolving universe, contradicting Hoyle's perfect cosmological principle (see chapter 15), which endorsed an unchanging universe.

From the numbers of quasars we see in the early universe, we predict that supermassive black holes in the present-day universe must be ubiquitous. After all, black holes only grow; they don't go away once they've formed. (We'll see in chapter 20 that black holes can eventually evaporate due to quantum effects, but for supermassive black holes, this process is slow indeed, and is quite negligible over the billions of years we're discussing here.) The fact that we don't see these black holes shining as quasars in nearby galaxies today simply tells us that they are currently quiescent, without gas falling into them. The supermassive black hole in the center of our Milky Way, whose presence we inferred from the motions of stars in its vicinity, is just one such example.

Looking for black holes in the centers of other galaxies is a challenge. If the black hole is not being fed by gas coming in from an accretion disk, there will be no quasarlike emission for us to see. However, we can use the Doppler shifts of stars near the centers of galaxies to infer the presence of a massive gravitating object. This is done most easily for nearby galaxies, for which we can resolve the central regions, where the gravity of a black hole will dominate the motions of stars.

Astronomers have now searched in detail for black holes in about 100 galaxies. In essentially every case that they had the sensitivity to detect it, they did find evidence for a supermassive black hole in the center. As far as we can tell, essentially every large galaxy with a significant bulge (i.e., ellipticals and most spirals) hosts a black hole. Our Milky Way, with a black hole of a mere 4 million solar masses, is a relative wimp; the most massive black holes among the nearby galaxies are several *billion* times more massive than the Sun (as we saw for M87). Moreover, the larger the elliptical galaxy (or the bulge of the spiral galaxy), the more massive the black hole will be; the mass of the black hole is typically about 1/500 of the mass of the bulge of stars in which it sits.

The tremendous luminosities of quasars make them much brighter than galaxies. Thus a distant quasar is much brighter, and therefore easier to see, than a galaxy at the same distance. What is the most distant quasar we can see in the universe? Again, because of the finite speed of light, the light we see from such a distant quasar left it when the universe was much younger than it is today. When we look at objects at large distances in astronomy, we are looking into the past: our telescopes are time machines.

In chapter 15, I described the Sloan Digital Sky Survey, which has taken images of the sky and measured redshifts for 2 million galaxies. It has also obtained spectra of more than 400,000 quasars. From this sample, we know that quasars were most common between 2 and 3 billion years after the Big Bang; this is when the supermassive black holes found in big galaxies today are thought to have gained most of their bulk. Two billion years after the Big Bang, about 12 billion years ago, corresponds to a redshift of 3. That is, the spectral lines in the quasars appear with wavelengths 4 (i.e., the redshift + 1) times the wavelength they would have without the expansion of the universe. Redshift, in this case, is not a subtle phenomenon but a big effect!

Edwin Hubble found a linear relationship between redshift and distance for galaxies. At very large redshifts, this relationship is somewhat more complicated; it turns out that a quasar at redshift 3 is now about 20 billion light-years from Earth. How can this be, if the Universe is only 13.8 billion years old? Remember that in the time since the light left the quasar until now, the universe has expanded fourfold (again, redshift + 1), carrying the quasar farther away, and this distance of 20 billion light-years corresponds to where it is now (we call this its *co-moving* distance).

Figure 16.5 shows the spectrum of the most distant quasar my colleagues and I found in the Sloan Digital Sky Survey. The very strong emission line at a wavelength of 9,000 Ångstroms (0.9 microns) corresponds to the transition from the second energy level to the ground state in hydrogen—the Lyman alpha line. Blueward of this emission line (i.e., in the blue direction, at shorter wavelengths), the spectrum drops to zero; this turns out to be due to absorption from hydrogen gas distributed in the volume between the quasar and us. The spectrum shows emission at near-infrared wavelengths and essentially nothing at shorter wavelengths, making this object appear tremendously red.

Thus the job of finding the highest-redshift quasars is straightforward: look in the Sloan Digital Sky Survey images for the reddest objects you can find. This is not as easy as it sounds; the survey includes images of about half a billion objects, and we had to make sure that an apparent red color for any given object had not resulted from some sort of rare processing glitch.

There is another challenge. We know from our studies of stars that the cooler the star the redder it appears. In 1998, as the first images from the Sloan Digital Sky Survey became available, my student Xiaohui Fan and I started a program of obtaining spectra of the reddest objects we could find in the data, to confirm their quasar nature and to determine their redshifts. We used the Apache Point Telescope (at the same observatory where the Sloan Digital Sky Survey telescope is located, in Sunspot, New Mexico). The telescope is

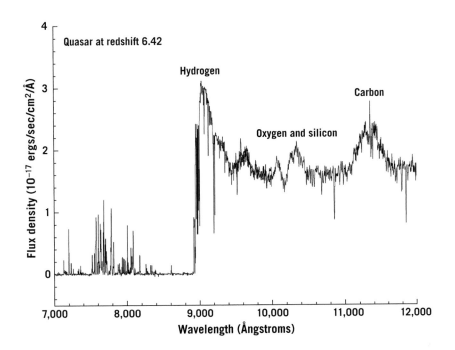

FIGURE 16.5. Spectrum of the quasar SDSS J1148+5251 at redshift 6.42. This quasar was discovered by Michael Strauss, Xiaohui Fan, and their colleagues in 2001, the highest-redshift quasar known from the time of its discovery until 2011. The light we are seeing from this quasar was emitted when the universe was less than 900 million years old. The strongest peak (emission line) in this quasar is due to emission from hydrogen atoms (the $n = 2$ to $n = 1$ transition; see figure 6.2), which has been greatly redshifted from its rest wavelength of 1,216 Ångstroms to 9,000 Ångstroms. The sharp drop in the spectrum below 9,000 Ångstroms is due to absorption from hydrogen gas between the quasar and us. *Credit:* Image by Michael A. Strauss using data in R. L. White, et al. 2003, *Astrophysical Journal* 126: 1, and A. J. Barth et al. 2003, *Astrophysical Journal Letters* 594: L95

remotely operable over the internet: rather than flying across the country, we could simply eat an early dinner at home, then drive into the office, where we would carry out our observations, sending instructions for moving a telescope 2,000 miles away.

As we started measuring spectra of these very red objects, we hit pay dirt almost immediately, but in an unexpected direction. Mixed in with an assortment of high-redshift quasars, we stumbled across some of the coolest (and thus lowest-mass) stars known, right here in our Milky Way. Indeed, these are the substellar objects discussed in chapter 8, with masses too low to burn hydrogen in their cores. These stars have temperatures of 1,000 K or even lower, and their spectra were quite unfamiliar to us when we first started finding these objects. I remember scrambling to look up the few papers that had described such cool stars at 3 in the morning, as we measured the spectra and struggled to understand them. In a single night of observing, we would take spectra both of the lowest-luminosity substellar objects known, at a distance of only 30 light-years, and enormously luminous quasars close to the edge of the observable universe. This is the most extreme illustration of the fact that with an astronomical image alone, we have no depth perception. The very nearby (in astronomical terms) and extraordinarily distant objects both appeared as very faint red points in our data, requiring detailed spectra to distinguish the two.

We continued pushing to ever-redder objects, as our techniques for removing glitches in the images improved. We broke the existing quasar redshift record (4.9 at the time we started our work) multiple times. Whenever we did so, we would call our colleague Jim Gunn (the project scientist for the Sloan Survey and a pioneer in quasar studies in his own right). Waking him out of a deep sleep (it was usually 3 a.m. or so, after all!), we'd say, "Jim, we broke the record again!" "Good work, boys," he would respond; "I always want to be woken up for this news!" And then he would go back to sleep.

The Lyman alpha hydrogen line apparent in the spectrum of our most distant object in figure 16.5 is normally found at a wavelength of 1,216 Ångstroms; here it has been redshifted all the way into the near-infrared part of the spectrum at 9,000 Ångstroms. The redshift is (9,000 Å – 1,216 Å)/1,216 Å, or 6.42, corresponding to a distance now of 28 billion light-years. This was the highest redshift quasar known when we discovered it in 2001. Perhaps even more impressive than its great distance is the fact that the light we see from this object left it about 13 billion years ago, when the universe was only about 850 million years old. If the CMB radiation comes from the universe's infancy, we're now probing a time when it was a toddler.

This brings up another cosmic mystery. As noted earlier, we can use a quasar's spectrum to estimate the mass of the black hole powering it. A typical value for the most distant quasars is about 4 billion solar masses, about as massive as the largest black holes we know of in the present-day universe. But remember that the smoothness of the CMB tells us that the very early universe was almost perfectly uniform. From this almost complete absence of structure, we have to form supermassive black holes, the densest conceivable objects, in only 850 million years. To make such a black hole, the universe had to form a first generation of stars, then explode them as supernovae, leaving behind stellar-mass black holes. These black holes then had to accrete matter at a tremendous rate to acquire such an enormous mass. Theoretical models suggest that this is barely possible under ideal conditions, implying that such high-redshift quasars should be rare. Indeed they are; after more a decade of searching, we've found only a few dozen quasars at the very highest redshifts.

The push for the most distant quasars continues: in 2011, our record was broken in spectacular fashion with the discovery of a quasar at redshift 7.08, using a survey that was sensitive to longer wavelengths (further into the infrared) than the Sloan Survey. Since the time when this quasar emitted the light we are seeing today, the universe has expanded by a factor of 8.08. Other teams are using the Hubble Space Telescope, the Subaru Telescope in Hawaii, and other telescopes to find galaxies at higher redshifts still. It remains unclear whether models for galaxy formation and black hole growth will be able to explain these and future discoveries, if the redshift records continue to be broken. There should be interesting times ahead!

The wonderful thing about astronomy is that every time we look at the heavens in a new way, we make fundamental new and unanticipated discoveries. The Sloan Digital Sky Survey, whose discoveries have been prominent in this chapter and chapter 15, is a good example of this. I am currently involved in planning for its successor, The Large Synoptic Survey Telescope, currently under construction on a mountaintop in the Chilean Andes. It will have a much larger light-gathering power than the Sloan Telescope, and in its 10-year survey lifetime, it will study the properties of faint galaxies and quasars, map the distribution of dark matter from the gravitational lensing distortions it causes in the shapes of galaxies, and discover hundreds of thousands of supernovae and other transient phenomena. The telescope will be making a movie of a quarter of the entire sky: 860 complete frames in 10 years. This will require us to process 30 terabytes of new data every day. The survey should discover hundreds of thousands of Kuiper Belt objects, and also spot Earth-approaching asteroids. But the most exciting discoveries are likely to be those we haven't even imagined yet, the "unknown unknowns," in Donald Rumsfeld's famous phrase.

PART III

EINSTEIN AND THE UNIVERSE

EINSTEIN'S ROAD
TO RELATIVITY

J. RICHARD GOTT

Einstein's name is synonymous with genius, as in "Hey, Einstein, get over here!" ("Hey, genius, get over here!") or "He's no Einstein," meaning "He's no genius." Einstein is famous for being a genius. Newton was also a genius. But around the world and throughout world history there have been other geniuses as well. Who is preeminent in English literature? Shakespeare! From his plays and poems, Shakespeare is often cited as the person in world history with the largest demonstrated vocabulary. His works contain a vocabulary of 31,534 different words. A statistical analysis of his works by Bradley Efron and Ronald Thisted suggests he must have actually known more than 66,000 different words. Shakespeare would take Newton on the Verbal portion of the SAT! But Newton would beat Shakespeare in the SAT's Math section, I suspect. Newton often gets the edge over Einstein, because in addition to his work in gravity and optics, he made important contributions in math, inventing differential and integral calculus. But Newton was also lucky, born at the right place at the right time—in Europe when they were talking about just these kinds of problems. Newton's mentor and his professor at Cambridge, Isaac Barrow, was interested in calculating the volumes of barrels and other such objects—a topic that integral calculus would tackle. Clearly, the time was ripe for discovering differential and integral calculus. In fact, the philosopher and mathematician Gottfried Wilhelm Leibniz invented differential and integral calculus independently in Europe. If you look at a world map, you see that Newton and Leibniz lived just a few hundred miles apart at roughly the same time. This is not simply a coincidence. Europe was talking about these ideas at that time.

The world of the late seventeenth century was primed for a great discovery, because Kepler had already quantified 600 pages of observations on the positions of the planets, as recorded by Tycho Brahe, and converted them into three simple laws of planetary motion that could be subjected to mathematical analysis. As Michael discussed in chapter 3, Newton used Kepler's third law to derive the $1/r^2$ force law for gravity. In similar fashion, in the twentieth century, experimental data on the wavelengths of the hydrogen Balmer series lines gave clues to a formula describing the energy levels in the hydrogen atom and paved the way for a quantum understanding of the atom by Neils Bohr and Edwin Schrödinger.

Time magazine picked Einstein as the most influential person of the twentieth century—the "Person of the Century." Gutenberg, Queen Elizabeth I, Jefferson, and Edison were each judged most important in their centuries by *Time*. Shakespeare just missed out, because *Time* selected Isaac Newton as its "Person of the Seventeenth Century."

Newton has a very nice life-sized statue of himself in Trinity College, at Cambridge University. William Wordsworth wrote a poem about the statue, calling it:

The marble index of a mind forever
Voyaging through strange seas of Thought, alone.

The statue has an inscription on it: *Newton Qui genus humanum ingenio superavit*. One translation of this is: "Newton, who in his genius surpassed the human race." For those like Neil who believe that Newton was the world's smartest person, here is some real evidence in favor of that—in marble. Einstein has a larger-than-life-sized statue in Washington, D.C., near the Vietnam Memorial, in front of the National Academy of Sciences. He is sitting down, and his statue is still 12 feet tall. Children come and play on his knees.

Now let me compare Einstein and Newton a bit more. I'm not going to contest Neil's contention that Newton is the greatest scientist ever. I want to give Newton his due. But I am going to argue that Einstein is someone who should compete for this title—someone in Newton's league.

What is Newton's most famous equation?

$$F = ma.$$

What is Einstein's most famous equation?

$$E = mc^2.$$

Which one of these two equations is more famous? Newton's equation, which we discussed in detail in chapter 3, says that more massive objects are harder to accelerate. Important for dynamics, but pretty simple. It's harder to get a piano moving than a harmonica. Einstein's equation says that a tiny bit of mass can be converted into an enormous amount of energy. It is the secret behind the atomic bomb. It tells us how the Sun shines. Which equation seems more important to you?

Newton has another famous equation: $F = GmM/r^2$ for the gravitational force between two particles of masses m and M. This is quite important. Einstein has another equation too: $E = h\nu$, where he found that light comes in particles of energy called *photons* with an energy equal to Planck's constant h times their frequency ν. Newton thought, to his credit, that light was made of particles, but you might say Einstein proved it. Light has a particle nature as well as a wave nature, a notion that is crucially important for quantum mechanics.

Both men invented things. Newton invented the reflecting telescope. All the big telescopes now are reflecting telescopes. The Hubble Space Telescope and the Keck telescopes are reflecting telescopes. Einstein invented the principle behind the laser. Every time you play a CD or a DVD, you are using Einstein's invention. Both men did some government work. Newton became Master of the Royal Mint. He invented the milling on the edges of coins that we still use today. This prevented thieves from scraping silver off the edges of silver coins and passing off the coins for full value. If they scraped off the milling, you could tell. Every time you pick up a quarter, you can see Newton's influence. Einstein's decisive role in world affairs is well known: he wrote a crucial letter to President Franklin D. Roosevelt, which led to the Manhattan Project and the atomic bombs that ended World War II. What Einstein did then was so important that we are still dealing with its effects today.

Einstein was such a famous character that people loved to tell anecdotes about him, which then added to the Einstein lore. One such story (perhaps apocryphal) goes like this: Einstein was talking to a man at the Institute of Advanced Study in Princeton. All at once the man reached inside his coat pocket and pulled out a small notebook and scribbled something down. Einstein asked, "What's that?" "Oh, this is my notebook," the man said. "I carry it with me everywhere, so if I have a good idea, I can write it down so I don't forget it." "I never had need for such a notebook," Einstein replied. "I only had three good ideas." So what were these good ideas, and how did Einstein get them?

The first was special relativity, which led to $E = mc^2$. The second was the photoelectric effect, the $E = h\nu$ equation, for which Einstein won the 1921 Nobel Prize in Physics. And the third was general relativity, Einstein's theory of curved spacetime to explain gravity. After he got the equations worked out, Einstein predicted that light would be bent, traveling in curved spacetime near the Sun, and he also predicted the amount of the bending. Stars seen near the Sun during a solar eclipse should appear in slightly displaced positions in the sky relative to pictures taken months earlier when the Sun was nowhere near those stars. The amount of deflection Einstein predicted (1.75 seconds of arc for stars near the *limb*, or outer edge, of the Sun) was twice what Newton would have predicted for particles traveling at the speed of light according to his theory. Sir Arthur Eddington led a British expedition to measure this. Einstein's prediction turned out to be right, and Newton's prediction turned out to be wrong. Today we believe Einstein's theory and not Newton's. Let's take a moment to appreciate that!

At the close of the twentieth century, I saw a program on its greatest moments in sports: Jesse Owens winning the 100 meters at the 1936 Berlin Olympics; Secretariat winning the Belmont Stakes by 31 lengths, to complete horse racing's Triple Crown; Mohammed Ali knocking out George Foreman in Zaire, to regain the heavyweight boxing championship of the world. What was the greatest play in science in the twentieth century? Imagine Newton and Einstein on a basketball court.

Newton's got the ball. He is dribbling the ball down court. And it's not just any ball, it's his theory of gravity—the proudest thing he ever did! Einstein comes along, steals the ball, shoots it up, and, *swish*, it's in the basket! This is the greatest play in science in the twentieth century.

I want to explain how Einstein got his great ideas. Einstein was good in school. He got good grades in science. Those stories you may have heard that Einstein got all bad grades in school—forget them. He was introduced to science at the age of 4, when his father showed him a compass. Einstein was quite taken with it, and this set him on a career in science. Einstein taught himself differential and integral calculus when he was about 12 years old. Smart fellow. And when he was 16, he started to think about the most exciting physical theory of his day—Maxwell's theory of electromagnetism. Maxwell put together all the different laws of electricity and magnetism.

Electric charges can be either negative or positive. Opposite charges attract each other and like charges repel each other with a $1/r^2$ force. Two positive charges repel each other, two negative charges repel each other, but a positive and a negative charge attract each other. This is *Coulomb's law*. It causes static electricity. Charges create electric fields, filling the space around them, and if you are a charge, the electric field acts to accelerate you. The electric field causes that $1/r^2$ electric force. It causes static cling in your clothes in winter. But moving charges create a magnetic field in addition, and a magnetic field can affect you if you are a moving charge. If a charge is not moving, the magnetic force on it is zero, but if it is moving and there are magnetic fields, there will be a magnetic force on the charge. These ideas had been worked out in several more physical laws. *Ampère's law* tells you how moving charges (e.g., a current in a wire) create a magnetic field, and if you know the magnetic fields and the electric fields at a given point, you can calculate the electric and magnetic forces on a moving charge at that location. *Faraday's law* describes how a changing magnetic field creates an electric field. And it was known that there are no "magnetic charges"; that is, one never finds an isolated north (or south) magnetic pole with a magnetic field spreading out from it. The *law of charge conservation* states that the total number of charges (number of positive charges minus the number of negative charges) stays constant. For example, if you had 10 positive charges and 9 negative charges in a region, the total charge was +1. A positive and negative charge could combine and eliminate each other, leaving 9 positive charges and 8 negative charges, but the total number of charges would remain +1.

Maxwell looked at the known laws of electromagnetism and showed that they were inconsistent with the law of charge conservation. To rectify this, he showed that a new effect needed to be added: a changing electric field creates a magnetic field. He put all these effects together into a set of four equations: *Maxwell's equations*. (You sometimes see physics students wearing them on T-shirts!)

Maxwell's equations included a constant, c, which was related to the ratio of the strength of electric to magnetic forces. If you had a swarm of charges moving with velocities v, the ratio of magnetic to electric forces they created was of order v^2/c^2, where c was a velocity. He then did experiments in the lab where he compared magnetic and electric forces to determine what the constant c was, and he got a very high value. He estimated the constant c was 310,740 km/sec. Maxwell also found a highly interesting solution to his own equations: it was an electromagnetic wave that traveled through empty space with velocity c.

The magnetic and electric fields were perpendicular to the velocity of the wave. The wave was sinusoidal, and the electric and magnetic fields oscillated at your location as the sinusoidal wave passed by you. Thus, the electric and magnetic fields were both changing. The changing electric field created the magnetic field, and the changing magnetic field created the electric field, and they bootstrapped themselves along with the wave, moving forward through empty space at a velocity c = 310,740 km/sec.

Eureka! Maxwell recognized that velocity—it was the velocity of light! Light must be electromagnetic waves! It was one of the great moments in science. How did Maxwell know the velocity of light? It was because astronomers—I want to speak up for astronomers here—had measured the velocity of light! In 1676, Danish astronomer Ole Rømer noticed that successive eclipses of Jupiter's moon Io by Jupiter were more closely separated in time when Earth was approaching Jupiter but were more widely spaced in time when Earth was moving away from Jupiter. Looking at those satellites orbiting Jupiter was like looking at a giant clock face. When we approach Jupiter, we observe the clock running fast, whereas when we move away from Jupiter, we observe the clock running slow. Rømer correctly attributed this to the finite velocity of light. As we approach Jupiter, the distance to Jupiter shrinks, and light beams from successive eclipses have less and less distance to travel to get to us, speeding their arrival. This effect is like a Doppler shift with light beams from successive eclipses being crowded together. He deduced that it must take light approximately 11 minutes to cross the half-diameter of Earth's orbit. It actually takes about 8 minutes, so Rømer was pretty accurate. When Earth is closest to Jupiter, the Jupiter clock is about 8 minutes fast, and when we are farthest away, the Jupiter clock is about 8 minutes slow. As discussed in chapter 8, Giovanni Cassini, in 1672, measured the parallax distance to Mars, which allowed one to deduce the radius of Earth's orbit. Using Rømer's data and knowing the approximate radius of Earth's orbit, Christiaan Huygens was able to estimate the speed of light: he got 220,000 km/sec (only about 27% low relative to the actual value of 299,792 km/sec).

In 1728, another astronomer, James Bradley, used a different method to measure the speed of light. Imagine a star directly overhead. Its light comes straight down onto you like rain. If you drive in a car, the rain on the windows comes down at a slant, because you are moving. Earth is moving at 30 km/sec in its orbit around the Sun. It's like moving in a car. If you point your telescope straight up, the light will fall down and hit the side of your telescope rather than reach the eyepiece at the bottom—because you are moving. To see the star, you will have to tilt your telescope to match the slant of the rain you are seeing in your moving vehicle, Earth. How much? It has to be slanted by about 20 seconds of arc. When you observe the same star 6 months later, it will be shifted 20 seconds of arc in the other direction. Bradley was able to measure that effect, called *stellar aberration*. The slope of this tilt is v_{Earth}/v_{light}, which Bradley found to be about 1 part in 10,000. Thus he could deduce that the velocity of light was about 10,000 times faster than the 30 km/sec orbital velocity of Earth, or 300,000 km/sec. So, in 1865, when Maxwell predicted that his electromagnetic waves traveling through empty space should have a velocity of about 310,740 km/sec, he recognized it as corresponding to the speed of light, which astronomers had already measured (300,000 km/sec). Within the plausible errors of his prediction (due to errors in his measurements of electric and magnetic forces) and the astronomical observational errors, the two numbers agreed. Light was electromagnetic waves. Maxwell recognized that electromagnetic waves could have wavelengths much shorter or longer than those of visible light. We know the shorter ones today as ultraviolet rays, X-rays, and gamma rays, while the longer ones are known as infrared, microwaves, and radio waves. In 1886, Heinrich Hertz proved the existence of electromagnetic waves by transmitting and receiving radio waves across a room. Maxwell's was the most exciting scientific theory of Einstein's day, and Einstein was very excited about it too.

Einstein did the following thought experiment in 1896, when he was just 17 years old. He imagined traveling away from the town clock at the speed of light. As he looked back at the clock, it would seem frozen at noon, because the light showing it at noon was traveling right along with him. Did time somehow stop if you traveled at the speed of light? He imagined looking at the light beam traveling alongside him. He would see static waves of electric and magnetic fields like furrows in a field; they were not moving relative to him. He was traveling along at the same speed as the wave, so it would look static to him. But such a *stationary* wavelike configuration of electric and magnetic fields in empty space was not allowed by Maxwell's field equations. What he was seeing out the window of his imagined spaceship seemed impossible. Einstein figured there was a paradox here—something must be wrong. It took him 9 years to figure out how to fix it.

What Einstein did was very original. In 1905, he decided to adopt two postulates:

1. Motion is relative. The effects of the laws of physics must look the same to every observer in *uniform motion* (motion at constant speed in a constant direction without turning).
2. The speed of light through empty space is constant. The velocity of light *c* through empty space should be the same as that measured by every observer in uniform motion.

These two postulates are the basis of Einstein's theory of *special relativity*. It's called *relativity* because "motion is relative" (the first postulate), and *special* because the motion is uniform. The first postulate you have tested yourself. Have you ever been on a jet plane traveling at 500 miles per hour (in a straight line without turning), with the shades pulled down so you can see some bad movie? It seems just like you are still sitting on the ground. In the moving plane, it seems just like you are at rest. Right now we are orbiting the Sun at 30 km/sec, and yet it seems like we are at rest. This first postulate is the relativity principle: that only relative motions are important, and that you cannot determine an absolute standard of rest. Newton's law of gravity obeys this postulate. It said that the acceleration (change of velocity) of two particles depended on their separation and had nothing to do with their velocities. Thus the solar system would work the same way if the Sun were stationary with the planets orbiting around it, or if the whole kit and caboodle were moving along at 100,000 km/sec. It wouldn't matter to Newton whichever was the case. You cannot tell by any gravitational experiment in the solar system whether the whole solar system is moving or not. In fact it is moving, going around the center of the galaxy at about 220 km/sec. Newton's theory obeyed the first postulate, and Einstein thought Maxwell's equations should obey this postulate too. All the laws of physics should obey this postulate.

The second postulate is peculiar. It means that if I see a light beam pass me, I must measure its speed to be 300,000 km/sec. But if another person comes running past me at 100,000 km/sec and looks at the same light beam, he must *not* measure it to be going at 200,000 km/sec, as you might think. He must see it going 300,000 km/sec, just like I do. It's crazy!

It doesn't make any common sense. Velocities should add. In fact, the only way it can make sense is if his clocks are ticking at a rate different from mine and if his measurements of distance are also different from mine. Remarkably, what Einstein did was to believe these two postulates and throw common sense right out the window! If this were a chess game, we would call this a "genius move" (denoted this way: !!), the kind of move that forces a checkmate 17 moves later. Einstein was going to assume these two postulates were true, prove theorems based on thought experiments derived from the postulates, and see what he got. If those theorems were then checked with observations and turned out to be correct, then that would be evidence that the postulates were true. This was amazing. No one had ever done anything quite like it before. Einstein's postulates were falsifiable.[1] If Einstein's theorems gave answers that were contradicted by observations, his theory would be proven wrong. If the theorems agreed with observations, while it would not prove the postulates themselves, it would certainly provide evidence supporting them.

Why did Einstein believe the second postulate? It was because the velocity of light was a constant in Maxwell's equations, related to a ratio of magnetic to electric forces you could measure in the lab. Maxwell calculated that light waves traveled through empty space at about 300,000 km/sec. If you saw a light beam passing you at any other speed (say, at 200,000 km/sec), you would be able to deduce that you were moving at 100,000 km/sec—you could deduce you were moving. That would violate the first postulate. In 1887, Albert Michelson and Edward Morley, in a famous experiment, tried to measure the velocity of Earth moving around the Sun by bouncing light beams off mirrors in their laboratory. Effectively, they measured differences in the speed of light relative to their lab for light beams traveling parallel and perpendicular to the velocity of Earth. They achieved enough precision to be sensitive to the 30 km/sec velocity of Earth around the Sun. Amazingly, they got a result of zero for the velocity of Earth, as if Earth were stationary and light beams in all directions traveled at the same speed relative to their lab. But we know Earth is moving—we see stellar aberration. It was quite puzzling. But their result is exactly what Einstein's second postulate would have predicted. You would always measure the speed of light to be the same whether Earth were moving or not, and therefore, if you believed the second postulate, you would have predicted that Michelson and Morley should have gotten a result of zero.

So, Einstein is going to believe his two postulates and prove theorems based on them. Here is one result. You can't build a rocket ship that travels faster than light. Why is that? Suppose I shine a laser beam toward a wall in my living room; it hits the wall. I am allowed to think I am at rest. But if you built a rocket that was going faster than light, and tried the same experiment on board the rocket, you would get a different result. If you sat in the middle of your rocket ship and directed your laser beam toward the front end of your ship, it would never get there. Any athlete can tell you that you cannot catch a runner who is faster than you and has a head start. The light beam from the laser can't catch the front end of the rocket, because the front end of the rocket is traveling faster (faster than light) and it has a head start. Clearly, if you did this experiment on the rocket ship, the laser beam would never reach the front end of the rocket, and you would know that you were moving (faster than the speed of light, in fact). But wait—that's not allowed by the first postulate. Since you are going at constant speed without turning, you must not be able to prove that you are moving. You must get the same results that I get in my living room. From this, it follows that you must not be able to build a rocket ship that travels faster than the speed of light. A strange result, but if you believe the two postulates, you must believe this result also. If you go slower than the speed of light, the laser beam eventually catches the front end of the rocket. It might take a very long time, but if your clocks were ticking slowly, for example, it might work out fine. Traveling slower than the speed of light is okay, but you can't build a rocket that travels faster than light. We have tested this in our particle accelerators, where we make particles like electrons and protons go faster and faster, nudging them ever closer to the speed of light but never quite reaching it.

Here is another result. Imagine a "light clock," in which a light beam bounces vertically between two mirrors, say, one on the ceiling and one on the floor; each bounce represents a tick of the clock. Light travels at 300,000 km/sec, or about 1 foot per nanosecond. One nanosecond is one billionth of a second. If we separate the two mirrors vertically by just 3 feet, the clock will tick once every 3 nanoseconds (figure 17.1).

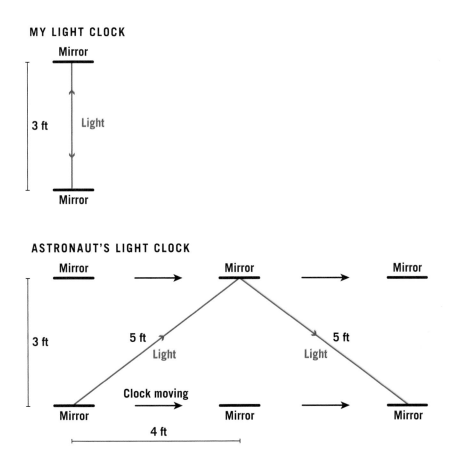

MY LIGHT CLOCK

Mirror

3 ft · Light

Mirror

ASTRONAUT'S LIGHT CLOCK

Mirror · Mirror · Mirror

3 ft · 5 ft Light · 5 ft Light

Clock moving

Mirror · Mirror · Mirror

4 ft

FIGURE 17.1. Light clocks. My light clock ticks once every 3 nanoseconds. A similar light clock is carried by an astronaut moving at 80% of the speed of light relative to me. Light moves at a constant velocity of 1 foot per nanosecond. I see the light beams in the astronaut's clock traveling on long diagonal paths 5 feet long, and therefore I see the astronaut's clock ticking only once every 5 nanoseconds.

Credit: Adapted from J. Richard Gott (*Time Travel in Einstein's Universe*, Houghton Mifflin, 2001)

It's a very fast clock, like a grandfather clock, only much faster. The light beam will bounce up and down, up and down, between the two mirrors. It hits a mirror once every 3 nanoseconds. This is my light clock. Now imagine an astronaut who passes me going left to right at 80% the speed of light, holding a similar light clock (figure 17.1). That's slower than the speed of light, so she can do that. From the point of view of the astronaut, she sees her light clock ticking normally with the light beam going up and down, ticking every 3 nanoseconds according to her. But if I look through the window of her space ship, I see her light clock moving along at 80% the speed of light and I see her light beam traveling on a diagonal path. The light beam starts at the bottom, but by the time it has moved up 3 feet, the upper mirror has moved from left to right by 4 feet. The light beam travels on a diagonal path that is 5 feet long. We have a 3-4-5 right triangle—3 feet vertically, 4 feet from left to right, and 5 feet along the diagonal hypotenuse. It solves Pythagoras's theorem $3^2 + 4^2 = 5^2$. While relative to me the light beam moves 5 feet diagonally from bottom left to top right, the astronaut moves 4 feet from left to right. Thus, she is traveling at 4/5 or 80% of the speed of light relative to me. Since I must observe the light beam to be traveling at 1 foot per nanosecond (according to the second postulate), I must say it takes 5 nanoseconds to go from bottom left to top right on the diagonal path of 5 feet I observe. I must see it take another 5 nanoseconds to come diagonally back down, arriving 8 feet to the right of where it started. Thus, I must say her clock ticks only once every 5 nanoseconds rather than once every 3 nanoseconds. I must see her clock ticking slowly (at 3/5 of the rate mine does).

Now for the interesting part. I must observe the astronaut's heart to be ticking slowly (also at 3/5 of the rate of mine), or she would notice that her light clock was ticking slowly relative to her heart, and she could deduce that she was moving, which is not allowed by the first postulate. Any clock she has on board must also be ticking slowly, at the 3/5 rate, or else she would be able to tell that she was moving. If she has a *muon* (an unstable elementary particle heavier than the electron) that is decaying, it must decay more slowly. She must age more slowly. She eats dinner more slowly. And . . . she . . . talks . . . more . . . slowly. Every process on the rocket ship goes more slowly.

How much more slowly depends on the astronaut's velocity v: if I age 10 years, a similar calculation using the light clock[2] shows the astronaut ages 10 years times $\sqrt{[1 - (v^2/c^2)]}$. For velocities that are small compared with the speed of light, such as we encounter in everyday life, this aging factor will turn out to be nearly exactly 1. If v/c is small relative to 1, then (v^2/c^2) will be really tiny relative to 1; something really tiny subtracted from 1 leaves something still about equal to 1, and the square root of 1 is 1—all of which means that this factor does not appreciably change the astronaut's aging. That is, the astronaut would also age 10 years, and I wouldn't notice any difference between her aging and mine. That's why we don't ordinarily notice that moving clocks are ticking slowly. However, if the astronaut is moving at a speed close to the speed of light—say, at 99.995% the speed of light—then $v/c = 0.99995$ and $\sqrt{[1 - (v^2/c^2)]}$ is only 0.01. You can check that on a calculator. While I age 10 years, I observe the astronaut aging only 1/10 of a year. At velocities approaching the speed of light, the slowing of time on the spaceship can be very dramatic.

We believe this formula, because we have checked it experimentally. Physicists took atomic clocks on plane trips around the world, going east so that the velocity of the plane added to the rotational velocity of Earth, and they observed that those atomic clocks came back slow (by about 59 nanoseconds) relative to atomic clocks left on the runway—just as Einstein would have predicted. Muons in the lab decay with a half-life of 2.2 microseconds—meaning half of them decay in 2.2 microseconds. But muons that are traveling toward Earth at nearly the speed of light (as cosmic rays) decay much more slowly, in accord with Einstein's formula. We believe this formula is right, because we have tested it many times. This is a funny universe, operating in surprising ways, but it seems to be the universe in which we live. Einstein's two postulates seem to be true. We'll see in the next chapter that these postulates also lead to the conclusion that $E = mc^2$, and that was verified in the atomic bomb. These are some truly remarkable results. The results are remarkable, because the postulates are remarkable. The more all these theorems check out, the more we may trust that the postulates are true.

18

IMPLICATIONS OF
SPECIAL RELATIVITY

J. RICHARD GOTT

Einstein's theory of special relativity revolutionized our ideas of space and time. It implied that time could be regarded as a fourth dimension—added to the three dimensions of space. Interestingly, it was Einstein's teacher Hermann Minkowski, using Einstein's work in special relativity, who developed this geometric picture of space and time, publishing his results in 1907. Einstein immediately adopted this view. We live in a four-dimensional universe. What do I mean by that? We say that the surface of Earth is two-dimensional. It takes two coordinates, latitude and longitude, to locate a point on Earth's surface. If you know your latitude and longitude, you know your location on the surface of Earth. But the universe is four-dimensional, meaning it takes four coordinates to tell you where you are. If I want you to come to a party, I will have to tell you where in latitude and longitude on Earth's surface to go. I must also tell you the altitude. You wouldn't want to show up on the fourth floor, if the party was on the twelfth floor! And I must tell you at what time you should arrive. If you come at the wrong time, you will miss the party just as surely as if you went to the wrong floor. Every event, such as a New Year's Eve party on the fifty-fourth floor, at 5th Avenue, and 34th Street, requires four coordinates to locate it: two coordinates to tell you where to go on Earth's surface, the altitude, and the time of the event. Since four coordinates are needed, we know that we live in a four-dimensional universe.

We can use this idea to draw spacetime diagrams. You have undoubtedly seen a picture in a book of Earth orbiting the Sun. The Sun is a big white dot in the center, and Earth's orbit is shown as a dashed circle surrounding it (because Earth's elliptical orbit is nearly circular). Earth can be shown as a small blue dot at the 12 noon position on the circle, representing its position on January 1. If we wanted to show Earth circling the Sun, we could have a sequence of pictures, with Earth working its way counterclockwise around the circle. By February 1, it will have reached about 11 o'clock on the circle, by March 1, it will reach 10 o'clock, and so forth. You could make a movie of this by making each picture in the sequence a frame of the movie. As the frames were played, you would see Earth circling the Sun.

Now imagine taking that film, cutting it up into individual frames, and stacking those frames on top of one another to form a vertical stack. Each frame would represent an instant of time, and frames that were higher in the stack would represent later times. In this way, you could make a spacetime picture of how Earth orbits the Sun. Time is the vertical dimension of the stack—the future is toward the top and the past is toward the bottom. The two horizontal directions represent two dimensions of space (as you would see them in a two-dimensional picture of Earth's orbit around the Sun). The Sun is not moving—it's always in the center; therefore, all these images of the Sun form a white rod extending vertically up the stack. In each frame, however, Earth has moved to a new position as it keeps advancing counterclockwise in its orbit around the Sun, which makes Earth in the stack appear as a blue helix winding around the white rod. The radius of this blue helix is 8 light-minutes—the radius of Earth's orbit. Vertically, the helix winds around the Sun once a year. The blue helix winding around the vertical white rod represents a spacetime diagram. We can add the orbits of Mercury, Venus, and Mars to the diagram by adding helixes for them that also wind around the vertical rod representing the Sun. This diagram is three-dimensional; I am leaving out one of the spatial dimensions to allow you to visualize the diagram. You couldn't visualize this diagram if it were four dimensional—you can only see three dimensions. We are showing the diagram in 3D in this book using a stereo pair (figure 18.1). You can either enjoy it as two photographs of a three-dimensional model from slightly different viewpoints or follow the instructions (appearing in the text near figure 4.2) for using both eyes to see it in its full three-dimensional glory.

The vertical white rod is called the *worldline* of the Sun—its path through space and time. It's white because, as we learned in chapter 4, the Sun is white (not yellow). The blue helix is the worldline of Earth—its path through space-

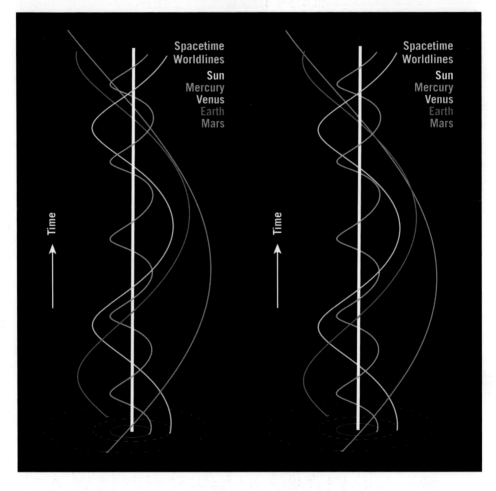

FIGURE 18.1. Spacetime diagram of the inner solar system. Time is vertical, and two dimensions of space are shown horizontally. This is a three-dimensional picture, so we have produced a cross-eyed stereo pair. Follow the same instructions for stereo viewing as for figure 4.2. The worldline of the Sun is the vertical white line in the middle. Earth, orbiting counterclockwise, circles first in front of the Sun and then passes behind it later (further up in the diagram). Mercury, Venus, Earth, and Mars have successively larger orbital periods and therefore successively more loosely wound helixes. *Photo credit:* Robert J. Vanderbei and J. Richard Gott

time. Notice how blue helix passes alternately in front of and behind the vertical worldline of the Sun. The orange helix winding tightly around the Sun is the worldline of Mercury. It orbits the Sun once every 88 days. The grey helix is Venus, and the red helix is Mars. The farther out a planet is, the less tightly its helix is wound around the Sun. If you think four-dimensionally, you should not think of Earth as a sphere, but as a long piece of spaghetti, a helix wound around the Sun. Earth has an extension in time.

You have a worldline too. It starts at your birth, snakes through all the events of your life, and ends at your death. Your worldline is about 1 foot, front to back;

about 2 feet wide; 6 feet tall; and if you are lucky, maybe 80 years in duration. These are spacetime diagrams where motionless worldlines intertwine in a static four-dimensional spacetime sculpture.

We can draw spacetime diagrams of some of the thought experiments that Einstein proposed on the concept of simultaneity. Suppose I am sitting in the center of my lab, and it is 30 feet wide. I'm an Earthling. My lab is stationary with respect to Earth, and I am stationary in the center of my lab. In the space-time diagram, the horizontal coordinate represents space, and the vertical coordinate represents time. Because I am advancing in time but not moving in space (left to right, or right to left), my worldline goes vertically upward. The front of my lab is not moving; it likewise has a vertical worldline, as does the back of my lab. The worldlines of the back of the lab, me (the Earthling), and the front of the lab are three parallel vertical lines. The future is toward the top, and the past is toward the bottom. The front of my lab is the vertical line to the right, and the back of my lab is the vertical line to the left. To make the hori-zontal and vertical scales, I am going to use units of feet and nanoseconds. Light moves through empty space at 1 foot per nanosecond. In the diagram (figure 18.2), light beams will be diagonal lines tipped at 45° with respect to vertical.

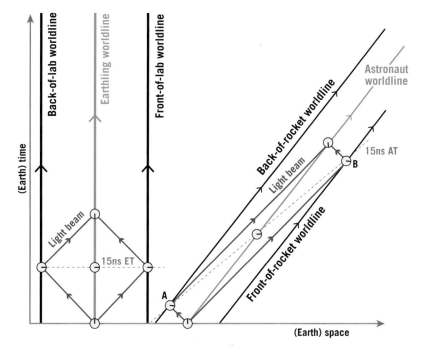

FIGURE 18.2. Spacetime diagram of my lab and an astronaut's rocket.
Credit: Adapted from J. Richard Gott (*Time Travel in Einstein's Universe*, Houghton Mifflin, 2001)

Let's say at time $t = 0$ (Earth time, ET; shown by the little clock on my worldline with its hand pointing vertically) I send out two laser beams, to the right and to the left, which hit mirrors on the front and back walls, respectively, of my lab. The worldlines of these two light beams are diagonal lines tipped at 45°. They reach the front and back of the lab simultaneously (traveling equal distances of 15 feet) at time 15 nanoseconds (ET). Two little 60-nanosecond clocks each reading 15 nanoseconds (ET) are shown at the points where the two light beams intersect the front and back walls of the lab. The Earthling's (my) worldline also has a little clock on it that reads 15 nanoseconds (ET). A horizontal dashed line connects those three little clocks reading 15 nanoseconds. That horizontal line connects simultaneous events according to me. After the laser beams bounce off the mirrors at the front and back of my lab, they return to me. And they both arrive back at the same time, 30 nanoseconds after the start. When the laser beams arrive back to me, my clock reads 30 nanoseconds, because the laser beams traveling at the speed of light have traveled out 15 feet and back 15 feet, traveling a total distance of 30 feet in 30 nanoseconds. So far, so good.

But then following Einstein's argument, consider an astronaut traveling at 80% of the speed of light in a rocket ship (going from left to right). The astronaut's worldline must therefore be tilted. For every 4 feet to the right he moves, he moves 5 nanoseconds upward in time. He is traveling at 4/5 (or 80%) of the speed of light. The front of his rocket is moving at the same speed and has the same tilt, as does the back of his rocket. The worldline of the back of the rocket, the worldline of the astronaut, and the worldline of the front of the rocket are all parallel. They are not moving relative to one another. Now that astronaut sitting at the center of his rocket sends laser beams toward the front and back of his rocket, just as I do in my lab. I measure the length of his rocket to be 18 feet. I will say more about that later. The light beam he sends to the left hits the back of his rocket, which starts out 9 feet away (that's half of 18 feet). I'm watching this experiment through a window of the rocket. I see the back of the rocket move 4 feet to the right in 5 nanoseconds, while the laser beam travels 5 feet to the left in those same 5 nanoseconds. Now 4 feet plus 5 feet is 9 feet, so it takes 5 nanoseconds for the astronaut's laser beam to hit the back of the rocket, closing that original distance of 9 feet. The astronaut's laser beam hits the back of the rocket at 5 nanoseconds Earth time (ET), according to me. The laser beam going to the left and the rocket going to the right are closing on each other and so collide quickly, according to me (that is, from my viewpoint).

The laser beam that the astronaut sends out to the right has to catch up with the front end of the rocket, which is moving away, and therefore it takes longer to catch up and meet it, according to me. While the laser beam travels 45 feet in 45 nanoseconds (ET; according to Einstein's second postulate, light travels at a constant speed of 1 foot per nanosecond), the front of the rocket travels just 36 feet (or 4/5 of 45 feet). In 45 nanoseconds, the light beam travels 45 feet, while the rocket travels 36 feet, or 9 feet less, but the front of the rocket had a 9-foot head start. Thus, the astronaut's laser beam hits the front of his rocket 45 nanoseconds (ET) later. This means that I observe the laser beam that is headed toward the back of his rocket hitting the back *before* the other laser beam headed toward the front hits the front. The events in which the laser beams hit the front and back of his rocket are not simultaneous, according to me.

What does the astronaut see? The astronaut is traveling at constant speed in a constant direction; by virtue of Einstein's first postulate, the astronaut is entitled to think of himself at rest. He does think he is at rest. He sits in the center of his rocket, which is also at rest with respect to him, and sends out laser beams toward the front and back of his rocket. Since he sits in the middle of his rocket and his rocket is not moving, he must think that the two laser beams traveling at the speed of light must take equal amounts of time to get to the front and back. From his perspective, he must see the two events, in which one laser beam hits the back of the rocket and the other laser beam hits the front of the rocket, as simultaneous events. I (the Earthling) do not think they are simultaneous events: I see the lasers hit in succession—the back of the rocket gets hit first, then the front. We do not agree on which events are simultaneous. This is counter to common sense, but it is a direct result of the postulates of special relativity.

Interestingly, when Einstein formulated this thought experiment, he did not use an astronaut on a rocket with mirrors at the front and back of his rocket, but a man on a train with mirrors at the front and back of the train. In 1905, the fastest vehicles we had were trains—about 120 miles per hour!

I slice spacetime differently from the astronaut. Think of four-dimensional spacetime as a loaf of bread. I slice the loaf in slices like those of American bread. Set a loaf of American bread on its end, so the individual slices are horizontal. These horizontal slices show individual instants of Earth time (ET), and each slice of bread contains simultaneous events according to me. The astronaut slices spacetime differently. Let us call him Jacques (he is French). He slices the loaf of spacetime on a slant, like French bread. His slanted slices measure instants of *astronaut time* (AT). Jacques and I do not agree on simultaneous events, that is, on which events we say are in the same slice. We slice the loaf differently, but we see the same loaf. According to Einstein, the things that are real are those that are observer-independent. Space and time as separate entities are not real. I say the present is a horizontal American-bread slice, but Jacques says the present is a slanted French-bread slice. Since he is moving with respect to me, we do not agree on what the present is. We therefore disagree on which events lie in the past and the future. But we can agree on the spacetime loaf. It is the entire four-dimensional spacetime that is real.

Now let's return to my view of Jacques's rocket. After Jacques's laser beam reflects off the mirror at the front of his rocket, it takes only 5 nanoseconds to get back to him, according to me. I see the light beam and the astronaut closing on each other. It only takes 5 nanoseconds for the light beam to move back 5 feet, while the rocket moves forward 4 feet, to close the distance of 9 feet. According to me, the front laser beam takes a total of 45 nanoseconds + 5 nanoseconds = 50 nanoseconds to go out and back. The laser beam that has hit the back mirror takes 45 nanoseconds to catch up with the astronaut. To go out and back takes 5 + 45 = 50 nanoseconds Earth Time (ET), according to me. I thus see both laser beams return to the astronaut at the same time. He must also see them returning at the same time, because they return at the same time and the same place.

I see 50 nanoseconds elapse between when he sent the laser beams out and when they returned. I see him moving at 80% of the speed of light (v/c = 0.8), so I must see his clocks ticking at 60% (or $\sqrt{[1 - (v^2/c^2)]}$) of the speed of my clocks. If I see 50 nanoseconds elapse, I must see the astronaut age only 30 nanoseconds. When the astronaut sees his laser beams return, he must say that *that* event occurs at 30 nanoseconds astronaut time (AT), because he is 30 nanoseconds older when they get back. The laser beams must have hit the front and back of the rocket simultaneously at 15 nanoseconds AT. Note the French-bread-tilted slice labeled "15 ns AT." This connects simultaneous events according to the astronaut. The astronaut thinks he is at rest, and the situation looks to him exactly like what I see in the lab on Earth. Since the laser beams go out and back in 30 nanoseconds according to him, he must deduce that his rocket is 30 feet long.

The events where the astronaut's two laser beams hit the front and back of the rocket are events that I see as separated in space by 50 feet and in time by 40 nanoseconds. Using the speed of light (1 foot per nanosecond) to compare distances in space with distances in time, I see these two distant events as being separated by more distance in space than distance in time. Any two events that I see separated by more distance in space than distance in time have what we call a *spacelike separation*. There is always some astronaut traveling at high speed (but lower than the speed of light) who will see those two events as simultaneous. He or she will see those events as having a separation in space but no separation in time. Einstein showed that what the two observers can agree on is the square of the separation in space of the two events minus the square of the separation in time of the two events; we refer to this quantity as ds^2. Using units where the speed of light is 1 (i.e., where 1 foot = 1 nanosecond), I find the separation in space of the two events to be 50 and the separation in time of the two events to be 40, so I calculate ds^2 to be $50^2 - 40^2 = 2,500 - 1,600 = 900$. The astronaut Jacques, however, sees the time difference between the two events to be 0 and the spatial separation between the two events to be 30 (remember he judges his rocket to be 30 feet long); but when he calculates ds^2, he gets $30^2 - 0^2$, or 900, just as I do. We may disagree on both distances and times, yet surprisingly, we will still agree on some important things.

Consider now the separation between the astronaut's sending of the light signal and its arrival at the back of his rocket. The separation I measure in space between these two events is 5 feet, and the time between these two events I measure to be 5 nanoseconds. So I calculate $ds^2 = $ (separation in space)^2 – (separation in time)^2 to be $5^2 - 5^2 = 0$. The astronaut measures a separation in space of 15 feet between the two events and a separation in time of 15 nanoseconds between the two events, so he calculates $ds^2 = 15^2 - 15^2 = 0$, just as I do. Events connected by a light ray (called a *null separation*) always have $ds^2 = 0$ as seen by any observer. Einstein's second postulate states that all observers must see a light beam traveling at a constant speed of 1 in these units (1 foot per nanosecond); thus, the separation in space must equal the separation in time, and ds^2 must be zero. Indeed, the minus sign in the ds^2 formula connected with the difference in time is designed to guarantee that the second postulate is always obeyed.

The Pythagorean theorem tells you that in a plane with an (x, y) Cartesian coordinate system, if two points are separated by distances dx and dy, then their (separation in space)^2 = $dx^2 + dy^2$. The square of the hypotenuse of a right triangle is equal to the sum of the squares of the other two sides. In three-dimensional space with x, y, and z Cartesian coordinates, the Pythagorean theorem generalizes to (separation in space)^2 = $dx^2 + dy^2 + dz^2$. That's high-school Euclidean solid geometry. But Einstein is saying that ds^2 = (separation in space)^2 − (separation in time)^2. Substituting, we find $ds^2 = dx^2 + dy^2 + dz^2$ − (separation in time)^2. But separation in time is just dt. So substituting for that, we have: $ds^2 = dx^2 + dy^2 + dz^2 − dt^2$. So that's the difference between the dimension of time t and any one of the three dimensions of space (x or y or z): there is a minus sign in front of the dt^2. It is that little minus sign that makes all the difference. That minus sign makes the time we know different from an ordinary dimension of space—all just to make the speed of light a constant.

Whew! That's a lot of arithmetic—but it gets us to an important point, the difference between time and the dimensions of space.

Remember that I began by noting that I measured the astronaut's rocket to be 18 feet long. So I say that his rocket is shorter than the astronaut thinks it is (which is 30 feet). I think his rocket is only $\sqrt{[1 − (v^2/c^2)]}$ as long as he does. Our clocks do not agree and our rulers do not agree—once again ensuring that we observe the speed of light always to be 1 (foot per nanosecond). How can we differ on the width of the worldline of his rocket? It results from our taking different "slices" through it. I am measuring its width at a particular instant of Earth time (ET), and he is measuring its width at a particular instant of astronaut time (AT). I am taking a horizontal American-bread slice through his rocket's worldline, and he is taking a tilted French-bread slice. To use a different metaphor, it is as if I were to saw through a tree trunk horizontally and then say "the trunk is 6 inches wide." If someone else saws through it on a slant, he may conclude it is 10 inches wide, but the trunk itself is the same. We are just making different cuts through it. The astronaut and I are simply taking different cuts through the rocket's worldline.

Why is this important? Take an extreme case where an astronaut passes by me on Earth traveling at 99.995% the speed of light: then the magic factor $\sqrt{[1 − (v^2/c^2)]}$ is 1/100. I see the astronaut fly out to the star Betelgeuse, 500 light-years away. I will see him take about 500 years to get there. After all, he is traveling at nearly the speed of light, and Betelgeuse is 500 light-years away—so it should take him about 500 years (ET) to get there. I observe him to age only 1/100 × 500 years, or 5 years, during the trip. I see his clocks ticking very slowly, because he is moving so fast. Everything he does looks slow to me—I see him take 100 hours to finish breakfast! When he reaches Betelgeuse, he will indeed be only 5 years older.

How does the trip look to him? He thinks he is at rest, and he sees Earth and Betelgeuse fly past him at 99.995% the speed of light. First he sees Earth pass him—*whish*, and then 5 years later, he sees Betelgeuse pass him—*whish*. Earth and Betelgeuse are basically at rest with respect to each other, on parallel worldlines. The Earth + Betelgeuse system looks to him like a long rocket with Earth at the front end and Betelgeuse at the back end. Since this rocket is moving at nearly the speed of light past him, and it takes 5 years to pass, he must deduce that the length of the Earth + Betelgeuse rocket is 5 light-years. Thus he must deduce that the distance between Earth and Betelgeuse is only 5 light-years. He thus judges the distance between Earth and Betelgeuse to be 1/100 of the distance I see. He sees my lengths compressed: he sees them to be 1/100 as long as I do. The factor of length compression he observes, $\sqrt{[1 - (v^2/c^2)]}$, must be the same as the factor by which I see him aging more slowly. This is certainly one of the most remarkable results of special relativity, beautiful in its symmetry and ironclad logic.

The fact that different observers have different ideas of simultaneity explains a paradox. Suppose my original astronaut Jacques, traveling at 80% of the speed of light, is instead a pole vaulter and carries a 30-foot-long pole with him pointing in the direction he is going. I will see his pole as only 18 feet long as it passes me. Suppose I have a barn 30 feet wide. Its front door is open, and its back door is closed. Jacques comes in the open front door; when he is in the center of my barn, I can close the front door, and his 18-foot-long pole will be trapped inside my 30-foot-wide barn. Then I open the back door and let him go on out the back. But how does it look to Jacques? He must think he is at rest holding a 30-foot-long pole. He sees my barn traveling toward him at 80% of the speed of light; he must think it is only 18 feet wide. When he is in the middle of my barn, he must see his 30-foot pole sticking out both the front and the back ends of my 18-foot-wide barn. Both doors can't be closed around it simultaneously trapping it inside. It looks like a paradox. But here is the answer. I close both doors around the pole at the same time—simultaneously according to me. But those events are not simultaneous to Jacques. He slices spacetime differently, on a slant. He sees me close those two barn doors at different times, one after the other, according to him. Because he never sees both barn doors closed simultaneously, he can see his pole sticking out both ends as he passes through the barn with both doors open.

It is a tribute to Einstein that he was able to work through all his thought experiments correctly. No one had ever tried doing thought experiments based on postulates in the way Einstein did. It was one of the most original features of his work.

Now we come to another apparent paradox, the famous *Twin Paradox*. In this paradox, the first twin—we will call her Eartha—stays home on Earth, while her twin sister Astra voyages to Alpha Centauri, 4 light-years away, at 80% of the speed of light, then turns around and comes back at 80% of the speed of light. Eartha sees Astra going at 4/5 of the speed of light, so she sees Astra taking 5 Earth years to get out to Alpha Centauri, and 5 Earth years to get back. When Astra gets back, Eartha is 10 years older. Because Eartha sees Astra moving at 80% of the speed of light, according to our formula $\sqrt{[1 - (v^2/c^2)]}$, Eartha must see Astra aging slowly, at 60% of the rate Eartha ages. When Astra returns, Eartha expects Astra to have aged only 6 years. So far so good. But what does Astra see? Since motion is relative, why doesn't Astra think Eartha has gone away and come back at 80% of the speed of light, and why doesn't she expect Eartha to be the younger one when Eartha returns? The answer is that Astra has accelerated during her trip; she has slammed on the brakes at Alpha Centauri to stop and start back. All her stuff would have hit the front windshield of her spacecraft. She has changed her velocity; she has reversed its direction. She no longer obeys the first postulate's requirement for an observer to be *in uniform motion in the same direction without turning* (figure 18.3.)

During the first half of her trip, Astra is traveling away from Earth, and slices of Astra Time (AT) are tilted like slices of French bread. Just as she is arriving at Alpha Centauri, her clock reads 3 years (AT), telling her how much she has aged. But the line of simultaneous events "3 AT" is tilted, so it intersects Earth only 1.8 years after the start. As she is arriving at Alpha Centauri, Astra thinks she is simultaneous with Eartha 1.8 years after the start. Astra says she has aged 3 years in the time Eartha has aged only 1.8 years. Now 1.8 years is 60% of 3 years. Therefore, Astra sees Eartha aging slowly, since Astra thinks of herself at rest and sees Eartha receding from her at 80% of the speed of light.

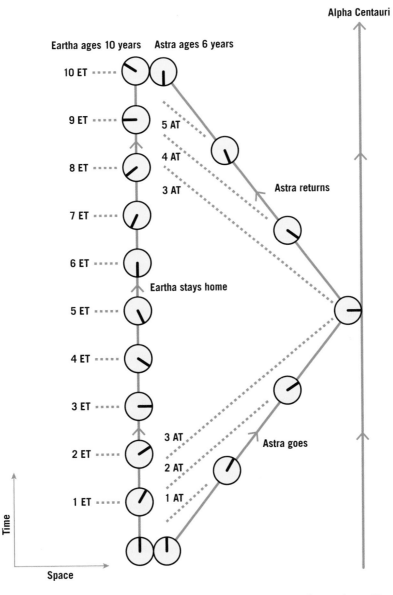

FIGURE 18.3. The Twin Paradox spacetime diagram of twins Eartha and Astra. Eartha stays home. Her worldline is straight. Astra goes to Alpha Centauri and returns—her worldline is bent. Astra ages less than Eartha does. Clocks show time measured by each in years. Dashed lines show Eartha Time (ET) and Astra Time (AT). *Credit:* J. Richard Gott

At this point, Astra thinks Eartha has aged less. But wait! Now Astra slams on the brakes, stops, and reverses course. Astra's worldline is bent at this point. She has changed velocity, and therefore her notion of simultaneity changes radically as well. Just as she is leaving Alpha Centauri, her clock still reads "3 AT," but now, because she is moving in the opposite direction, the slice

"3 AT" marking simultaneous events is tipped in the opposite direction and intersects Earth 8.2 years after the start. Once she is on the way back, Astra thinks that her departure from Alpha Centauri is simultaneous with Eartha being 8.2 years older than at the start. During the trip back, Astra sees Earth age another 1.8 years while she, Astra, ages another 3 years. That makes Eartha a total of 8.2 + 1.8 = 10 years older than at the start, while Astra is 3 + 3 = 6 years older than at the start. So Astra agrees with Eartha, as she must, that Astra is younger than Eartha when they meet again. Eartha has traveled on a straight worldline, whereas Astra's worldline is bent. This is the solution to the Twin Paradox. The idea of simultaneity is very important here.

The Twin Paradox enables you to visit the future. If you want to visit Earth a thousand years from now, all you have to do is get on a rocket ship and travel at 99.995% of the speed of light out to the star Betelgeuse, 500 light-years away. Your clock ticks 1/100 as fast as one on Earth. It will take you 500 years to get there, according to clocks on Earth. But you will only age 5 years. Come back at 99.995% the speed of light and you will age another 5 years on the way back. But when you get back, you will find Earth 1,000 years older. You will have time-traveled into the future. A trip such as this would be a lot more expensive than the current NASA budget (!), and of course the technology for building such a fast spacecraft doesn't exist yet, but we know it is possible under the laws of physics. We send protons in our particle accelerators at speeds faster than this, so we know such speeds are possible. It's just a matter of money and engineering—NASA, take note.

You might complain that the high acceleration at the turn-around point would kill you. But it turns out that you can arrange the same trip with only a comfortable 1 g acceleration, such as you experience on the surface of Earth. Your feet would be pressed to the floor, because your rocket was accelerating, just as they are pressed to the floor now because of gravity. Your trip would take you longer this way, but it would be comfortable. You would accelerate outward toward Betelgeuse for 6 years and 3 weeks of spaceship time, reaching a peak velocity of 99.9992% the speed of light. At that point, you are halfway to Betelgeuse. Then you decelerate at 1 g for another 6 years, 3 weeks of spaceship time to bring yourself to a halt at Betelgeuse. Accelerate back toward Earth for 6 years, 3 weeks, and finally decelerate for another 6 years, 3 weeks to bring yourself to a stop at Earth. You will age 24 years, 12 weeks during the trip, but when you get back, Earth will be 1,000 years older. You just have to invest a little more of your time (24 years versus 10 years) to do it comfortably. It took Marco Polo 24 years to make his famous visit to China and return to Europe. You would just need to invest as much of your time in your trip as Marco Polo did in his, and you could visit the future. You could visit Earth a millennium from now.

Gennady Padalka, a Russian cosmonaut, is our greatest time traveler to date. By orbiting Earth at high speed for a total of 879 days during visits to the Russian space station Mir and the International Space Station, he aged 1/44 of a second less than he would have if he had stayed home. (This calculation also includes some smaller general relativity effects due to his high altitude.) When he returned, he found Earth 1/44 of a second to the future of where he expected it to be. He has time traveled 1/44 of a second into the future. I know you're laughing. It's not a big trip, but it is a trip into the future. I was once interviewed on National Public Radio, and they asked why it was so easy to travel in space and so difficult to travel in time. I replied that the truth is we have not gone very far in space either! Einstein showed us that when comparing distances in space with distances in time, we should use the velocity of light. Thus, astronomers know to say that Alpha Centauri is 4 light-years away, because it takes light 4 years to come from it to us. The farthest our astronauts have gone is to the Moon. The Moon is only 1.3 light-seconds away. Humans have traveled as far as 1.3 light-seconds in space, and have time traveled 1/44 of a second into the future. These are roughly comparable.

Interestingly we have actual identical twin astronauts today to illustrate the twin paradox. Mark Kelly has spent 54 days in low Earth orbit, while his identical twin brother Scott Kelly has spent 519 days in low Earth orbit. Because Scott spent more time traveling at high speed in low Earth orbit, he is now about 1/87 of a second younger than his twin brother Mark.

I have pointed out that if we sent an astronaut to the planet Mercury and she lived there for 30 years before returning to Earth, she would be about 22 seconds younger than if she had stayed home. Clocks on Mercury tick more slowly than those on Earth both because Mercury circles the Sun at a faster speed (a special relativity effect) and because Mercury is deeper in the Sun's gravitational field (a general relativity effect).[1]

In 1905, Einstein showed that time travel to the future was possible. This is just 10 years after H. G. Wells proposed the idea, in 1895, in his book *The Time Machine*. In Newton's laws of physics, you could forget it—everyone agreed on time, everyone agreed on what "now" was, and time travel to the future was impossible. But Einstein showed that observers did not always agree on what was happening "now"; time was flexible—moving clocks ticked more slowly. Einstein gave us an entirely new picture of the universe, a universe with three dimensions of space and one dimension of time.

Now I am going to derive Einstein's famous equation $E = mc^2$. Suppose you had a laboratory with a particle moving slowly from left to right inside it with velocity v much, much less than c (i.e., $v \ll c$). Newton's laws will apply, and if the particle has a mass m, it will have, according to Newton, a momentum $P = mv$ pointed toward the right. The particle gives off two photons each of energy $E = h\nu_0$ in opposite directions: one to the right and one to the left. We are using Einstein's famous equation for the energy of photons, where h is Planck's constant and ν_0 (Greek letter nu) is the frequency of the photons as measured by the particle. The particle loses an amount of energy $\Delta E = 2h\nu_0$, equal to the energy the particle sees carried off by the two photons. Einstein showed that photons carry not only energy but momentum. The momentum of a photon is equal to its energy divided by the speed of light c. The particle sees the two photons carry away equal amounts of momentum but in opposite directions, making the total momentum carried off by the two photons zero as seen by the particle. The particle "thinks" it is at rest (by the first postulate), and it gives off two equal photons in opposite directions. By symmetry, a particle at rest that gives off two equal-frequency photons in opposite directions stays at rest. The recoils from the two photons on the particle cancel out. The particle's worldline remains straight: it does not change in velocity (figure 18.4).

Now consider what happens to those two photons. The one going to the right will eventually slam into the right wall of the lab. It hits the wall, and the wall is pushed a tiny bit toward the right. Einstein showed that a photon carries a momentum equal to its energy divided by the speed of light. This is the effect of radiation pressure: the wall absorbs the momentum of the photon, and this pushes the wall to the right. An observer sitting on the right wall will see the photon headed to the right hitting the right wall with a frequency that is higher than the emitted frequency, because the particle is approaching the right wall. This is an instance of the Doppler effect, which you will recall from previous chapters. In contrast, an observer sitting on the left wall of the lab will see a redshifted photon traveling to the left hit the left wall with a lower

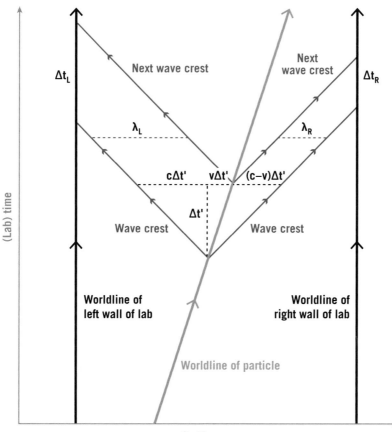

FIGURE 18.4 Spacetime diagram of $E = mc^2$ thought experiment. Stationary walls of the lab have vertical worldlines. The particle moves from left to right with velocity v, its worldline is tipped. It emits a photon to the left (whose wave crests move at 45° to the upper left) and an equivalent photon to the right (whose wave crests move at 45° to the upper right). The lab time between the particle's emission of the two sets of wave crests is $\Delta t'$, shown by the vertical dashed line. In that time the first leftward wave crest moves a distance $c\Delta t'$ to the left, while the particle moves a distance $v\,\Delta t'$ to the right as shown. The wavelength (distance between wave crests) of the leftward-moving photon is shown: $\lambda_L = (c + v)\Delta t'$. The wavelength of the rightward-moving photon is shorter: $\lambda_R = (c - v)\Delta t'$ due to the Doppler shift. *Credit:* J. Richard Gott

frequency than emitted, because the particle is going away from him. A higher frequency (bluer) photon carries a larger momentum than a lower frequency (redder) photon does. So, the right wall receives a harder kick (to the right) than the left wall receives (to the left). The two kicks do not cancel out, and the lab receives an overall kick to the right. The lab has received some momentum. There must be conservation of momentum, as Newton supposed (otherwise one could construct various unphysical levitating devices!), and therefore, this momentum must come from someplace. The only place it can come from is the particle itself.

Now the velocity of the particle is $v \ll c$, so the momentum of the particle should be given by Newton's formula: mv. Since the lab has gained momentum, the particle must have lost momentum. But the particle's worldline is not bent—it remains straight (see spacetime diagram in figure 18.4). Its velocity does not change. If the particle's momentum mv decreases while its velocity v remains the same, its mass m must have decreased. It gave up some energy (in the form of two photons) and it lost some mass. Some of its mass was turned into energy! Whoa! That's a remarkably bold conclusion. What is the relation between the amount of energy given off and the amount of mass lost? This just requires you to calculate the Doppler shifts of the two photons. The total rightward momentum gained by the walls of the lab is $2hv_0(v/c^2)$. I show the full calculation of these in appendix 1. The energy given off by the particle in the form of the two photons is $\Delta E = 2hv_0$, so the total rightward momentum gained by the walls is $\Delta E(v/c^2)$. The factor v/c^2 comes from a factor v/c due to the Doppler shifts and a factor of $1/c$ due to the ratio of momentum to energy carried by photons. The total rightward momentum gained by the walls, $\Delta E(v/c^2)$, in turn, must equal the momentum lost by the particle: $(\Delta m)v$. So we have $\Delta E(v/c^2) = (\Delta m)v$. Divide both sides of the equation by v. (The velocity of the particle cancels out!) We get $\Delta E/c^2 = \Delta m$. Now multiply both sides of the equation by c^2. That gives $\Delta E = \Delta mc^2$. Get rid of the Δ signs. The answer is $E = mc^2$.

In the thought experiment, the particle loses some energy by emitting two photons, and it loses some mass. A particle that loses mass emits energy. The emitted energy is related to the mass that is lost by the formula $E = mc^2$. It's as simple, yet as powerful, as that. The c^2 appears in the equation because all the Doppler shift and momentum calculations involve light, and c is its speed.

As you know, c is a very large number (300,000 km/sec in ordinary units), enabling a tiny amount of mass to be converted into a huge amount of energy. Newton's laws show us that the kinetic energy of a truck is $\frac{1}{2}mv^2$, where m is the mass of the truck, and v is its velocity. That's accurate as long as $v \ll c$. A truck going at a velocity of 100 miles per hour has a velocity of 0.045 km/sec (that's only $0.00000015c$). If two trucks each going 100 miles per hour collide head on, all that kinetic energy—2 times ($\frac{1}{2}mv^2$)—will be released in a giant explosion. Pieces of the trucks will fly all over. But now suppose a truck made of matter would hit a truck made of antimatter. These two trucks would annihilate each other, converting all their mass into energy—an extreme case.

That would cause an explosion releasing $2 \times (mc^2)$ worth of energy, which is a lot bigger than mv^2 for the normal trucks. How much bigger? By a factor of $2/(0.00000015)^2 = 89$ trillion! That matter–antimatter explosion would be 89 trillion times as energetic as the explosion caused by two normal trucks hitting each other at 100 miles per hour. There is a tremendous amount of energy locked up in the mass of ordinary matter.

That is the secret of the atomic bomb. Uranium or plutonium atoms can fission to create decay products that weigh slightly less than the original atoms, releasing an enormous amount of energy. In the Sun, four hydrogen nuclei fuse to become one slightly lighter helium nucleus, releasing energy. This is what has powered the Sun for the past 4.6 billion years. Chemists have measured accurate masses for different elements, which show slight differences in the mass per nucleon in different elements. Consequently, one can calculate how much nuclear energy could be produced by fusing light elements or by splitting heavy ones. Iron has the lowest mass per nucleon—there is no getting nuclear energy out of it, as we discussed in chapter 7.

Einstein realized, along with other physicists, that his equation implied that atomic bombs, from splitting atoms, could be made, and he wrote the crucial letter to President Franklin D. Roosevelt on August 2, 1939, urging him to build an atomic bomb before Hitler did. Thus, the Manhattan Project was born, and American and European refugee physicists developed a working atomic bomb. As the Americans later learned, Germany did have an atomic bomb program just as Einstein had feared, but it was ineffectual and did not succeed. Germany had already surrendered by the time the first U.S. atomic bomb was tested in New Mexico. But ultimately, two atomic bombs were dropped on Japan. Japan surrendered a short time later, ending World War II. The devastation was horrendous: approximately 200,000 people were killed by the bombs and their aftermath, including radiation exposure. Robert Oppenheimer, who led the Manhattan Project, later said the first test of the atomic bomb reminded him of lines from the *Bhagavad Gita*: "now I am become Death destroyer of worlds." President Truman took full responsibility for the decision to drop the bomb. He felt its use was necessary to end World War II as swiftly as possible. But Truman said, "I realize the tragic significance of the atomic bomb." Years later, in Truman's private library, a book on the atomic bomb was found where he had underlined words from Horatio's last speech in *Hamlet*: "Let me speak to the unknowing world. So shall you hear of carnal, bloody and unnatural acts, of accidental judgments, casual slaughters, of deaths put on by cunning and forced cause, and, in this upshot, purposes mistook fall'n on the inventors' heads." After the war, Einstein devoted himself to the cause of nuclear disarmament.

By thinking about traveling near the speed of light, which was far from practical in his day, Einstein nevertheless discovered a principle that would change the course of history. Einstein's work in his miracle year of 1905 would catapult him into the first rank of scientists, along with Marie Curie and Max Planck, but his greatest work was yet to come.

19

EINSTEIN'S GENERAL THEORY OF RELATIVITY

J. RICHARD GOTT

Einstein's greatest scientific achievement was general relativity, his theory of curved spacetime to explain gravity, which has replaced Newton's theory of gravity.

Einstein was considering the following problem. Drop a heavy and a light ball simultaneously—they fall and hit the floor at the same time. Galileo knew this. What would Newton say? He would say that the gravitational force between the ball and Earth was $F = Gm_{ball} M_{Earth}/r_{Earth}^2$. He would also say $F = m_{ball}a_{ball}$, so the acceleration, a_{ball}, of the ball was the force on the ball divided by its mass. Combining the equations, we get $a_{ball} = GM_{Earth}/r_{Earth}^2$. The mass of the ball cancels out. The acceleration of the ball is independent of its mass, which means that heavy and light balls must fall at the same rate. Newton would say that the heavy ball had a larger gravitational force pulling it toward Earth. But he would say that it was harder to accelerate because of $F = ma$, and this would just compensate for the bigger force such that the acceleration of both balls was exactly the same. That's quite a coincidence, and is a statement that the mass we use in the gravitation formula (the *gravitational mass*) and the mass we use in the $F = ma$ formula (the *inertial mass*) are identical.

Einstein thought about the problem differently. He considered what would happen if you were in an accelerating spaceship in interstellar space where there was no gravity. (Like the accelerating matter-antimatter powered interstellar spaceship Neil discussed in chapter 10.) If you drop the two balls, they just float weightless next to each other. Then, because the spaceship is firing its rockets and accelerating upward, the floor of the spaceship cabin just accelerates upward and hits the two balls floating there. The balls automatically hit the floor at the same time. They just float in place, and it's the floor that comes up and hits them. Simple. Now it is not a coincidence that the two balls hit at the same time. Imagine dropping the two balls again on Earth. This time try to imagine them just floating together in place and the floor coming up to meet them. People knew that on an accelerating spaceship it would seem like you were back home on Earth. But Einstein said, if the experiment on an accelerating spaceship looked just like gravity, then it must be gravity. He called this the *Equivalence Principle*. He called this his happiest idea, and 1907 was the year it came to him. If the two different phenomena looked exactly the same, they must be the same. This was very bold.

Einstein had used this reasoning before. A charge moving past a magnet was accelerated by a magnetic field, but a stationary charge experienced the same acceleration when a magnet was moved past it. In the second case, according to Maxwell's equations, the acceleration was produced by an electric field generated by the changing magnetic field. Einstein deduced that the two phenomena must be identical, and that it was only the relative motion that was important. This meant that the notions of electric and magnetic fields as separate entities needed to be replaced by a notion of one *electromagnetic* field. In the same way, Einstein found that our ideas of space and time as separate should be replaced with the idea of four-dimensional spacetime. Often a great breakthrough in science occurs when someone realizes that two different things are actually the same. Newton realized that the same force that makes the apple fall is what keeps the Moon in its orbit. Aristotle knew that gravity made an apple fall to Earth, but he assumed that something different, something celestial, kept the Moon in its orbit. Newton realized that the two phenomena were the same.

Einstein had great faith in his idea of the Equivalence Principle. If you dropped a heavy and a light ball, they just floated together in free fall and Earth's surface accelerated upward to hit them at the same time. The only trouble was that it didn't seem to make sense. How could Earth's surface be accelerating upward everywhere if it is not getting bigger? If it were expanding like a balloon, it could be plowing into those balls we drop, but Earth is not getting any bigger, so the idea appears to make no sense. This could only make sense if one had curved spacetime, where the laws of Euclidean geometry do not apply.

Let's discuss curvature. Figure 19.1 shows a globe of Earth. Its surface is curved, and therefore the laws of Euclidean plane geometry don't apply on its surface. Euclid tells us that, on a plane, the sum of angles in every triangle is 180°. On the globe, the straightest line you can draw is a great circle route—it's the shortest distance between two points. A *great circle* is a circle on the globe whose center is at the center of the globe. Earth's equator is a great circle. A meridian of longitude is a great circle. The shortest distance from New York City to the North Pole traces the meridian of longitude connecting New York City to the North Pole. On the globe we can make a triangle connecting the North Pole with two points on the equator that are 90° apart in longitude, and we will form a triangle (made up of great circle routes) that has three 90° angles, for a total of 270°.

If you go south from the North Pole, when you get to the first point on the equator, you will have to turn 90° to start going west along the equator. Then you have to turn 90° again when you get to the second point on the equator, to head due north and return to the North Pole. When you arrive, you will see

FIGURE 19.1. Triangle with three right angles on a sphere.
Photo credit: J. Richard Gott

that the two sides of the triangle meeting at the Pole make another 90° angle, because they are two meridians of longitude separated by 90°. You have traced a triangle with three right angles, impossible in Euclidean plane geometry. The surface of the sphere is curved and does not behave like a flat Euclidean plane.

Imagine drawing a circle on the globe centered at the North Pole. Make the radius of the circle measured along the surface of the globe equal to the distance from the Pole to the equator (that's 1/4 of the circumference of Earth). The circumference of your circle (centered on the North Pole) will be the equator. The equator has a length equal to the circumference of earth, and so the radius of the circle you have drawn must be 1/4 of the length of the circumference. Therefore, in this case, the circumference of your circle is 4 times its radius or less than 2π times the radius you would expect from Euclidean geometry. Again, we find that the curved surface of the sphere does not obey the laws of Euclidean plane geometry.

Einstein thought about a spinning phonograph record. If an ant stood on the phonograph record, it would have to hold on tightly to stay on. It would need to produce a centripetal acceleration (by holding on tightly) to keep it on the record, and it would feel a "gravitational" force pulling it outward. Certain carnival rides can give you this effect; they are like being inside a spinning tin can, and you feel a g-force pushing you against the cylindrical wall. You can even lift your feet off the floor. In both cases, the spinning record and the spinning carnival ride, accelerated circular motion mimics gravity, just like an accelerating spaceship does. We expect the phonograph record to be flat. But Einstein knew that, because the outer edge of the phonograph record is moving rapidly, measuring rods placed on it would have a different length if measured by observers sitting on the rotating edge of the record than they would if measured by someone sitting still at the center. The record's circumference as measured by observers on the rotating record should differ from 2π times the radius, the value we expect from Euclidean plane geometry. The geometry of a rotating phonograph record would be non-Euclidean (it would have curvature) precisely because it was rotating, Einstein deduced, and gravity would be simulated. If such simulated gravity *is* gravity (that's Einstein's Equivalence Principle), then curvature of spacetime could itself create gravity.

If I'm in New York City and I want to go to Tokyo, I should go on a great circle route, the shortest possible path. I can stretch a string tight between the two cities on a globe. The great circle route goes through northern Alaska

FIGURE 19.2. Great circle route on a globe, connecting New York City and Tokyo.
Photo credit: J. Richard Gott

(figure 19.2). Get a globe and try it yourself. That's the route an airplane would take. It is also the straightest possible path between the two cities. You can demonstrate that by taking out a little toy truck and driving it on the globe between the two cities. The wheels are aimed straight ahead on the toy truck; if it heads out in the right direction on the path to Tokyo, it can simply drive straight ahead on the great circle route, pass through northern Alaska, and arrive at its destination. We call this straightest possible path a *geodesic*. Start a truck traveling west on the equator, drive it straight ahead, and you will trace out the entire equator. Starting out in any direction, if you drive straight and never turn your steering wheel, you will follow a geodesic route. If you look at a flat Mercator map of Earth, the geodesic great-circle path linking New York and Tokyo looks curved. Because both cities are at about 40° latitude, it might seem from the Mercator map that the best way to get to Tokyo would be to go west along a circle of latitude. But that path is actually longer on the globe. It is also not straight. A circle of latitude is a small circle on the globe; its circumference is less than that of the equator, and its center (inside Earth) lies north of Earth's center. It is not a great circle. The U.S. border with Canada out west (past the Great Lakes) is a segment of such a small circle. If you drove a truck along that border going east, you would have to turn your steering wheel slightly to the left as you drove along to stay on the path. On a flat map of Earth, depending on what coordinate system it has, a straight geodesic line can end up looking curved.

Throw a basketball up into a basket, and it will arc up and then down as it goes into the basket. It seems to take a curved path—a parabola. Its path may appear to be bent by several feet. It is bent just like the geodesic route from New York to Tokyo is bent on a Mercator map. Einstein's idea was that objects in free fall like the basketball would travel along geodesics in curved spacetime, along the straightest possible trajectories they could possibly follow (as long as they were not acted on by any other forces, like electromagnetic forces). The marching orders for a particle were simple— just go straight ahead. The particles did not add up a bunch of forces from different masses all pointing in different directions, as Newton would have it. They just flew straight ahead. Spacetime was curved, and this curvature created gravity. Recall the spacetime diagram in figure 18.1 with the Sun's worldline as a vertical rod and Earth's worldline as a helix winding around it. It is actually a rather tall helix. It is 8 light-minutes in radius and has a 1 light-year-tall segment between turns. Einstein's idea is that the mass of the Sun slightly curves the spacetime around it, so that the helical worldline of Earth is actually following the straightest possible trajectory through that curved spacetime, like the truck driving straight to Tokyo. Earth's worldline may look bent in the coordinate system of figure 18.1, but actually, it is the straightest possible geodesic path in the curved spacetime. If you knew what that curvature looked like, you could calculate the geodesic path that Earth would take around the Sun.

This is how Einstein is going to explain gravity. Newton would say that if you took two masses and set them at rest in interstellar space, they would start accelerating toward each other because of their gravitational attraction, and they would eventually hit each other. Newton would say this is because they exert forces on each other at a distance, and these forces pull them together. Einstein says that the two masses cause the spacetime around them to be curved. In that curved geometry, the two particles simply travel on the straightest possible trajectories they can, which brings them together.

Suppose you have two trucks located some distance apart on the equator, both headed north (figure 19.3 at bottom). They start out on parallel trajectories, neither approaching nor separating from each other initially, but they don't stay on parallel trajectories, because Earth's surface is curved. Let the two trucks both travel straight north on adjacent meridians of longitude (which are geodesic trajectories). They are traveling parallel initially, both headed north, but as they keep heading north, going straight along separate meridians of longitude, they will find themselves drifting toward each other. Eventually they will collide at the North Pole.

Einstein is saying that the masses of particles cause a curvature of spacetime like the curvature of Earth. The direction "north" represents the direction of time toward the future. The lines of longitude they travel along represent the worldlines of the two particles. These straightest possible worldlines of the two particles are

FIGURE 19.3. Trucks traveling due north, drawn together by the curvature of the globe, hit at the North Pole. *Photo credit:* J. Richard Gott

drawn together because of the curvature of spacetime. Notice that if you started off the two trucks on parallel paths on a flat desktop, they would continue to travel parallel to each other, and their geodesics would remain the same distance apart. Gravitational attraction is caused by curvature of spacetime in Einstein's theory.

Mass and energy cause spacetime to curve—but how? Einstein started to work on this idea. He asked one of his mathematician friends, "Am I going to have to learn about Riemann curvature tensors?" The friend said, "Yes, I'm afraid you are." Bernhard Riemann had worked out the theory of curvature in many dimensions. Riemann was starting on the equivalent of a PhD thesis, under the supervision of Carl Friedrich Gauss. Gauss was a great mathematician who had worked out the theory of curvature (Gaussian curvature) on two-dimensional surfaces, such as the surface of Earth. Gauss told Riemann to suggest three possible thesis topics. Riemann's third favorite topic was curvature in higher dimensions. Gauss said "work on that." Riemann did, and it was a tour de force. Riemann showed that to understand curvature in many dimensions, you needed something we now call the *Riemann curvature tensor*: $R^\alpha_{\beta\gamma\delta}$. In four dimensions, this was a mathematical monster containing 256 components.[1] Luckily, many of these components were the same, effectively reducing the number to only 20 independent components—still a lot. This is the mathematical creature Einstein had to master. Einstein wanted to come up with field equations for the gravitational field that were precisely analogous to Maxwell's field equations for electric and magnetic fields. How do energy and mass curve spacetime? What geometries are possible? He wanted his theory to answer these fundamental questions, but it also had to agree approximately with Newton's theory for low velocities and small amounts of curvature, because Newton's theory worked pretty well under these conditions.

Einstein worked on this problem from 1907 to 1915. It required very difficult mathematics. He had many false leads. But he never gave up. Finally, in late 1915, he arrived at the correct field equations. Here they are (in appropriate units where Newton's constant G and the speed of light c are set equal to 1), just so you can see what they look like: $R_{\mu\nu} - \frac{1}{2}g_{\mu\nu}R = 8\pi T_{\mu\nu}$. The right side of the equation represents the "stuff" (mass, radiation, etc.) at a location in spacetime, and the left side of the equation tells how spacetime is curved at that location.[2] The stuff in the universe tells spacetime how to curve. Einstein had gotten rid of Newton's mysterious "action at a distance." The stuff of the universe (matter, radiation) at a location caused spacetime to curve in a certain way *at that location*. Particles, and planets also got their marching orders locally—they just went straight ahead in the curved spacetime. Deriving these equations was very tough going. At first, Einstein thought the correct equations were $R_{\mu\nu} = 8\pi T_{\mu\nu}$. He was missing a term. Interestingly, these equations were correct for empty space. Empty space is devoid of stuff, so $T_{\mu\nu} = 0$ in empty space, Einstein reasoned. Einstein thus deduced that $R_{\mu\nu} = 0$ for empty space as well. But if $R_{\mu\nu} = 0$ for empty space, R (which is calculated from the components of $R_{\mu\nu}$) would also be zero, and the correct field equations of 1915 with the extra term $-\frac{1}{2} g_{\mu\nu} R$ would also be satisfied, because the extra term would be zero in the case of empty space as well. Even though Einstein had the wrong field equations at first, luckily they were correct for the case of empty space. A week later, he figured out that he needed to add this extra term $-\frac{1}{2} g_{\mu\nu} R$ to ensure local energy conservation. Local energy conservation demands that the only way the total mass-energy in a room can go up is if something comes in the door. This is a very nice property for the equations to have. It's just like when Maxwell found he had to add a term to his equations to produce charge conservation, an extra term that in Maxwell's case led directly to the famous result that light was electromagnetic waves.

Einstein did some calculations using his field equations. He calculated the curvature expected in the empty space around the Sun. He could then calculate the geodesic representing the helix for a planet's worldline. He found that, in general, planets in curved spacetime did not follow simple elliptical orbits as Kepler predicted, but ellipses that were *precessing* (i.e., slowly rotating). They did not retrace the same ellipse over and over; instead the ellipse for each planet slowly rotated. For most planets far from the Sun, the effect was tiny, but for Mercury, which orbited closest to the Sun where the curvature was greatest, the effect was measurable. Einstein calculated that its elliptical orbit would precess, or rotate, by 43 seconds of arc per century. Eureka! That was equal to the unexplained precession in the orbit of Mercury, something astronomers had measured, which Einstein knew about and which Newton couldn't explain.

Einstein was so excited doing this calculation that he said it gave him palpitations of the heart. His equations had given the right answer—43 seconds of arc per century—nature had spoken. He did this calculation on November 18, 1915. At that time he was using the incorrect field equations $R_{\mu\nu} = 8\pi T_{\mu\nu}$ but luckily, in the specific case of the empty spacetime around the Sun, they were actually perfectly good.

On the same day, he calculated the bending of light beams passing near the Sun. He calculated the geodesic path that light would take in the curved empty spacetime around the Sun. The answer he got said that a light beam from a distant star passing near the limb of the Sun on its way to Earth would be deflected by 1.75 seconds of arc. This was twice the amount Newton would have calculated if he had thought that light was made up of little massive bullets traveling at 300,000 km/sec. Newton would have calculated a deflection of 0.875 seconds of arc. But maybe light was *not* made of massive particles, so in Newton's theory it was also possible that light was not deflected at all. Einstein had no choice, however: light had to travel on geodesics and had to be deflected by 1.75 seconds of arc. This deflection could be observed. How could you observe stars near the limb of the Sun? You had to wait for a solar eclipse, when the Moon just blocked out the bright light from the solar surface. You could measure the star positions on a photographic plate during the eclipse, and then measure again 6 months later, when Earth was on the other side of the Sun and the Sun was far away from those stars, and compare the two photographs for differences in positions. Close to the edge of the Sun, the stars should be shifted by 1.75 seconds of arc, according to Einstein's equations. Einstein proposed this as a test to be done during a solar eclipse.

Einstein was lucky in this regard. Earlier, before he had his field equations, he had made a qualitative argument using the Equivalence Principle's accelerating spaceship argument. A straight horizontal light beam in interstellar space would look bent in the accelerating spaceship, because a straight horizontal light beam would eventually hit the floor as the spaceship accelerated upward to hit it. By this analogy, he argued, light ought to be bent by gravity. The argument correctly accounted for the curvature in time, but left out the curvature in space required by the full field equations, and so Einstein only got half of the correct answer. He got a deflection of 0.875 seconds of arc, just as Newton would have done. Einstein published this and suggested people look during the eclipse of 1914. But World War I broke out, and the expeditions never made the observations. Lucky for Einstein. By 1915, he had the correct answer of 1.75 seconds of arc for curved spacetime, and this differed from what Newton would predict. If a deflection of 1.75 seconds of arc were observed, Einstein would be right and Newton would be proven wrong. If 0.875 seconds of arc deflection were observed, Newton would win and Einstein would be wrong. If no deflection were observed, Einstein would be wrong, but Newton could still be right, because he might say that mass attracts mass, but mass does not attract light. Newton would still be in business, in that event. Here was a decisive test. Einstein's calculation of the precession of Mercury was a postdiction. It explained an already known experimental fact, unaccounted for by Newton. But in this case, he was making a prediction—much more dramatic.

Two British expeditions were mounted to observe the May 29, 1919, solar eclipse. One observed from Sobral, Brazil, and the other from Príncipe Island off the coast of Africa. Sir Arthur Eddington reported the results at the combined Royal Society and Royal Astronomical Society meeting in London on November 6, 1919. From Sobral a deflection of 1.98 ± 0.30 seconds of arc was observed, while from Príncipe a deflection of 1.61 ± 0.30 seconds of arc was observed. Both results agreed with Einstein's value of 1.75 seconds of arc to within the observational errors of ±0.30 seconds of arc, and both disagreed with Newton. Nobel Prize winner J. J. Thompson, discoverer of the electron, chaired the meeting, and pronounced: "This is the most important result obtained in connection with the theory of gravitation since Newton's day . . . the result [is] one of the highest achievements of human thought."

The next day Einstein was in the *London Times* under the headline "Revolution in Science." Two days later, he was in the *New York Times*. This was the moment Einstein moved up from being one of the greatest scientists of his day to being the world-famous person you know. This is the moment he joined the company of Isaac Newton.

The light bending results of Eddington were soon independently confirmed with higher accuracy by W. W. Campbell and R. Trumpler, observing a 1922 eclipse from Australia. They found a deflection of 1.82 ± 0.20 seconds of arc, again consistent with Einstein's prediction of 1.75 seconds of arc.

Einstein said of his travails, while struggling from 1907 to 1915 to work out his theory: "But the years of anxious searching in the dark for a truth that one feels but cannot express, the intense desire and the alternations of confidence and misgiving until one achieves clarity and understanding, can be understood only by those who have experienced them."[3]

20

BLACK HOLES

J. RICHARD GOTT

This chapter is about the most mysterious objects in the universe, black holes. One of the first exact solutions obtained for Einstein's equations of general relativity corresponded to a black hole. An exact solution to Einstein's equations is a spacetime whose geometry has a curvature at each point that solves the equations locally at each point. Of particular interest is the solution for the geometry of the empty space around a point mass. This is called a solution to the *vacuum* field equations, because they apply in empty space. These were exactly the equations Einstein was trying to solve when he worked out the orbit of Mercury and the light bending around the empty space near the Sun. But this solution was difficult to find, because one had no idea what the geometry of the solution would be like, so Einstein settled for an approximate solution. In his approximate solution, spacetime was approximately flat, just as in special relativity, but with small *perturbations* (departures from flatness). The equations for the small perturbations were easier to solve, because one knew that one had a flat geometry to start with and with that as a starting point, the equations for the small corrections were easier to solve. Since the velocities of objects orbiting the Sun are small with respect to the speed of light, the geometry around the Sun is only slightly curved. Thus, Einstein's approximate solution is quite accurate, as are his values for Mercury's orbit and the bending of light near the Sun. Perhaps Einstein thought that solving the equations exactly would be too difficult. In any case, he contented himself with an approximate solution.

The first person to find an exact solution to Einstein's field equations for the empty space around a point mass was the German astronomer Karl

Schwarzschild. What he found was the solution for a black hole, that is, a point source of mass in otherwise empty space. When Einstein published his work on general relativity, he estimated that there were only 12 people in the world who could understand it. Karl Schwarzschild was one of them. In 1900, Schwarzschild had written a paper about the possible curvature of space. This was even before special relativity. He had reasoned that space might be positively curved, like the surface of a sphere, or even negatively curved, like the surface of a Western saddle. He wanted to know how big that radius of curvature had to be, given the then-current astronomical observations. He was someone who was already willing to think about the curvature of space. When Einstein's paper came out, Schwarzschild was very receptive: he understood it and, just as important, he was able to deal with the difficult math involving Riemann curvature tensors. He had all the tools he needed to do something new and original with it. Schwarzschild was able to solve the problem, because he came up with a clever coordinate system in which to solve these complex equations, which took advantage of the fact that the problem had spherical symmetry and was unchanging in time. This exact solution to Einstein's vacuum field equations for the empty space around a point mass turned out to map the exterior of a black hole.

While serving in World War I, Karl Schwarzschild contracted a rare skin disease, which ultimately proved to be fatal; he was sent home sick in 1916 and at that time he learned of Einstein's paper and found his solution. He sent the solution to Einstein, saying that during the middle of the war he had been pleased to "spend some time in the garden of your ideas." Schwarzschild died a few months later.

Finding this exact global solution to the vacuum field equations was very much like making a patchwork coat. At each point in spacetime, you are sewing together pieces, where locally there are different curvature terms that sum up to zero. The equations are telling you the rules by which you can stitch the pieces together. You just keep sewing and adding little pieces. But ultimately, you must come up with a global solution—a patchwork coat—that satisfies the rules at every point. This is quite difficult. Karl Schwarzschild was the one who first managed to do this for the curved space around a point mass.

Karl Schwarzschild's son, Martin Schwarzschild, was our longtime colleague at Princeton (see figure 8.3). He was also an astronomer who made many important contributions. In particular, Martin figured out that a star like the

Sun would eventually become a red giant. He definitely followed in his father's footsteps. Martin never really got to know his father, who died when Martin was only 4 years old. Interestingly, Karl fought in World War I on the German side, whereas his son Martin fled Germany when Hitler came to power, and fought on the American side against Germany in World War II.

To understand black holes, let's first go back to Newtonian gravity. If I take a ball and throw it up in the air, what will happen? It will go up and then fall back down. There is even a saying about this: "What goes up must come down." The only trouble with this saying is that it's wrong. Ignoring air resistance, if you throw a ball up fast enough, at greater than Earth's escape velocity of 25,000 miles per hour, it will escape the gravitational field of Earth and never come back. Apollo astronauts had to travel at nearly this speed to get out to the Moon. Newtonian theory has a formula for the escape velocity: $v_{es}^2 = 2GM/r$, where G is Newton's gravitational constant, M is the mass of Earth, and r is the radius of Earth. Now suppose I got an enormous trash compactor and crushed Earth to a smaller size, wadding it up like a ball of paper and crushing it into a smaller radius. What would happen to the escape velocity? Earth's mass would be the same, but its radius would be smaller, making the escape velocity from its surface rise. Eventually, if I crushed Earth to small enough size, the escape velocity would become equal to the velocity of light c. How small is this? I could just set $v_{es}^2 = c^2 = 2GM/r$ and solve for r. I would get $r = 2GM/c^2$. We call this radius the *Schwarzschild radius*, in Karl's honor. For Earth's mass, the Schwarzschild radius is 8.88 millimeters. This is about the size of a large marble. If you crush Earth to a size smaller than this radius, the escape velocity would be greater than the speed of light and nothing, not even light, could escape. Einstein showed that nothing can travel faster than the speed of light—if you crush Earth to a radius inside its Schwarzschild radius, it's never going to re-emerge: it forms a *black hole*. We call it a "black hole," because no light from inside it can ever get out. The mass will continue to collapse to still smaller size, where gravity will pull it together even more strongly, increasing the escape velocity even more. Inside the Schwarzschild radius, gravity wins over all other forces, and the mass collapses to a point, a *singularity* of infinite curvature at the center. General relativity would say that this point is of zero size, but we believe that quantum effects will ultimately smear this out to a size of perhaps 1.6×10^{-33} cm, called the *Planck length* (we'll see where this number comes from in chapter 24). This is much smaller than an atomic nucleus. We are left with a point mass at the center, of essentially zero size, surrounded by empty curved spacetime.

If you were to venture inside the Schwarzschild radius, could you ever get back outside? No. You would have to travel faster than light to do so, and Einstein showed that was impossible.

The Schwarzschild radius of a black hole is proportional to its mass. The bigger the mass is, the bigger the Schwarzschild radius will be. The truth is that it would be very hard to crush Earth to be within its Schwarzschild radius. But after they run out of their nuclear fuel, massive stars have dense cores that are in danger of falling inside their Schwarzschild radii. When the Sun dies, it will become a red giant and then shed its envelope, leaving a white dwarf core about the size of Earth. If the core of a dying star is more massive than 1.4 solar masses but less than 2 solar masses, the white dwarf star will collapse to form a neutron star with a radius of about 12 kilometers. A neutron star is only about a factor of 2–3 larger than its Schwarzschild radius, and therefore is close to dangerous ground. If you try to make a neutron star with a mass larger than about 2 solar masses, it is unstable to collapse and collapses inside its Schwarzschild radius, where gravity takes over completely and a black hole is formed. A 10-solar-mass black hole, such as may form when a very massive star collapses at the end of its life, has a Schwarzschild radius of 30 kilometers. A 4-million-solar-mass supermassive black hole, such as we find in the center of our galaxy, has a Schwarzschild radius of 12 million kilometers (a bit less than 1/10 of an AU). One of the largest black holes we have ever found is at the center of the giant elliptical galaxy M87. It has a mass of 3 billion solar masses and thus has a radius of 9 billion kilometers. That's twice the radius of our entire solar system out to the orbit of Neptune.

Let's imagine taking a trip inside a Schwarzschild black hole of 3 billion solar masses. Suppose we have a professor and a graduate student; the professor wants to know what happens inside a black hole, so he sends the graduate student to investigate. The professor stays outside the black hole, firing his rocket to stay at a fixed radius of, say, 1.25 Schwarzschild radii. The professor feels an acceleration caused by his rocket to keep at that fixed radius and not fall in. As long as the professor stays outside the black hole, bad things do not happen to him. To investigate the black hole, however, the brave graduate student just free-falls in. AHHHHHH! As the graduate student free-falls in, he sends radio signals outward, back to the professor, to tell the professor how things are going. The first part of his message says, "THINGS." The radio signal moves outward at the speed of light.

The graduate student falls in farther, and the radio signal reaches the professor. The professor receives the first word of the message: "THINGS." Meanwhile the graduate student falls in still farther. He sends the second word of his message: "ARE." This word is sent just outside the Schwarzschild radius. It travels outward at the speed of light, but it is going to take a long time to climb back out and reach the professor. The professor has to fire his rockets just to stand still and avoid falling in, and so he is actually accelerating away from the horizon, so it takes the signal "ARE" a long time to catch up with him.

Meanwhile, the graduate student crosses the Schwarzschild radius. Is this good? No. Will the graduate student ever be getting back out to join the professor? No, unfortunately not. Yet as he crosses this point of no return, there is no special road sign there marking it. Nothing funny happens to the graduate student here. The graduate student doesn't know that anything bad has happened. Everything looks normal to him. In fact, you might even be crossing inside the Schwarzschild radius of some enormous black hole right now, and from your room where you are reading this, you wouldn't even know it. Locally a tiny piece of spacetime looks approximately flat, and therefore you have no hint as to what the global solution looks like from local measurements. Just as the graduate student crosses the Schwarzschild radius, he sends the third word of his message: "GOING." The second word of the message, "ARE," is still on its way out to the professor. So far, the professor has only received "THINGS." Now the graduate student falls inside the Schwarzschild radius. The signal saying "GOING" continues outward at the speed of light. But it is like a kid running up on a down escalator and making no progress. At the Schwarzschild radius, the escape velocity is the speed of light; the radio signal traveling outward at the speed of light just stays at the Schwarzschild radius, making no progress. The signal "ARE" continues to make its way outward.

As the graduate student continues to fall farther inside the Schwarzschild radius, something begins to happen. The graduate student is falling in feet first. His feet are closer to the center than his head. Because gravity is a $1/r^2$ force, the mass at the center is pulling his feet more strongly than his head, with the force on his midsection being intermediate. His head and feet are being pulled apart by this tidal force. It's like being stretched on a rack. In addition, the graduate student's left shoulder is being drawn radially inward toward the center, while the right shoulder is also being drawn radially inward toward the center. His shoulders are being wedged together as they are being drawn closer on lines that point toward the center of the black hole. It is like being crushed in an Iron Maiden. So he sends out the last word of his message: "BADLY." "THINGS ARE GOING BADLY."

As he gets closer to the center, the forces grow larger and larger. He is being crushed from the sides and stretched head to toe—being turned into a piece of spaghetti. This is called *spaghettification*. That really is the technical term astronomers use for this process! Eventually the graduate student is ripped apart, crushed, and deposited in the central point. The mass of the central point is now 3 billion solar masses plus a little! The Schwarzschild radius moves just a little bit outward. The signal "ARE" is still working its way out to the professor. The signal "GOING" is still running in place at the Schwarzschild radius. The signal "BADLY" is going outward at the speed of light but is like a kid running up on a down escalator, where the escalator is going faster than the kid can run. Although running up, the kid is being drawn downward. The signal "BADLY," although running outward, is sucked backward into the center, where it is crushed and deposited on the point singularity as well.

Finally, after a long time, the signal "ARE" is received by the professor. The professor has received the message "THINGS A . . . R . . . E." He never receives the rest of the message: "GOING BADLY." "GOING" remains stuck at the Schwarzschild radius, and "BADLY" has been sucked into the pointlike singularity at the center along with the graduate student. "BADLY" is the news of an event occurring inside the Schwarzschild radius. That signal is never going to reach the professor, and he never finds out what happens inside that radius. The professor never sees any events that occur inside the Schwarzschild radius, which is why it is called the *event horizon*: the boundary of the region containing all the events the professor can see. The professor simply can't see past the event horizon. In the same way, on Earth you can't see past the horizon when you look out; it marks the limit of what you can see. Any observer who stays outside the event horizon of the black hole can never see any events that occur inside the event horizon.

If the professor ever wonders what happened to the poor graduate student, he can turn off his rocket motor, which has kept him hovering outside the black hole, and then he will free fall in himself. When he crosses the event horizon, he will see that signal "GOING," which is still stuck there. As he rides down the "escalator," he will see the signal "GOING" running past him at the speed of light. Light will always pass him at 300,000 km/sec. But then the professor will fall into the center of the black hole and be killed as well.

For a 3-billion-solar-mass Schwarzschild black hole, the graduate student would have 5.5 hours of free-fall time as measured on his watch before he hit the center and was killed. Luckily for him, the spaghettification process, from the moment the tidal forces begin to hurt him until he is completely ripped apart and killed, only takes up the last 0.09 seconds of his trip. So at least it is a quick end.

We might also like to know what the curved geometry of the exterior of the black hole looks like. I was once asked to appear on the *McNeil/Lehrer Newshour*, because astronomers using the Hubble Space Telescope had just discovered evidence that the big black hole in M87 existed, and they wanted Kip Thorne and me to explain this to the viewers. I made up a little demonstration. If you cut a plane through the center of the black hole, you might expect that plane to be a flat, two-dimensional surface like a basketball court, with the Schwarzschild radius a circle like the free-throw circle. The singularity would be a point at its center. But that would be wrong. This two-dimensional slice through the black hole is actually curved. It looks like the horn of a trumpet pointing upward (figure 20.1.) The third dimension here is purely to allow us to show you the curvature of two-dimensional funnel surface. The third dimension is not real here. Forget the space above and below the funnel, the only thing that is real is the funnel shape itself. At large distances, the horn of the trumpet flattens out, so that it begins to look flat like a basketball court. Far from the hole the curvature is weak. The extended trumpet horn slopes ever more dramatically downward toward the hole as you approach it. The slope becomes vertical at the Schwarzschild radius. The Schwarzschild radius marks the circumference of the trumpet at its narrowest point. That's why we call it a black *hole*—it really is a hole. In fact, in the coordinate system Karl Schwarzschild invented, the coordinate r is called a *circumferential radius*, because $2\pi r$ is the circumference at that point. This circumference lies within the surface of the funnel. You can think of the funnel as a series of smaller and smaller circles reaching a minimum circle at the bottom (where the circumference is equal to 2π times the Schwarzschild radius). The Schwarzschild radius is the radius of the hole at the bottom of the funnel. (Ignore the flange base at the bottom in figure 20.1—it just holds up the model of the funnel.)

FIGURE 20.1. Black hole funnel. The geometry around a black hole is not flat like a basketball court, but curved like a funnel. The funnel becomes vertical at the Schwarzschild radius, indicated by the red band showing the circumference: 2π times the Schwarzschild radius. An astronaut can fall straight in. When he passes the Schwarzschild radius (the red band), that is the point of no return. Ignore the base that holds the funnel up. Also, ignore the inside and outside of the funnel, it is only the funnel shape itself that is real. *Photo credit: J. Richard Gott*

For my TV demonstration, I used a trumpet-horn-shaped funnel. I set it up with the rim of the bell at the top and the narrowest circumference of the funnel at the bottom (see figure 20.1). Astronomers had detected gas orbiting the black hole in M87 at high velocity. I illustrated this by throwing marbles sideways into the funnel and letting them spin around as they spiraled slowly downward before disappearing through the hole at the bottom. Gas likewise orbits the hole, with the gas farther in orbiting faster such that gas rubs on gas, causing friction. The friction heats the gas, making it glow. We can see this radiation, because it is emitted outside the event horizon. Meanwhile this energy production causes the gas to lose energy and spiral into the hole. This is the power source for quasars: gas spiraling into a supermassive black hole. We see the hot gas while it is spiraling inward toward the event horizon, but we do not see it once it has passed that horizon. My demonstration showed all these things. I thought it was pretty good, and was ready to go for filming the news segment. Then I showed it to my daughter, 7 years old at that time, and she said, why not drop an astronaut in? She went to her room and came back with a cute little one-inch-tall *Apollo* astronaut wearing his spacesuit and holding a tiny American flag—a toy I didn't know she had. If you are spiraling around the black hole, like the marbles, you will spiral around as you slowly descend into the black hole, or you can just fall straight in like the graduate student. I put the toy astronaut at the upper edge of the funnel and just let it slide straight in—disappearing into the hole at the bottom. Perfect. A black hole is a hotel where you check in but you don't check out. The path of the astronaut falling straight in is a curved radial line going straight down into the funnel (it is a geodesic). When I let go of the astronaut, he falls straight down along this line in my model, so it made a good illustration. When a TV crew comes to film you, they usually take hours, filming lots of shots, but for national TV news this usually gets edited down to a short clip. After filming all my elaborate demonstrations with circling marbles spiraling in, what do you think they chose to show in the end? Just the little astronaut falling straight in, of course! Now you know what the geometry of a black hole's exterior looks like: it looks like a funnel, with a hole at the bottom.

The Schwarzschild solution, found by Karl Schwarzschild in 1916, showed the shape of this funnel. But Schwarzschild's coordinate system, clever though it was, broke down at the Schwarzschild radius. His solution showed the geometry outside the Schwarzschild radius, but it didn't show what happens inside. It was like having a map of the world that only showed the Northern Hemisphere—but nothing south of the equator. People thought the exterior solution was all that there was. Finally in the mid-1960s, my colleague Martin Kruskal of the applied math department at Princeton, and George Szekeres of the University of New South Wales, independently found a way to extend the coordinates to cover all the interior of the black hole solution. We can look at a spacetime diagram of the solution, now called a *Kruskal diagram* (figure 20.2).

This two-dimensional diagram shows one dimension of space horizontally with time shown vertically—the future is toward the top of the diagram. This diagram has the property that light beams travel in straight lines tipped at 45°. The speed of light is constant, and the constant slope of 45° shows that. Let's illustrate the coordinates by returning to the professor and his ill-fated graduate student. We start by drawing the worldline of the professor in black (see figure). It is not straight, because the professor is accelerating, firing his rocket to stay at 1.25 Schwarzschild radii from the black hole. The professor stays outside the black hole. The worldline is vertical at the point midway up and then bends toward the right. In flat spacetime, this would be a worldline

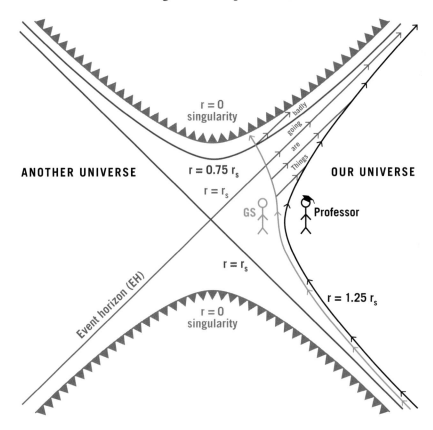

FIGURE 20.2. Kruskal diagram. Spacetime diagram that shows the geometry both outside and inside the Schwarzschild (nonrotating) black hole. The future is toward the top. The diagram represents the curved empty space around a point mass that has lasted forever. Our universe is to the right. The worldlines of a professor and a graduate student (GS) are shown. The professor stays safely outside the black hole at 1.25 Schwarzschild radii (1.25 r_S.) The grad student falls into the black hole and hits the singularity at $r = 0$. The event horizon (EH) runs along a line where the radius is equal to the Schwarzschild radius ($r = r_S$). *Credit:* J. Richard Gott

representing a particle that was at rest at the midpoint and accelerated off to the right, picking up velocity. The full worldline of the professor is a hyperbola. It bends over, so that in the far future, it is going upward at about 45° as it approaches the speed of light. Remember the Equivalence Principle, in which an accelerating observer in flat spacetime is like a stationary observer (the professor) in a gravitational field. The professor's worldline is bent like a hyperbola in the Kruskal diagram.

A horizontal line extending to the right from the point at the center of the X where the two 45° lines cross represents a snapshot of a radial ray extending from the hole at the bottom of the funnel straight out of the funnel, pictured at a single instant of time. (The other dimension of the funnel, the circumferential direction, is left out of the diagram.)

The grad student's worldline (labeled "GS") is shown in green. He travels along with the professor at early times, near the bottom of the diagram, where both worldlines track each other side by side until the graduate student leaves the professor at the vertical midpoint of the professor's worldline. The graduate student is free falling; his worldline falls into the black hole, while the professor accelerates off to the right. The event horizon (labeled "EH," and $r = r_s$, i.e., a radius equal to the Schwarzschild radius) is a line tipped at 45° that is asymptotic to the professor's worldline in the far future. It never touches the professor's worldline. It is tipped at 45°, because the light beam (in this case a radio wave) with the signal "GOING" can travel along it. The grad student's worldline crosses the diagonal event-horizon line, just as he emits the photon signal "GOING." The professor will never receive that signal. The professor's worldline marks the set of points on the diagram where r is equal to 1.25 Schwarzschild radii. The light (radio) signals "THINGS" and "ARE" are two lines tipped at 45° that were emitted before the graduate student crossed the event horizon; those two signals do intersect the worldline of the professor. He receives those signals. You can see now why it takes the signal "ARE" such a long time to reach the professor.

Where are the points lying at 0.75 times the Schwarzschild radius? They are twisted around to form a hyperbola that looks like a smile hovering above the diagonal event-horizon line. At the far right, it approaches the diagonal event-horizon line EH from above but never touches it. The *singularity* at $r = 0$ is also a hyperbola-shaped smile lying above the $r = 0.75$ Schwarzschild radius one. The grad student's worldline hits this horizontal smile. We put teeth on this grin, for these are jaws that will eat the grad student. The spacetime is so warped that the singularity, which you might expect to be a vertical line over to the far left, has been twisted around until it lies in the future. In fact, once the grad student crosses the event-horizon line, this hyperbola looms in the graduate student's future. He can't avoid it any more than you can avoid next Tuesday. No matter how he fires his rocket, he can't go faster than the speed of light, and he must go upward at more than a 45° angle. Once he has passed that event horizon, the hyperbola representing the singularity looms above him, spanning more than ±45°, and his worldline has to hit it. He is doomed. Likewise, the light signal "BADLY" that he sends out at 45° toward the right after he crosses the event horizon will hit the jaws of the singularity at $r = 0$ as well.

We can complete the Kruskal diagram to obtain the complete point-mass solution. This represents a point mass that began in the infinite past and lasts into the infinite future, in an otherwise empty universe. The diagonal event-horizon line EH is joined by another diagonal line going in the other direction to form a giant X in the center of the diagram. This X divides the spacetime into four regions. The outside of the black hole, where the professor lives, is to the right of the X. That's our universe. Above the X is the inside of the black hole, where the singularity looms in the future at the top. Below the X is an initial singularity labeled $r = 0$, looking like a frown at the bottom in the past. To the left is another universe like ours. It is connected to ours by a *wormhole* in the middle. If we were to make a horizontal slice through this spacetime in the middle, we would have a slice at a given instant of time. Its geometry is like two funnels joined at their narrowest point. Starting on the far right, the funnel has a large circumference, representing a large radius far from the hole. Going toward the left, the funnel gets narrower and narrower until it has a circumference of $2\pi r_{\text{Schwarzschild}}$ at the event horizon at the center of the X. It then fans out as again to large radius to make another universe on the left side of the X. The two funnels are joined to form a wormhole. Far away from the hole, the funnels flatten out to look like basketball courts, and they extend to infinity. Imagine a basketball court on the second floor of a building with a curved funnel leading downward into a hole at the center of the court (like the hole on a golf green). This funnel begins to open out again and fans out to make a pretty flat ceiling on the floor below the floor containing the basketball court. The basketball court represents our large universe, and the ceiling of the floor below represents another large universe connected to ours by the small hole smoothly connecting the court surface to the ceiling below it. The two large universes are connected by a wormhole at the instant represented by the horizontal line through the diagram. But you can't use this wormhole to travel from one universe to the other. That's because the arms of the X are tipped at exactly 45°. To cross from the region to the right side of the X (our universe) into the region to the left side of the X (another universe), you would have to have a worldline with a slope inclined at more than 45° to the vertical. That would mean you were traveling at greater than the speed of light, and that's not possible. But you could in principle meet an extraterrestrial from the other universe inside the black hole in the upper (future) quadrant. You could even shake hands. You might say to each other, "Boy, are we in trouble," before you both died as you hit that smiling singularity in the future. You would hit that singularity in a finite time.

The singularity in the past, at the bottom, is rather like the Big Bang singularity at the beginning of our universe. This part of the solution is called a *white hole*. It is a time-reversed version of a black hole—like a movie of a black hole run backward. A particle can be created in the white hole singularity at the bottom and have its worldline come out into our universe. If a particle can fall into a black hole, it can come out of a white hole.

The black holes that we might encounter now have not been around forever. In a realistic case, a black hole might be formed from the collapse of a star. In the spacetime Kruskal diagram, imagine that the surface of the collapsing star lies just below the graduate student's feet: just below the grad student's feet when he is with the professor and just below the grad student's feet as he falls in. This represents the situation when the surface of the star maintains a radius of 1.25 Schwarzschild radii for a long time and then free falls inward just below the grad student's feet as he free falls in. The worldline of the surface of the star is thus parallel to that of the grad student and just to the left of it. Beneath the falling graduate student is the interior of the star, beneath its surface, where the density of matter is greater than zero and the vacuum solution of the Kruskal diagram does *not* apply. Just ignore that part of the diagram to the left of the graduate student's worldline—no wormhole, no other universe, no white-hole singularity at the bottom. These are not formed when a star collapses to form a black hole. But the part of the diagram to the right of the graduate student's worldline is in the vacuum region and it accurately depicts what is happening. The grad student does get crushed as his worldline hits the singularity at $r = 0$.

If you lived inside the star (in an air-conditioned little room), you would find yourself crushed when the volume of the star shrinks to zero and its density blows up to infinity. A curvature singularity awaits your worldline in the future as well: you hit $r = 0$ when the size of your star collapses to zero.

Here's some advice: just stay outside the Schwarzschild radius, and you will be okay. You can happily orbit outside the black hole event horizon. If the Sun were to collapse and form a black hole, Earth would stay in its current orbit outside. You could see the black hole, which would appear as a black disk in the sky. It would be surrounded by gravitationally lensed images of stars behind it (figure 20.3).

FIGURE 20.3. Simulated view of a Schwarzschild black hole. It looks like a black disk in the sky, surrounded by gravitationally lensed images of background stars. You can see two images of the galactic plane whose light is bent around opposite sides of the black hole on the way to your eye.

Photo credit: Andrew Hamilton (using Milky Way background image adapted from Axel Mellinger)

In 1963, Roy Kerr discovered an exact solution to Einstein's field equations for a rotating black hole (one having angular momentum). It has a more complicated geometry inside its event horizon, which we discuss in chapter 21. But its event horizon marks the point of no return, just as in the Schwarzschild black hole. Kerr's solution was brilliantly vindicated on September 14, 2015, when astronomers from the Laser Interferometer Gravitational-Wave Observatory (LIGO) witnessed the formation of a 62-solar-mass rotating Kerr black hole from the collision of a 29-solar-mass black hole and a 36-solar-mass black hole. The two had formed a tight binary, and spiraled inward, losing energy due to emission of gravitational radiation. By studying these gravitational ripples in the geometry of spacetime, astronomers were able to deduce the masses of the black holes involved. Since the two black holes had orbital angular momentum as they circled each other, it was not surprising that a rotating black hole formed at the center. The ringing oscillations of this final black hole that formed and then settled down matched exactly those expected from the decay of perturbations to a Kerr black hole. Astronomers could even determine that the Kerr black hole had approximately 67% of the maximum angular momentum allowed for a black hole of its mass. The whole collision, including the emission of gravitational waves, could be simulated on a supercomputer that solved Einstein's equations to calculate the geometry of spacetime. The agreement between the computer simulation and the observed gravitational wave ripples shows that Einstein's equations work even when spacetime is highly curved—a very important result.

In 1974, Stephen Hawking surprisingly and famously discovered that a black hole actually radiates thermal radiation: energy can, and does, escape from a black hole. How did this discovery come about? Princeton graduate student Jacob Bekenstein was talking with his PhD thesis advisor John Archibald Wheeler. Wheeler was the one who coined the name "black hole." And it's a good name! Black holes are holes, and they are black—they emit no light. As Neil has said, astronomers prefer simple names for things—"it's black and it's a hole, so why not just call it a 'black hole'?" Wheeler was the godfather of black hole research and helped revive interest in general relativity in the 1960s. He got people interested in working on the subject and with Charles W. Misner and Kip Thorne wrote an influential textbook on it whose early proofs I studied as a graduate student. When Kruskal discovered his diagram, he sent it to Wheeler to ask his opinion of it and went on vacation. Wheeler read the paper and thought it was so important that he wrote it up himself and sent it immediately to the *Physical Review* journal with Kruskal's name alone on it! When Kruskal got back from vacation, he found his paper had already been sent in to the journal.

Wheeler invited his student Bekenstein in for a talk. He brought out a cup of hot tea and mixed some cold water into it. Wheeler said, "I have just committed a crime: I have increased the entropy (disorder) in the universe, and I can't take it back, because I can't unmix the tea and water." Bekenstein knew that the entropy in the universe always increases with time. If you break a vase, the disorder in the universe increases. We don't often see pieces bounce up and assemble themselves into a vase. In fact, when you see a film showing this by running the film backward, you laugh, because you know that it is unlikely to occur. There is a certain chance that something like that would occur, but it is very small. Statistically, we expect to see the disorder in the universe increasing with time—this principle is called the *second law of thermodynamics*. People like order; it is a shame to break a beautiful vase into pieces. Following this logic, any increase in entropy, like mixing the tea and water, can be considered a crime. Wheeler went on: "But now I can hide the evidence of my crime by throwing the lukewarm tea and water mixture into a black hole. It will increase the mass of the black hole: it will now have the mass it did before, plus that of the tea and water, but not by any more than if I had thrown the tea and the water separately into the black hole. I have gotten back the result I would have had if I had not mixed the tea and water in the first place, and that seems to violate the second law of thermodynamics. Think about it!"

Bekenstein took Wheeler's idea seriously and thought about it. The resulting paper struck me as particularly brilliant. Bekenstein noted that Hawking had proved a theorem saying that the total area of all the event horizons in the universe always goes up with time if the mass density everywhere is nonnegative, which seemed reasonable. When mass is added to a black hole, the mass of the black hole increases, and its Schwarzschild radius increases. The surface area of the event horizon, which is $4\pi r_{\text{Schwarzschild}}^2$, goes up as well. If two black holes collide, as they did in the case discovered by LIGO, they form a black hole that has an event horizon whose total area is larger than the sum of the areas of the event horizons of the two initial black holes. In the LIGO case, for example, calculation shows the area of the event horizon of the final rotating 62-solar-mass Kerr black hole is at least a factor of 1.5 greater than the sum of the areas of the event horizons of the initial 29- and 36-solar-mass black holes. To Bekenstein, this phenomenon of the total area of the event horizons always going up with time sounded to him like entropy, which of course also is always going up with time.

Bekenstein did a thought experiment: he lowered a particle on a string as gently (nearly reversibly) into a Schwarzschild black hole as possible, and calculated how much the area of the black hole went up. He noted that this corresponded to a loss of one bit of information, namely, the information about whether that particle existed or not. Because a loss of information in his thought experiment is equal to a specific increase in entropy, he was able to calculate the relationship between the number of bits of information lost, to the increase in the area of the black hole horizon. He found the loss of one bit of information corresponded to a tiny increase in area—of order $(1.6 \times 10^{-33} \text{ cm})^2 = hG/2\pi c^3$ (where these are our old friends, Planck's constant h, Newton's gravitational constant G, and the speed of light c). We'll see this distance of 1.6×10^{-33} cm, called the *Planck length*, again, in chapter 24. It's the scale where spacetime geometry becomes uncertain due to Heisenberg's uncertainty principle in quantum mechanics. When Wheeler dropped his mixed cup of lukewarm tea and water into the black hole, he increased the area of its horizon and its entropy. The entropy in the universe still went up appropriately because the black hole had an entropy that increased as the mixed cup fell in. Black holes had a large but finite entropy, Bekenstein concluded.

Interestingly, Bekenstein's work puts a limit on the amount of information your 6-inch-diameter hard drive can store—10^{68} bits = 1.16×10^{58} gigabytes. Try to pack more information than that inside its diameter, and it will become so massive that it collapses and forms a black hole. (To follow the reasoning in detail, see appendix 2.) Bekenstein's argument also puts limits on the number of bits of information you can cram into the finite radius of the observable universe, and therefore on the number of different visible universes of our size and energy one can have—namely, 10^(10^124), the big number Neil gave you in chapter 1. So, Bekenstein's paper has a lot of applications.

But Hawking, unlike me, thought Bekenstein's paper was wrong. If you added a finite amount of energy to a black hole and that increased its entropy by a finite amount, this implied by a simple thermodynamic argument that it had a finite temperature. Hawking thought that must be incorrect. Black holes do not glow like an object of finite temperature would glow. Black holes are black—they have a temperature of zero.

Roger Penrose had shown that in the special case of a rotating black hole, a particle could decay into two particles in a region just outside the black hole event horizon, and one particle could fall inside the event horizon in a counter-rotating way so that it lowered the angular momentum of the black hole, while the second decay particle sailed out with more energy than the total energy the initial particle possessed. In a rotating black hole, some of the hole's mass is tied up in its rotational energy, and at the end the black hole is rotating more slowly, so its total mass is smaller than it was before. Tapping the rotational energy of the black hole provides the energy to power the high-energy escape of the second decay particle. In this process, the area of the rotating black hole's event horizon goes up a little. Demetrios Christodoulou, another of Wheeler's students, investigated these questions, putting limits on how much energy could be extracted from the rotating black hole. In the Soviet Union, Yakov Zeldovich had applied this idea to electromagnetic waves. He presented a heuristic argument that an electromagnetic wave impinging near a rotating black hole could be amplified, gaining more energy, like the Penrose escaping particle. This looked like stimulated emission, the laser effect discovered by Einstein. By that logic, the rotating black hole should have some spontaneous emission as well, slowly losing rotational energy by emitting electromagnetic waves. Alexei Starobinski calculated these effects for waves for a rotating Kerr black hole.

As recounted by his student, Don Page,[1] Hawking wanted to put these ideas on a firmer foundation. Hawking set out to apply quantum mechanics in curved spacetime—to calculate the creation and annihilation of particles in the curved Schwarzschild spacetime to discover whether the nonrotating black hole really emitted any radiation. Hawking found, quite to his surprise, that particles were created—the black hole emitted thermal radiation. The black hole had a finite temperature after all! Hawking used the fact that, in the vacuum of empty space, particle pairs are always being created, falling back together, and anni-hilating again. These are called *virtual pairs*. They are always popping in and out of existence. Heisenberg's uncertainty principle of quantum mechanics says that the energy of a system is significantly uncertain over a short enough time period. Thus the energy needed to create an electron and a positron (you need both; the total electric charge still needs to be conserved), can be "borrowed" from the vacuum for a short amount of time. Thus an electron-positron pair can be created near each other out of the vacuum and can fall back together and annihilate again after a short period (of order 3×10^{-22} seconds). But in the black hole case, the electron can be created slightly inside the event horizon and the positron slightly outside the event horizon. The electron created inside the event horizon can't get back outside to recombine with the positron on the outside. The electron falls into the black hole, and the positron escapes. The electron created inside the event horizon has a gravitational potential energy that is negative and larger in magnitude than its rest mass energy from $E = mc^2$. Thus, its total energy is less than zero and when it falls in, it robs the black hole of some energy and therefore of some of its mass. This makes up for the mass and energy of the emitted positron. There is a quantum vacuum state (now called the *Hartle-Hawking vacuum*) around the black hole of slightly negative energy density, which violates the positive energy assumption on which the Hawking area-increase theorem was based. In this case, the area of the event horizon goes down slightly as the positron escapes. Alternately, it can be an electron that escapes while a positron falls in. The same effect can produce pairs of photons, where one photon created just inside the horizon falls in and the other created just outside the horizon escapes. Hawking found that black holes give off thermal radiation (now called *Hawking radiation*). This causes black holes to shrink, and eventually evaporate. This thermal radiation has a characteristic wavelength (λ_{max}) about 2.5 times the size of the Schwarzschild radius of the black hole. For a 10-solar-mass black hole, this means it is giving off 75-kilometer-long radio waves—far too feeble to be detected; this thermal radiation is very low temperature, 6×10^{-9} K (with very few positrons and electrons in the mix). This is why Stephen Hawking has not garnered a Nobel Prize yet. If the radiation were strong enough to have been detected by now, he would surely have gone to Stockholm by now. No one, I think, doubts that the radiation exists; but the radiation is predicted to be extremely weak. Black holes of stellar mass or larger are actually absorbing more radiation from the cosmic microwave background (CMB) than they are emitting. Only in the far future will the microwave background redshift and cool enough to allow the evaporation process to proceed.

It takes black holes a long time to evaporate. A 3×10^9-solar-mass black hole like the one in M87 currently should be emitting thermal radiation with a temperature of about 2×10^{-17} K—mostly in the form of photons and gravitons. According to calculations by Don Page, a 3×10^9-solar-mass black hole will take 3×10^{95} years to evaporate. Today it is accreting more radiation from the CMB than it is emitting in thermal radiation. Its loss of mass will not really start until the CMB temperature has dropped below 2×10^{-17} K. That should occur about 700 billion years from now. Ultimately, due to evaporation, it will finally shrink to a size of about 10^{-33} cm and then go out of existence in a blaze of ultra-high-energy gamma rays. It is thought that the information lost when the black hole formed eventually leaks back out in the Hawking radiation that is emitted as it evaporates, but in a scrambled (disordered) form.

The details of how this evaporation affects the interior of the black hole are still being hotly debated. Some physicists believe that the antiparticles (or particles) just inside the event horizon paired with the Hawking particles (or antiparticles) being emitted outside the horizon can form a *firewall*, a wall of hot photons, just inside the event horizon that will kill any astronaut falling in. This effect might become important only after the black hole has evaporated more than half its mass, something that would only occur in the far future. The details depend on the quantum vacuum state that forms around the black hole.

James Hartle and Hawking found a quantum vacuum state that did not blow up on the event horizon and in which an infalling astronaut would not get burned up as he passed through to the inside. When a particle and an antiparticle (such as a positron and an electron) are created out of the vacuum, their quantum states are entangled. The two particles have angular momentum and spins that are opposite. If you measure the spin of one relative to a particular direction, you instantly know that the spin of the other relative to that same direction is the opposite. This remains true even as the particles separate to large distances. This effect puzzled Einstein, who called it "spooky action at a distance." It was one of the things that troubled him about quantum mechanics. In a recent paper, Juan Maldacena and Leonard Susskind, two of the leading experts in the field, have argued that quantum entanglement between the particles being emitted and their partners on the inside of the horizon can keep the astronaut cool as he falls through, just as Hartle and Hawking intended. They argue that the particle and its antiparticle are connected by a tiny microscopic wormhole. They are essentially touching each other through the wormhole while being separated by a large distance in regular space. The wormhole is like a hole in a dining room tabletop that allows an ant to get from the top surface of the tabletop to its underside. Yet the two wormhole openings, or mouths, are separated by a large distance if one must follow a path along the table's big surfaces. An ant would have to crawl a long way to get from the upper wormhole mouth to the lower wormhole mouth this way. She would first need to crawl along the top of the table to reach its side edge; then she would have to crawl around the side, underneath the tabletop, and then along the bottom surface of the tabletop until she reached the lower wormhole mouth. That traveling ant would say the upper and lower mouths of the wormhole were widely separated, whereas an ant zipping through the wormhole would realize that they were actually quite close to each other. This could solve Einstein's "spooky action at a distance" problem. The particle and antiparticle are always close to each other through the wormhole. Interestingly, Wheeler had already commented that electric field lines converging on a wormhole mouth could look like an electron (on the underside of the tabletop), but when they emerged and fanned out on the top of the table, they would look like a positron. Thus, he argued that particles and antiparticles could be linked by a wormhole like that occurring in a black hole, such as we encountered in the Kruskal diagram connecting two universes (in that case called an *Einstein–Rosen bridge*). Einstein's paper on spooky action at a distance was co-authored with Nathan Rosen and Boris Poldowski. Thus, Maldacena and Susskind argued, the Einstein, Rosen, Poldowski paradox of spooky action at a distance can be resolved using a microscopic Einstein–Rosen bridge! Surprisingly, Einstein and Rosen (and everyone else) missed the connection! If this picture is correct, it looks safe for the graduate student to pass inside the event horizon, as Hawking had originally supposed. This example points out some of the deep connections that Hawking's work has illuminated.

I remember well the excitement when Hawking came to Caltech to tell us of his discovery that black holes would evaporate. Kip Thorne, one of the world's experts on black holes, introduced him. Nobel laureate Murray Gell-Mann was in the audience. Thorne assured us all of the revolutionary importance of this research. I agree—it is the most important result in the theory of general relativity since Einstein's day. You've heard of Stephen Hawking—this is how he became world famous. Some of these exciting events have been recounted in the 2014 movie *The Theory of Everything*, which earned Eddie Redmayne an Oscar for his compelling and accurate portrayal of Hawking.

21

COSMIC STRINGS,
WORMHOLES,
AND TIME TRAVEL

J. RICHARD GOTT

Since I work on time travel in general relativity, the neighborhood children think I have a time machine in my garage. Once I attended a cosmology conference in California, and I happened to wear a turquoise sports coat. A colleague of mine, Robert Kirshner, then chair of Harvard's astronomy department, came up to me and said, "Rich, you must have bought this coat in the future and brought it back, because they haven't invented this color yet!" Ever since then, this has been known as the "Coat of the Future," and I have worn it when giving talks about time travel.

I start my usual talk about time travel by entering wearing this turquoise sports coat and carrying a brown briefcase. I hide the briefcase in a cabinet and make a hasty exit. I return wearing a tee shirt. I explain to the audience that I have another meeting to go to and that I have arranged for a guest speaker to give the talk for me, and I exit again. I return a second time wearing the turquoise sports jacket and tell everyone that it is the "Coat of the Future." I explain that I couldn't give the talk, because I had a meeting to go to at the same time; but since I have a time machine, after my meeting I could simply go to the future, buy the coat of the future then, and come back in time to give the talk as my older self!

At that point I notice that I forgot to bring the notes for my time travel talk. What to do? Since I have a time machine, I realize I can get them the next day (after my talk) and come back in time to deposit my briefcase containing my notes somewhere in the classroom ahead of time. I look around but don't see them. So, I must have hidden them. Is there anywhere around to hide them? Maybe in the cabinet. I open the cabinet, find the briefcase, and open it. Yes! My time travel notes are inside.

Let's see what's going on here by tracing the worldlines on a spacetime diagram. Space is shown horizontally and time vertically with the future toward the top. The classroom where I'm giving the talk is a vertical band in the center. Here is what my worldline looks like (see figure 21.1).

In the spacetime diagram, I am outside the room and wearing a white tee shirt. I come into the room briefly and mention that I am not really able to give the talk, because I have to go to a meeting. I leave, go to the meeting, and then proceed into the future, where I buy the "Coat of the Future." Now my worldline becomes turquoise in color. I come back in time and re-enter the room where I then give the talk. After the talk is over, I have to go back in time to just before the talk to bring the time travel notes into the room. I will enter the room and then quickly leave before my younger tee-shirt-wearing self comes in. I will then continue to live the rest of my life on into the future. I have a complicated worldline.

But what about the worldline of the briefcase? I have the briefcase in my possession just after finding it in the cabinet. If I simply hold onto it, I can take it along with me, looping back in time and delivering it to the room earlier, where it stays until I find it in the cabinet. The worldline of the briefcase is a circular loop (colored orange). The briefcase's worldline is odd, because it has no beginning and no end. My worldline has a beginning when I am born and an ending when I die, but the briefcase's worldline is a closed loop. The briefcase is what we call a *jinn* particle. This term is named after a jinn, or genie, who appears out of nowhere.

The briefcase never leaves my sight. The briefcase never visits a briefcase factory. Physicists who work with time travel to the past have to deal with jinn particles when they consider quantum effects. What if my briefcase gets a scuff mark on it as I take it with me after the lecture? Igor Novikov has pointed out that such wear and tear undergone by a jinn particle would have to be repaired at some point to return it to its original condition—my briefcase is no excep-

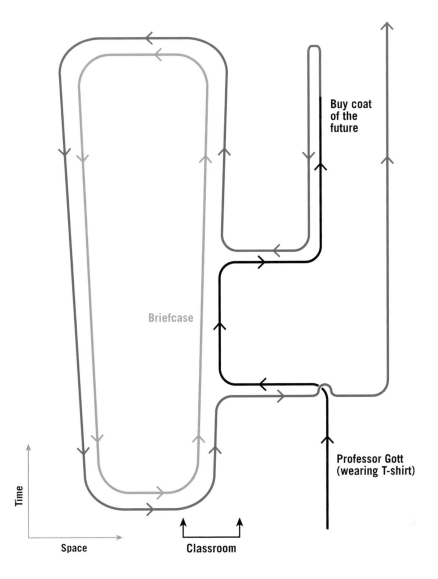

FIGURE 21.1. Spacetime diagram of Professor Gott's time travel talk. *Credit:* J. Richard Gott

tion. This does not violate the laws of entropy, because the briefcase is not an isolated system; energy coming from outside is used to repair the briefcase.

Information can also be a jinn. Imagine that I go back in time to 1915 and give Einstein the correct field equations of general relativity. He could then write them up and publish them. Where did the information come from? I learned it by reading his paper, and he learned it from me—a circular worldline.

Jinn particles are possible under the laws of physics—they are just improbable—and the more massive and complex the jinn particles are, the more improbable they become. We could have had the same story if I had found a paper clip on the floor in the lecture hall and carried it with me instead of the briefcase and gone back in time to place that paper clip on the floor in the spot I found it. Then the paper clip is a jinn, and it would have been simpler and less massive than the briefcase. Simpler still, I could even have found an electron and taken it back in time to place it in the lecture hall . It is just more improbable to find an object as large and complex as a briefcase, and particularly lucky to find one containing the exact notes I need for my talk. I think such complicated jinn are possible but very unlikely to occur.

Time travel to the past happens when you have a worldline that loops back into the past. The usual state of affairs is captured by figure 18.1 of the worldline of Earth and other planets as helixes around the worldline of the Sun. Nothing is going faster than the speed of light, and the worldlines all proceed toward the future. Figure 21.2 shows the situation when you have time travel to the past. The time traveler's worldline loops backward in time to visit an event in his own past.

The time traveler starts at the bottom in the past and comes upward until he encounters a worldline of his older self, who says, "Hi! I'm your future self! I've traveled back in time to say hello!" He replies, "Really?" and goes on to loop back into the past. He then encounters his younger self and says, "Hi! I'm your future self! I've traveled back in time to say hello!" His younger self replies "Really?" The time traveler experiences this scene twice, once as his younger self and once as his older self, but the scene only happens once. You can think of this as one four-dimensional sculpture with worldlines on it. It never changes: that's what the picture looks like. If you want to know how it would be to experience it, then follow a worldline around and see what other worldlines would approach you.

This brings us to one of the ways to address the famous *grandmother paradox*: What if I go back in time and accidently kill my grandmother before she gave birth to my mother? Then she would not give birth to my mother, and my mother could not give birth to me, which means I wouldn't exist; therefore, I can't go back in time to kill my grandmother, which in turn means she's okay and gives birth to my mother, who gives birth to me after all. It's a paradox. The conservative solution to the grandmother paradox is that time travelers cannot change the past. They were always part of the past. You might have gone back in time and had tea and cookies with your grandmother as a young

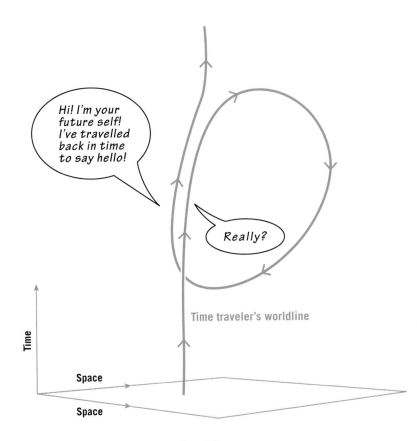

FIGURE 21.2. Spacetime diagram of time traveler's worldline.
Credit: Adapted from J. Richard Gott (*Time Travel in Einstein's Universe*, Houghton Mifflin, 2001)

girl, but you could not have killed her, because she gave birth to your mother, who gave birth to you. The solution must be self-consistent. Kip Thorne, Igor Novikov, and their collaborators constructed a set of thought experiments involving time-traveling, colliding billiard balls to show that it always seems possible to find self-consistent solutions free of paradoxes.

You don't have to worry about changing history: no matter how hard you try, you don't change anything. If you went back to the *Titanic* to warn the captain about the iceberg, the captain would ignore your warning just as he ignored all the other iceberg warnings, because we know the ship went down. You would find it impossible to change events. The time travel in the film *Bill and Ted's Excellent Adventure* is built on that same principle of self-consistency.

The alternative solution to the grandmother paradox is the Everett *many-worlds theory* of quantum mechanics. Physicists disagree about this, but let's first consider how it works. In the many-worlds theory, many parallel worlds can coexist like railway lines in a railway switching yard. We see one history—like riding down one railway track. The events we see are like stations we pass: Here is World War II, . . . here are people landing on the Moon, and so on. But there are many parallel worlds. There is a world where World War II never happened. This is based on Richard Feynman's many-histories approach to quantum mechanics: he found that to calculate the probability of an outcome in any future experiment, you must consider all possible histories that could have led up to it. Some people think this is just one of the weird rules of how to do calculations in quantum mechanics, but proponents of the many-worlds model think that all those histories are real, and they interact with one another. David Deutsch has argued that a time traveler could go into the past and kill his grandmother as a young girl. That would cause a new track to branch off. In that branching history, there would be a time traveler and a dead grandmother. The track where the time traveler was born and where his grandmother lived is a separate track that still exists. He still remembers the part of his history on that track before he switched to a new one. Both tracks exist.

We now have two solutions to the grandmother paradox, each of which solves it: the conservative one, which is a single, self-consistent, four-dimensional sculpture that does not change, and the more radical many-worlds theory of quantum mechanics. Either solution works.

Now if we return to our picture of the time traveler's worldline looping back into the past, we can notice one thing wrong with it. Light travels at a slope of 45° in this diagram. As the time traveler loops over the top, to start returning to the past, at some point the slope of his worldline relative to the time axis must be greater than 45°. That means that at some point, he must be exceeding the speed of light. Just as he rolls over the top, he is actually traveling at infinite speed. The notion that if you could travel faster than light, you could travel back in time is recognized in A.H.R. Buller's limerick:

> There was a young lady called Bright
> Who could travel far faster than light;
> > She set off one day,
> > In a relative way,
> And returned home the previous night.

The trouble with this is that Einstein showed, in his theory of special relativity, that you cannot build a rocket that travels faster than light. If you are always going slower than the speed of light, the slope of your worldline never tips more than 45° from the time axis, and you cannot circle back to the past. However, in Einstein's theory of general relativity, in which spacetime is curved, you can beat a light beam by taking a shortcut, either by going through a wormhole or (as we shall see) by going around a cosmic string. If you can beat a light beam, you—like Ms. Bright—can travel back in time.

Suppose you have a piece of paper, representing one dimension of space horizontally and the dimension of time vertically (figure 21.3). Your worldline is then a vertical green line on this sheet of paper. You are lazy, just staying at home, so your worldline runs straight up from the bottom of the paper to the top. With curved spacetime, however, the rules change. Let's bend the paper into a horizontal cylinder by taping the top to the bottom. Now your worldline is a circle going back into the past.

You are always going forward into the future, but you circle back into the past. The same thing happened to Magellan's crew. They traveled always west, west, west around the curved surface of Earth, and yet arrived back in Europe. This could never have happened if Earth's surface were flat. In the same way, the time traveler always travels toward the future, but if spacetime curves sufficiently, she can circle back to an event in her own past.

Various solutions in general relativity permit this. Before discussing them, let me describe *cosmic strings*. In 1985, I found an exact solution to Einstein's field equations for the geometry around a cosmic string. Alex Vilenkin of Tufts University had found an approximate solution, and I found an exact solution. William Hiscock of Montana State University also found the same exact solution independently, so we both share credit for that discovery. The solution tells us what the geometry around a cosmic string is like.

But what is a cosmic string? It's a thin (narrower than an atomic nucleus) high-energy-density thread of quantum vacuum energy under tension, something that might be left over after the Big Bang. Such strings are predicted in many theories of particle physics. We haven't found them yet, but we're certainly looking for them.

Physicists have learned that a vacuum (empty space—free of particles and photons) can acquire an energy from the presence of a field permeating space. This concept comes into play, for example, with the recently discovered Higgs field and its associated particle the Higgs boson. After the Higgs boson was discovered at the Large Hadron Collider, François Englert and Peter Higgs won

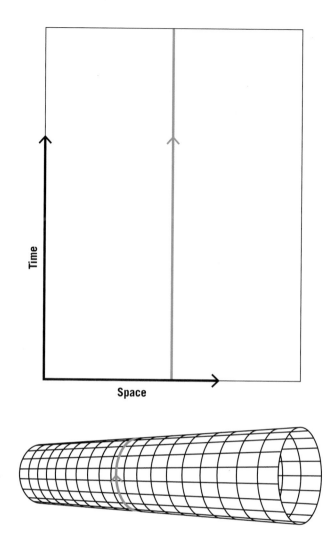

FIGURE 21.3. Curved spacetime allows a worldline to circle back into the past.

Credit: Adapted from J. Richard Gott (*Time Travel in Einstein's Universe*, Houghton Mifflin, 2001)

the 2013 Nobel prize in Physics for their theoretical work predicting its existence. As I discuss in chapter 23, we now believe that the very early universe had a high vacuum energy. When this vacuum energy decayed into normal particles, it is possible that some of it remained trapped in thin threads of high vacuum energy—the cosmic strings. It is like when a field of snow melts, and some snowmen are left standing at the end. Similarly, cosmic strings are made of vacuum energy left over from the early universe.

Cosmic strings have no ends: either they are infinite in length if the universe is infinite in extent, or they occur in closed loops. Visualize (infinitely long) strands of spaghetti, and SpaghettiOs. We expect to have both the infinitely long strands and the loops. Most of the mass in the network of cosmic strings is contained in the infinitely long ones.

As for the geometry of the space around the cosmic string, we need to ask: what would a cross-section through a plane perpendicular to the string look like? You might expect this to look like a piece of paper with a dot in the middle where the string went through. But a cosmic string is expected to be very massive—about a million-billion (10^{15}) tons per centimeter—and, therefore, it warps the space around it significantly. Instead of looking like a piece of paper with a dot in the middle of it, it looks like a pizza with a slice missing (figure 21.4).

You start with a pizza and simply remove one slice. Just eat it. Get rid of it. It's gone. Take the rest of the pizza and carefully grab the two edges of the

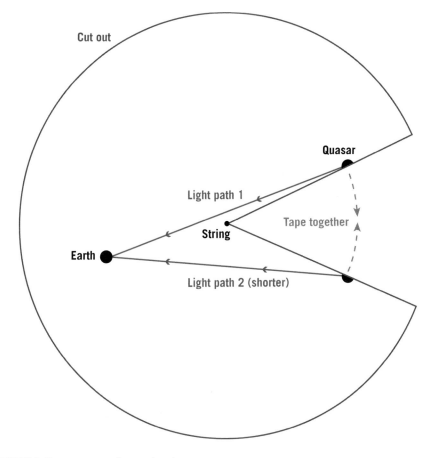

FIGURE 21.4. Geometry around a cosmic string.

Credit: Adapted from J. Richard Gott (*Time Travel in Einstein's Universe*, Houghton Mifflin, 2001)

pizza next to where the missing slice used to be. Draw these together so that you form the pizza into a cone. This is a cross-section of the geometry around the string. It is shaped like a cone. The string itself passes through the center of the pizza. The conical geometry shows that the circumference is not equal to 2π times the radius of the pizza. That's because there is a piece missing—the circumference is less than it would be if the pizza had all its slices. You can see that it does not obey the laws of Euclidean geometry for a flat plane.

The angular width of the missing slice is proportional to the mass per unit length of the string and, for cosmic strings that might be realistically produced in the early universe (grand unified models of particle physics predict that they are produced at an epoch when the unification of the weak, strong, and electromagnetic forces begins to break) this angle is actually rather small—maybe a half second of arc or less. This is very small, but it's nevertheless detectable.

In figure 21.4, the string is at the center, and you can see the missing slice where the two edges are taped together. Suppose I'm sitting on Earth and I'm looking at a quasar behind the string. Light can come to me via either of these two straight-line trajectories (light path 1 or light path 2) that pass on either side of the string. If you tape the sides of the missing slice together such that the piece of paper makes a cone, the two light paths bend around each side of the string. Their paths are being bent by gravitational lensing. It's the same effect that bends light passing near the Sun, as discussed in chapter 19. Yet their trajectories are as straight as they can be. I drew them with a ruler. When the paper pizza is taped together as a cone, one can drive little trucks steering straight ahead along either light path 1 or light path 2 from the quasar to Earth. Both paths are geodesics. Because two light beams can travel on straight-line paths from the quasar to Earth, we see twin images of the quasar on opposite sides of the cosmic string. We can look for a cosmic string by searching for pairs of quasar images appearing on opposite sides of the cosmic string in the sky, like pairs of buttons on a double-breasted suit. We haven't found any lensed by a cosmic string yet, but we are still looking.

One remarkable feature of this picture is that the two light paths can have different lengths. In figure 21.4, for example, path 2 is a bit shorter than path 1. So if I flew in my spaceship from the quasar to Earth on path 2 at 99.9999999999% of the speed of light, I could beat a light beam going along path 1, because the light beam had a longer distance to travel. I could beat a light beam by taking a shortcut!

Although we haven't seen a cosmic string yet, we have actually observed this sort of gravitational lensing phenomenon with a galaxy situated between us and a quasar. We see two images of the distant quasar QSO 0957+561 on opposite sides of a lensing galaxy in the sky. The warping of spacetime produced by the galaxy is bending the light in the same way as would occur for the cosmic string. In this case, the background quasar is varying in brightness; a team of astronomers led by Ed Turner, Tomislav Kundić, and Wes Colley, and in which I participated, was able to measure the same outburst in the quasar in both images and determine that there was a time delay between the two images of 417 days. That is a small fraction of the total light travel time of 8.9 billion years. If you want to know whether you can travel faster than light, in this case the answer is yes, you can! One light beam has beaten the other light beam by 417 days in a fair race through empty space, but only by taking a shortcut.

So looking for double images of quasars is one way to search for cosmic strings. So far, all the cases seem to be explained by galaxy lenses, but we would expect string-lensed quasars to occur more rarely, so this is not surprising. We keep looking.

Cosmic strings are under tension and are typically whipping around at velocities of about half the speed of light. Just as light beams are bent toward each other by passing on opposite sides of a cosmic string, two space ships at rest with respect to each other can be drawn toward each other after a cosmic string passes rapidly between them. The two spaceships pick up a velocity toward each other as the string passes between them. Now let one spaceship be Earth and the other spaceship be the CMB. As a string moves by, it causes a slight Doppler shift in the CMB in the distance behind it. If the string is passing from left to right between the CMB and us, this makes the CMB appear slightly hotter on one (the left) side of the string than the other. We are searching for such effects. Oscillating string loops, like vibrating rubber bands, can produce gravitational waves and we can search for these in the future as well, with space-based LIGO-type instruments. Thus we have a number of promising ways to look for cosmic strings.

How might one possibly make use of the shortcut effect shown by a single cosmic string? In 1991, I found an exact solution to Einstein's field equations in general relativity for two moving cosmic strings. In this solution, two parallel cosmic strings move past each other like the masts of two schooners passing in the night. String 1, which is vertical, moves from left to right and string 2, which is also vertical, moves from right to left. What does the geometry around two cosmic strings look like?

Not surprisingly, two slices are missing this time. A cross-section perpendicular to the two cosmic strings looks like a piece of paper with two missing slices, and you can fold it up into a little paper boat (figure 21.5). Laid out flat, we see the two missing slices, one originating at string 1 and extending upward on the page, and one originating at string 2 and extending downward on the page. (The two strings extend out toward you, perpendicular to the page.) Now there are two shortcuts. If you start on planet A in the figure, you can go to planet B on a straight-line path between the two cosmic strings labeled path 2. But there is a shorter straight-line path 1 that will get you to planet B faster, by going around string 1. Likewise, another shortcut, straight-line path 3, will get you from planet B back to planet A faster than going back by path 2. If you start at planet A and go to planet B on path 1 traveling at 99.9999999% of the speed of light, you can beat a light beam going directly to planet B along path 2. Path 1 is shorter than path 2, because a "pizza slice" is missing. That means you can depart planet A after the light beam going along path 2 departs planet A and yet arrive on planet B before the light beam arrives. Your departure from planet A and your arrival at planet B are therefore two events that have a *spacelike separation* along path 2: they are separated by more light-years in space than years in time. You are beating the light beam and therefore effectively going faster than the speed of light, because you've taken a shortcut. That means that some observer moving rapidly to the left—let's call him Cosmo—will judge those two events to be simultaneous. Because of his velocity (less than the speed of light), he slices spacetime on a slant, like French bread, and judges your departure from planet A to be simultaneous with your arrival on planet B.

Now move the upper half of the solution rapidly to the right, taking string 1 and Cosmo along with it. Now string 1 is not stationary but rather moving rapidly to the right, and since motion is relative, Cosmo is no longer moving to the left but is now just standing still, in the center. Cosmo sees you depart planet A at 12:00 p.m. Cosmo time, and sees you arrive on planet B at 12:00 p.m. Cosmo time. If you can do this trick once, you can do it twice. Slide the bottom half of the solution rapidly to the left carrying string 2 along with it at an equally high speed (but slower than the speed of light). You can depart planet B and travel along the shortcut path 3 and beat a light beam going to planet A along path 2. Your departure from planet B and your arrival back at planet A will be

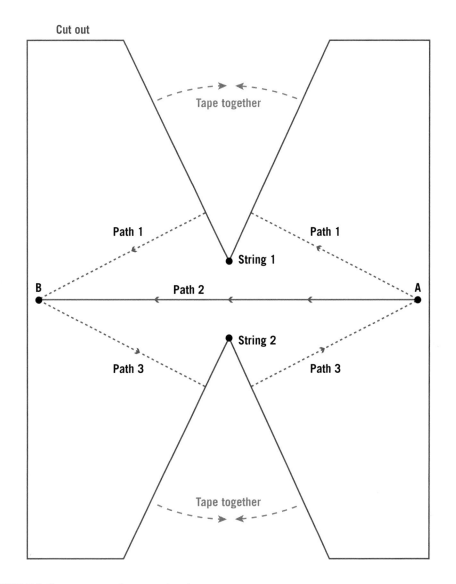

FIGURE 21.5. Geometry around two cosmic strings.
Credit: Adapted from J. Richard Gott (*Time Travel in Einstein's Universe*, Houghton Mifflin, 2001)

separated by more light-years of space than years of time. If the bottom half of the solution is moved rapidly enough (but still slower than the speed of light), then string 2 moves at nearly the speed of light as seen by Cosmo, and Cosmo will observe your departure from planet B and your arrival at planet A to be simultaneous. So, if he sees you depart from planet B at 12:00 p.m. Cosmo time, he sees you arrive back at planet A at 12:00 p.m. Cosmo time. But you departed planet A in the first place at 12:00 p.m. Cosmo time. Your departure from planet A and your arrival back on planet A are at the same time and place. You can get back in time to see yourself off and shake hands with your younger self! You have time traveled back to an event in your own past. That is real time travel to the past.

This is how it looks to you. You arrive at the spaceport on planet A. An older version of yourself arrives and says, "Hello, I've been around the strings once!" You will reply, "Really?" Then you depart on your spaceship traveling around string 1 and arriving at planet B traveling along path 1. Then you immediately depart from planet B, travel around string 2 and arrive back at planet A in time to meet your younger self. You say, "Hello, I've been around the strings once." You hear your younger self reply, "Really?"

Does meeting your younger self somehow violate energy conservation? After all, originally there was one of you and now at that meeting there are two of you. No, because general relativity has only local energy conservation. That means that the only way for the mass-energy in a room to go up is for something to come into the room. But as a time traveler, you are like anyone else who enters the room. The mass-energy goes up because you enter. So there is local energy conservation in these solutions.

It is important that the two strings pass each other going in opposite directions. Then all you need is a spaceship to travel around the strings, and you can come back to the time and place you started. Michael Lemonick wrote an article on my time machine for *Time* magazine; in it, he included a picture of me holding up two strings along with a small model spaceship.

Curt Cutler at Caltech discovered a very interesting property of my two-string solution. There was an epoch before which no time travel to the past occurred. When the strings are very far apart in the distant past, it takes a long time to circle them, and you always arrive back home on planet A after you started. But when the strings get close enough, when they are just passing each other, you can circle the strings and get back in time to visit an event in your own past. Such an event is in the time travel region. Figure 21.6 is a three-dimensional spacetime diagram of this.

Time is shown vertically, and two dimensions of space are shown in perspective horizontally. Because String 1 is moving to the right, its worldline is a straight line tilted toward the right. String 2 is moving to the left, and its worldline is a straight line tilted toward the left. The time traveler's worldline is also shown. She is moving slowly, so her worldline is nearly vertical until

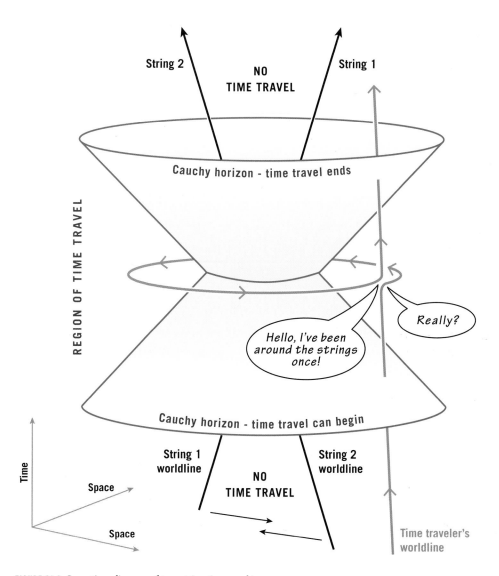

FIGURE 21.6. Spacetime diagram of two-string time machine.

Credit: Adapted from J. Richard Gott (*Time Travel in Einstein's Universe*, Houghton Mifflin, 2001)

she arrives on planet A. You can then see her depart at noon and circle the two strings, arriving back at noon. She says hello to her younger self. She then lives out the rest of her life, and her worldline is again nearly vertical. Cutler found that the region of time travel is bounded by a surface called a *Cauchy horizon*, which is shaped like two lampshades, one inverted on top of the other. Note that the time traveler approaching planet A starts out in the distant past in a region where no time travel to the past is possible. She then crosses a Cauchy horizon where time travel starts. After that point, she can see time travelers arriving from the future. For a while time travel is possible, but she eventually crosses the second Cauchy horizon where time travel to the past stops. After that, she will encounter no more time travelers arriving from the future. By then, the two cosmic strings are so far apart that any time traveler can no longer circle the strings and arrive back at the same time she started.

This answers Stephen Hawking's famous question: "So where are all the time travelers?" If time travel is possible, then why aren't famous historical events overrun with time-travel tourists from the future? Why don't we see time travelers from the far future with their video cameras and silver space suits in the film footage of the Kennedy assassination? The answer is that when you create a time machine in the future by twisting spacetime, a Cauchy horizon is created, and only at that point can you start to see time travelers from the future. But these time travelers cannot travel back before the time machine was created. If you create a time machine in the year 3000, then you can use it, in principle, to travel back from the year 3002 to the year 3001, but you can't use it to travel back before the year 3000, because that's when the time machine was created. We haven't seen such time travelers yet because we haven't built any time machines yet! This is also true of time machines built from wormholes and warp drives, which we discuss shortly. But that means that even if we inspect the past and find no time travelers from the future, it is still possible for us, at some future time, to pass through a Cauchy horizon and see time travelers from the future suddenly appear.

We expect cosmic strings to appear both as infinite strings, like the ones we have been discussing (as well as finite string loops). And because they are under tension, we would expect infinite cosmic strings to be whipping around at speeds of order half the speed of light. But, in practice, you would not expect to be lucky enough to find two infinite cosmic strings passing each other at the requisite speed to create a time machine. Grand unified cosmic strings would have to be moving at speeds of at least 99.99999999996% the speed of light (a bit slower than the speed of light but still very fast) in order to produce a time machine. But you could always find a loop of string and manipulate it gravitationally using massive spaceships in such a way as to cause it to collapse by a large factor due to its tension. A string loop is rather like a rubber band. By flying massive spaceships near it, you could manipulate it so that it snapped shut in just such a way that two very long, straight sections of the string could pass each other at high enough velocity to create a time machine. I was able to show (in my 1991 *Physical Review Letters* paper on the cosmic string time machine) that the string loop in this case is just at the point of collapsing inside a black hole that would form around it. That's not good!

I showed that this is likely to trap the time-travel region inside the black hole. Li-Xin Li and I would later find that the extra mass of your rocket circling the strings in the time machine would likely also help trigger the formation of a black hole around you.

The string loop would have two long straight segments of string passing each other in opposite directions at high speed, and thus the loop would have some angular momentum and therefore, the black hole that was formed would be a rotating black hole.

So let's discuss rotating black holes. As mentioned in chapter 20, an exact solution to Einstein's field equations for a rotating black hole (one having angular momentum) was discovered by Roy Kerr in 1963. What happens in the interior of the rotating black hole solution (inside the event horizon) was worked out by Brandon Carter. The Kerr solution has two critical radii: r_+, which marks the event horizon, and r_-, which is smaller and marks an inner, or *Cauchy horizon*.

At the center of the Kerr black hole one finds not a pointlike singularity but a *ring singularity*. Curvature only becomes infinite on this ring (actually, nearly infinite, because quantum effects would blur it out a little). If you hit the ring, the *tidal forces* (that Iron Maiden plus rack effect described in chapter 20) would kill you. But interestingly, the grad student who has fallen inside a rotating black hole can avoid hitting the ring singularity. It does not block his way to the future. The grad student first crosses inside r_+ (the event horizon) and then inside r_- (the Cauchy horizon). The ring singularity is inside the Cauchy horizon, and the grad student can see it the moment he crosses the Cauchy horizon. If the grad student jumps through the ring, like jumping through a hula hoop, he will enter an entirely new large universe (Universe 1). Carter showed that if the grad student goes through the ring into Universe 1 and circles the ring's circumference while on the other side in a specific way, he can actually jump back through the ring to our side before he entered. The grad student can do a little time loop into the past and say hello to his younger self just before he jumped through the ring initially. Of course no one on the outside of the black hole can see any of this, for it all happens inside the event horizon. Once the grad student crosses inside the Cauchy horizon, he enters a region where time travel to the past is possible—just as in figure 21.6. This Cauchy horizon marks the beginning of an epoch of time travel—an epoch entirely trapped inside the event horizon of the black hole. The grad student will never be able to get back to our universe to brag to his friends about his time-travel adventures. He can then travel on into the future. In the *spacetime diagram* for this, the ring singularity is off to one side and does *not* block the grad student's way to the future. He leaves the time travel region by crossing a second Cauchy horizon (again just as in figure 21.6) and can then pop out into yet another big universe like ours (Universe 2). He pops out of what we would call a *rotating white hole* into Universe 2. He can live out his life there, or jump back in the hole and voyage to additional universes in the future. This is like getting on an elevator in a multistory building. Imagine that you get in the elevator on the ground floor—that's our universe. The door closes and you go up—there's no getting back to the ground floor universe any more. You have left it in your past. The door opens again, and you see a new universe (Universe 1). You can exit the elevator by jumping through the ring singularity, and you can visit Universe 1. You can stay in Universe 1 until you die, or you can hop back into the elevator by jumping back through the ring again. If you do, you will go on up, and the door will open to the next universe (Universe 2). You can exit there and live there or just stay in the elevator and continue on into the future, just looking out the opening and closing elevator doors at new universes forever. But you will never get back to the ground floor universe (our universe). The Kerr solution indicates that all this occurs for a rotating black hole that has formed realistically in the finite past in our universe.

But we must also consider some caveats.

As in chapter 20, the professor stays safely outside the black hole. Photons sent by the professor that fall into the black hole can be received by the graduate student even after he crosses the event horizon. The professor could be sending the grad student messages, such as "good work" or "keep going and you will have a great thesis." The grad student will receive them all. Between his crossing the event horizon and his crossing the Cauchy horizon, two events that occur a finite time apart to the grad student—of order a few hours for a several-billion-solar-mass black hole—the grad student will see the entire infinite future history of our universe external to the black hole. News headlines will come at the grad student faster and faster. The grad student will get in principle an infinite number of newscasts from the outside in a finite amount of time before crossing Cauchy horizon, according to the Kerr solution.

This would be good for historians. If the grad student were curious about the future of our universe, he could find out the entire infinite future history of our universe in a finite time. But this is dangerous! Those speeded-up news reports would come in such rapid sequence because they are carried by very blue-shifted photons. The photons are blue shifted because they have fallen inside the black hole and have gained energy. The photons are blue-shifted by the same factor as the news reports they carry are speeded up. High-energy photons like this are gamma rays, and they can kill the grad student. The photons would become arbitrarily (approaching infinitely) blue-shifted as the grad student crossed the Cauchy horizon, and they would form a curvature singularity along the Cauchy horizon, blocking the way to the time travel region and other universes in the future.

But this singularity along the Cauchy horizon may be weak. Calculations by Amos Ori indicate that the tidal forces there may not tear your body apart. The tidal forces may build up to infinity, but stay there for only an infinitesimal amount of time. It would be like going over a speed bump. The grad student would get a jolt but could survive. The grad student might find his body was not stretched infinitely (spagettified) but just stretched an inch, rather like having a visit to the chiropractor's office. Another unknown is that the Cauchy horizon appears to be unstable: fluctuations at the Cauchy horizon could grow, sending the part of the solution beyond it off in new unpredictable directions. One thing in the grad student's favor is that we do not know the laws of quantum gravity—how gravity behaves at microscopic scales. This Kerr solution to Einstein's equations of general relativity does not consider quantum effects. We expect quantum effects to be important at microscopic scales and to smear out singularities. This may help the grad student through. But because we don't know the laws of quantum gravity, we really don't know for sure what would happen. When we get a grand unified theory of particle physics, we may be able to answer this question. Meanwhile, the rotating black hole holds some secrets still. One way to find out would be to jump in!

Now back to the string loop, which has just fallen inside a rotating black hole and made a time machine. The Cauchy horizon for time travel around the strings that Cutler found would become coincident with the Cauchy horizon in the rotating Kerr black hole that forms. Once you cross the Cauchy horizon, you are in the time travel region. We don't have exact solutions for the collapsing-string-loop case to guide us, but interestingly, in 1999, Sören Holst and Hans-Jürgen Matschull found an exact solution for an analogous lower-dimensional case (in flatland), where two particles (with conical exterior geometries—just like cosmic strings) pass each other at high speeds in a curved spacetime, creating a time machine trapped inside a rotating black hole!

For the string loop case, we have to consider several possibilities for what might happen. You might be able to circle the cosmic string loop and come back to shake hands with your younger self, but you would find yourself inside a black hole and, therefore, never able to get back outside to report your adventures. Then you might be killed by hitting a singularity. If you were really lucky, you might be able to pop out in another universe, but still you could never come back to see your friends. Even worse, you could be killed by hitting a singularity before being able to time travel in the first place. We don't know which of these possibilities would occur.

Stephen Hawking has proven that if a Cauchy time-travel horizon arises in a finite region and the matter density is never negative, a singularity should form somewhere on the Cauchy horizon. Basically it's a theorem that says it is hard to make a time machine out of normal material in your garage by only gently curving spacetime (without ever forming a singularity anywhere). In the case of the two infinite strings passing each other, the energy density is always nonnegative everywhere, but since the strings are infinite, the Cauchy horizon extends to infinity, and Hawking's theorem does not apply. But for the finite cosmic-string-loop solution, where we might imagine actually creating a time machine, you might think a singularity would form on the Cauchy horizon inside the black hole. This wouldn't necessarily block your way, but you would at least see it in the distance just as you crossed the Cauchy horizon. However, if this Cauchy horizon is trapped inside a black hole (as I suspect), and the black hole evaporates via Hawking radiation (as it must), then the quantum vacuum state outside the black hole has a slight negative energy density (causing the event horizon to shrink), in which case Hawking's theorem does not apply either. Thus it would not necessarily violate any theorems to form a time machine trapped inside a rotating black hole where you are not killed by a singularity before you can cross the Cauchy horizon.

The fact that the black hole evaporates in a finite time means that as you reach the Cauchy horizon, you do not see the entire future of our universe before you cross (but just what happens prior to the point the black hole event horizon shrinks to zero size by evaporation). Therefore, you are not hit by arbitrarily highly blueshifted photons from outside as you fall in. This is also helpful.

The Cauchy horizon is unstable, but we have fighter planes that are designed to be unstable so that they can be managed by the pilot to be very maneuverable, like balancing a tall pencil on its point on your finger tip by moving your finger rapidly back and forth to counteract it's tendency to fall. Jugglers do this with rods all the time. In principle, a supercivilization might be able to stabilize the Cauchy horizon by actively perturbing it in the right way.

If you wanted to travel back in time a year by circling the collapsing string loop once, (inside the black hole), this would require finding and manipulating a string loop with a mass equal to about half the mass of our galaxy. This is a project only a supercivilization could even attempt.

Would you be killed before doing the time travel? Would you survive to time travel to an event in your own past, all within a rotating black hole? To answer these questions, we will ultimately need to understand the laws of *quantum gravity*—how gravity behaves on microscopic scales. That's one of the reasons this problem is so interesting.

Moving cosmic strings are not the only time travel solutions to Einstein's equations of general relativity. The first one was a nonexpanding but rotating universe solution proposed by the famous mathematician Kurt Gödel in 1949. Even though our universe is expanding, not rotating, Gödel's solution showed that time travel to the past was permissible in principle in general relativity. If one such solution existed, there could be others. In 1974, Frank Tipler showed that an infinitely tall, rotating cylinder could permit time travel to the past. In 1988, Kip Thorne and his associates Mike Morris and Ulvi Urtsever proposed a time machine using a traversable wormhole. In general relativity, a wormhole is a short tunnel connecting two distant points in curved spacetime. A *traversable* wormhole is one that stays open long enough for you to get through it (unlike the wormhole in the Kruskal diagram we learned about in chapter 20). Such tunnels may exist, according to our understanding of general relativity, although they have not yet been discovered. One end of the tunnel might be near Earth, while the other end is at Alpha Centauri, 4 light-years away. Yet the tunnel might be only 10 feet long (figure 21.7).

If you sent a light beam from Earth to Alpha Centauri, it would take 4 years to get there. But jump through the wormhole, and you could be at Alpha Centauri only a few seconds later. In this way, you could beat a light beam to Alpha

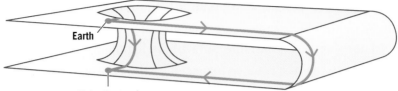

Earth

Alpha Centauri

Wormhole creates a shortcut from Earth to Alpha Centauri

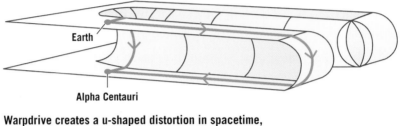

Earth

Alpha Centauri

Warpdrive creates a u-shaped distortion in spacetime, also creating a shortcut from Earth to Alpha Centauri

FIGURE 21.7. Wormholes and warpdrives.

Credit: Adapted from J. Richard Gott (*Time Travel in Einstein's Universe*, Houghton Mifflin, 2001)

Centauri by taking a shortcut through the wormhole. What does the opening, or mouth, of the wormhole look like? In the diagram, it is shown as a circle, but that diagram only shows two spatial dimensions. Actually the wormhole mouth looks like a sphere. It looks like one of those shiny reflecting balls one sometimes sees in a garden. This is correctly depicted in the movie *Interstellar*, for which Kip Thorne served as physics advisor. But don't expect to see the reflection of your Earthly garden in it. Instead you see a garden on a planet orbiting around Alpha Centauri. Jump into that ball on Earth, and you pop out in that other garden somewhere near Alpha Centauri.

Here's how we can make that wormhole into a time machine. Suppose you find such a wormhole on January 1, 3000. If you look through the wormhole, you will see Alpha Centauri, but at what time? If the two mouths (ends of the wormhole tunnel) are synchronized, you would find the clocks on Alpha Centauri also reading January 1, 3000. No time travel there. But now suppose you pull out a massive spaceship and gravitationally pull the wormhole mouth sitting near Earth on a 2.5-light-year journey out and back at 99.5% the speed of light. People on Earth would see that round trip journey taking just over 5 years, with the wormhole mouth arriving back on Earth on January 10, 3005.

Suppose an astronaut was sitting in the middle of the wormhole tunnel. You would see him aging ten times more slowly, because he was traveling at 99.5% the speed of light. During the trip, he would only age 5 years divided by 10, or just 6 months. When he got back, his clock would read July 1, 3000. But the wormhole tunnel is still only 10 feet long. Its length does not change during the journey, because its geometry is determined by the stuff inside the wormhole tunnel and that has not changed. Furthermore, the astronaut is at rest relative to the Alpha Centauri mouth, and the Alpha Centauri mouth is stationary with respect to Alpha Centauri, because nothing is moving it at that end. Thus the astronaut's clock must remain synchronized with Alpha Centauri. If you peer into the wormhole when it returns and you see the astronaut's clock reading July 1, 3000, when you look over his shoulder to the clocks behind him on Alpha Centauri, they must read July 1, 3000 as well. Therefore, just as the wormhole returns to Earth on January 10, 3005, you look through the wormhole and see the Alpha Centauri clocks reading July 1, 3000. You see your opportunity: you jump through the wormhole and find yourself on Alpha Centauri on July 1, 3000. Get in a spaceship and travel back to Earth at 99.5% of the speed of light. The trip through ordinary space will take a little more than 4 years. You will arrive back on Earth on July 8, 3004. But you started your trip on January 10, 3005, so you have arrived back before you started. You have traveled back in time. You can visit an event in your own past. You can shake hands with your younger self on Earth on July 8, 3004, before you started on your trip. Notice how you cannot use the wormhole to go back before the time machine was created, when that one wormhole mouth near Earth was taken on a trip. You can't go back before the year 3000, for example, because that was before the wormhole mouths were desynchronized.

The inspiration for this line of research started with Carl Sagan. He was writing a science fiction novel called *Contact*. Neil told you about the movie in chapter 10. For his plot, Sagan wanted his heroine, Jodie Foster in the movie, to jump in a wormhole and emerge near the star Vega, 25 light-years away. Carl wanted to be sure to get the physics right, so he called up his friend Kip Thorne. When Thorne and his associates investigated the physics of wormholes, they found that wormholes have to be propped open with some negative-energy stuff—stuff whose energy is less than zero, stuff that is gravitationally repulsive. Light converges on a wormhole, passes through the wormhole tunnel, and diverges on the other side. That is the hallmark of the repulsive effects of negative-energy stuff. Recall that there was a wormhole connected with the black hole in the Kruskal diagram, but you couldn't get through it to the other side. You couldn't get through to the other universe before hitting a singularity and being torn apart. But with negative-energy stuff, you could prop the wormhole open, enabling you to get through. But where to find negative-energy stuff?

Curiously, a quantum effect called the *Casimir effect* actually creates negative-energy stuff. If you put two parallel metal conducting plates close together, the quantum vacuum state between the two plates has a negative energy density. Pressure effects associated with the Casimir effect have been verified in the lab by M. J. Sparnaay and S. K. Lamoreaux. The Hartle–Hawking quantum vacuum state around a black hole also has a slight negative energy density, which allows the black hole to evaporate over time, decreasing the area of its event horizon. These two examples show that you can make negative-energy stuff. Thorne and his colleagues figured that if two spherical plates were placed in the wormhole tunnel back to back, blocking the tunnel, with only a 10^{-10} cm separation between them, the Casimir effect between the two plates could prop the wormhole open. You would open trap doors in the plates to pass through. (Since these solutions involve some negative-energy stuff, wormhole solutions can create a time machine in a finite region in a singularity-free manner, because Hawking's theorem about that, which I discussed earlier, does not apply.)

For the time machine proposed by Thorne and his colleagues, each wormhole mouth would weigh 100 million solar masses and have a radius of 1 AU. Building such a wormhole would be a massive project, only conceivable for some supercivilization to attempt. The only way to do this would be to find some microscopic quantum wormhole mouths 1.6×10^{-33} cm apart and 1.6×10^{-33} cm in diameter that are part of the quantum spacetime foam thought to exist at microscopic scales. Then you would have to move them apart and slowly enlarge them to 100 million solar masses each. This is not something you are going to build in your garage! But the recent work of Maldacena and Susskind suggests that microscopic wormholes connecting quantum entangled particles might give one at least a place to start.

The other famous time machine is the warp drive from *Star Trek*. This is a U-shaped distortion of space that also creates a shortcut through space, for example, to Alpha Centauri. There is no hole, just a U-shaped distortion (see figure 21.7). Physicist Miguel Alcubierre has looked at this from the point of view of general relativity and found that you need both some positive-energy stuff and some negative-energy stuff to make it work, but it is theoretically possible.

Amos Ori has recently proposed a toroidal (doughnut-shaped) time machine. Creative general relativity solutions involving time travel are still being discovered.

Stephen Hawking thought that some quantum effects, yet to be discovered, might always step in to prohibit time travel, even though general relativity allows it. He proposed his Chronology Protection Conjecture, suggesting that the laws of physics would somehow prevent time travel to the past. Of course, it was just a conjecture. He based it on some indications that the quantum vacuum state might blow up (become infinite) as one approached the Cauchy horizon and the region of time travel. Li-Xin Li and I found a counterexample, which had a different quantum vacuum state that did not blow up on the Cauchy horizon. Hawking's student Michael J. Cassidy found the same example from different reasoning. So it appears that in some situations, you may be able to time travel. Once again, to know for sure we will need to determine the laws of quantum gravity.

In 1895, when H. G. Wells published his novel *The Time Machine*, the known laws of physics, Newton's laws, had a universal time that everyone agreed on, and time travel to either the future or the past was forbidden. Yet just 10 years later, in 1905, Einstein would prove that time travel to the future was possible. Cosmonaut Gennady Padalka has already time traveled 1/44 of a second into the future(see chapter 18). In 1915, Einstein's theory of gravity, based on curved spacetime, permitted shortcuts allowing you to beat a light beam, thereby opening the door to time travel to the past. Currently, several solutions to Einstein's equations are known that allow time travel to the past in principle. Our current situation is the opposite of the one H. G. Wells found himself in when he wrote his famous book. Einstein's theory of general relativity, which has passed every test we have devised so far, is our best theory of gravity, and it does have solutions that allow time travel to the past in principle, even if the means required are ones only a supercivilization might attempt. We know how gravity behaves on macroscopic scales, but we also know that on microscopic scales, quantum effects must become important, and so we still need to develop a theory of quantum gravity. We must successfully marry general relativity and quantum mechanics in a workable theory to understand whether we can actually construct a time machine to visit the past. As we currently understand them, the laws of physics seem to allow time travel to the past, but the question remains open whether any laws of physics we will discover in the future will prevent such time travel.

I explored the ideas of special and general relativity as they relate to the possibilities of time travel in my book *Time Travel in Einstein's Universe* (2001). We do research on time travel to the past in general relativity, not in order to fabricate a time machine at present, but to discover clues about how the universe works. Time travel solutions test the laws of physics under extreme conditions. In chapter 23, I revisit time travel when considering the extreme conditions at the beginning of the universe.

22

THE SHAPE OF THE UNIVERSE
AND THE BIG BANG

J. RICHARD GOTT

To discuss the shape of the universe, let's first revisit the question of how many dimensions the universe has. As we have said, we live in a four-dimensional universe. You need four coordinates to locate any event: three dimensions of space and one dimension of time. In his theory of special relativity, Einstein showed that intervals between events (at least in flat spacetime) can be measured by $ds^2 = -dt^2 + dx^2 + dy^2 + dz^2$. That minus sign in front of the dt^2 term differentiates the dimension of time from any dimension of space and guarantees that all observers will agree that the speed of light is constant.

We can imagine a universe having a different number of space and time dimensions. A universe with two spatial dimensions and one time dimension would have intervals between events measured by $ds^2 = -dt^2 + dx^2 + dy^2$. People living in that universe would not know what the z coordinate was—they would not know about up and down. These people would be living in Flatland. A picture of Flatland (figure 22.1) shows a Flatlander standing in his house.

He has a doorway in the front, and he can even have a swimming pool in his back yard. But if he wants to go swimming, he has to go out his front door, climb over the roof, and dive off the roof into the pool. He has an eye: it has a lens in the front and a retina in the back. You may notice that we see his entire cross-section. We can see the complete interior of his body. We are in a position to give him a very good diagnosis of whatever might ail him, because we can see all his internal organs. He has a mouth, an esophagus, and a stomach, but no

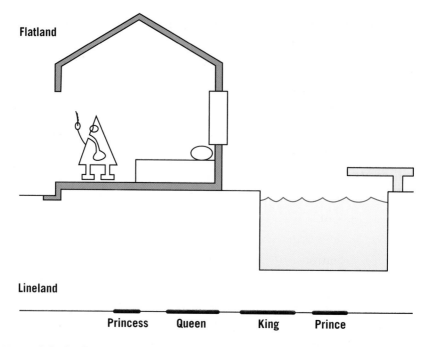

FIGURE 22.1. Flatland and Lineland.
Credit: Adapted from J. Richard Gott (*Time Travel in Einstein's Universe*, Houghton Mifflin, 2001)

alimentary canal that goes all the way through his body. If he did, he would fall apart into two pieces! He must digest his food in his stomach and regurgitate the remains. He is shown holding up a newspaper. Our newspapers are two dimensional—they are pieces of paper; but his is one dimensional, like a line. His newsprint consists of dots and dashes—Morse code. If he wants to get in bed, he just has to do a back flip into bed. How would his brain work? You can't build neurons (or wires) that cross in Flatland. But electromagnetic signals can cross each other in Flatland, so you would just use electromagnetic waves to replace neurons for sending signals from one cell to another.[1] In principle, a Flatlander might have a brain, but it would be much more difficult to arrange.

In 1880, Edwin Abbott wrote a wonderful book, *Flatland*, about creatures living in such a world of two spatial dimensions. The narrator was a square.[2]

What would happen if there were only one dimension of space and one dimension of time? This would be Lineland (also shown in figure 22.1). Every-thing would be on a single line. Then we would have $ds^2 = -dt^2 + dx^2$. People would be line segments. You could have a King and Queen and a Prince and Princess, but if you lived in Lineland, you could see only the people directly to your left and right. They would look like points. You had better like them, for you are never seeing anybody else. Intelligent life seems difficult to achieve in Flatland, and hopeless in Lineland.

We can also imagine spacetimes with more dimensions of space than we see. Suppose we add one additional dimension of space. Then we would have $ds^2 = -dt^2 + dx^2 + dy^2 + dz^2 + dw^2$. This is a spacetime with four dimensions of space and one dimension of time. It has one extra dimension of space (w). In 1919, Theodor Kaluza proposed that such an extra dimension existed. Why? Well, he found a remarkable thing. If you believed Einstein's equations of general relativity and applied them in such a five-dimensional spacetime, and the solution was uniform in the w direction, you would get something equivalent to Einstein's equations of general relativity in four dimensions (normal gravity), plus Maxwell's equations (as updated by Einstein using special relativity)! A miracle! Electromagnetism was equivalent to the action of gravity in an extra dimension. This would unify gravity and electromagnetism. It seems like too much of a coincidence that Einstein's general relativity with an extra dimension will automatically reproduce Maxwell's equations.

Attractive as this finding was, this theory had one big trouble: it didn't seem to make any sense. Why don't we see this extra dimension? In 1926, Oskar Klein came up with an answer. He had the idea that the extra dimension would be curled up like a soda straw. A soda straw is a cylinder, a two-dimensional surface. It is made from a two-dimensional piece of paper, after all. If creatures lived on the surface of a soda straw, they would have to be two-dimensional creatures, in other words, Flatlanders. It takes just two coordinates to locate yourself on the surface of a soda straw: a vertical coordinate to tell you how far up on the straw you are, and an angular coordinate to tell you your location around the circumference of the straw. But if the circumference is tiny and you look at the soda straw from a distance, it looks one dimensional, like Lineland. We only notice the macroscopic dimension of the straw—the dimension along its length. If the circumference of the straw is smaller than an atom, we won't see that circumference at all.

Thus Kaluza–Klein theory explains electromagnetism. Positively charged particles circle the soda straw in the counterclockwise direction, whereas negatively charged particles circle the straw in the clockwise direction; neutral particles like the neutron do not circle. If the soda straw is bent like a bow, then the clockwise and counterclockwise geodesics can bend differently in the macroscopic directions, because they have different starting velocities in the small extra dimension. This would explain how positively charged particles in an electric field could accelerate in the opposite macroscopic direction from negatively charged particles. Since their velocities in the small circumferential direction would be different, they would move on different geodesics. It also explains why charge is quantized. The wave nature of particles means that only an integer number (1, 2, 3, . . .) of wavelengths can circle the circumference of the soda straw. That means that the momentum of particles in the w direction (which depends on their wavelengths and is equal to their charge) must be an integer multiple of the charge on the proton or the electron. Given the observed size of the electric charge of the proton and electron, we can solve for the circumference of the soda straw: it is 8×10^{-31} cm. That is smaller than an atomic nucleus and explains why we don't see the extra dimension.

After he invented general relativity, Einstein dreamed of finding a grand unified theory of physics that would unify all the forces of nature. It is fair to say that Kaluza and Klein made some progress toward that goal: they unified electromagnetism and gravity. Electromagnetism was just gravity operating in a curled-up extra dimension. But the Kaluza–Klein theory had something in addition: the circumference of that straw might vary in time and from place to place. That was equivalent to having a scalar field that could vary from place to place in spacetime. A *scalar field* is a field that has a magnitude but does not point in any particular direction. Temperature is a scalar field. Wind velocity is a *vector field*, because it has a speed and points in a specific direction (north, for instance). In this case, the scalar field would be the magnitude of the circumference of the extra dimension at that point, and therefore, the magnitude of the electric charge of an electron at that location. If one wanted just general relativity and Maxwell's equations, that circumference would have to stay fixed and not vary, because we always observe electrons to have the same electric charge, wherever we find them. If the circumference did vary, that would cause the electric charge of an electron to vary, which is not observed. It was not clear what would cause the circumference of the straw to remain fixed. If it were fixed, as one might like, their theory gave no new predictions; it gave the same predictions as standard general relativity plus standard Maxwell's equations. Einstein was lucky—his theory of general relativity gave different predictions than Newton's theory did (concerning Mercury's orbit and light bending), and these could be tested. But Kaluza and Klein had no new predictions, so the theory could not be tested, and they got no Nobel Prize.

Today we know of four forces: strong and weak nuclear forces, electromagnetism, and gravity. The strong nuclear force is what holds the atomic nucleus together, and the weak nuclear force is important in some forms of radioactive decay. Steven Weinberg, Abdus Salam, and Sheldon Glashow won the 1979 Nobel Prize in Physics for unifying the weak force with electromagnetism. Their theory predicted that, just as the photon is the carrier of the electromagnetic force, cousins of the photon, the heavy W_+, W_-, and Z_0 particles, would be carriers of the weak force. These particles were discovered at the CERN particle accelerator (near Geneva); Carlo Rubbia and Simon van der Meer shared the Nobel Prize in 1984 for this work. Strong and weak nuclear forces and electromagnetism are all treated in the *Standard Model of particle physics*. Recently, researchers using the Large Hadron Supercollider in Europe discovered the Higgs boson, which was a prediction of the theory. The Higgs boson is the particle associated with the Higgs field, a scalar field that permeates space and gives the W_+, W_-, and Z_0 particles their mass. The Standard Model of particle physics has been very successful, but currently offers no explanation for dark matter or for the nonzero masses of neutrinos. Also, the strong, weak, and electromagnetic forces have not yet been unified with gravity.

Today our best hope for a grand unified theory that will bring together all four forces is *superstring theory*. This is based on the idea that elementary particles are not pointlike but rather are tiny lengths of string about 10^{-33} cm long. These strings are like the cosmic strings we have talked about, in that they have positive mass and a tension along their length. But instead of a microscopic thickness, the superstrings have zero thickness. Different vibration states in the string make for different elementary particles—quarks, electrons, and whatnot. Ed Witten has shown that the five different versions of superstring theory along with another theory called supergravity are actually limiting cases of one overarching theory, which he has dubbed *M-theory*. In M-theory, spacetime is eleven-dimensional, with ten dimensions of space and one dimension of time. It posits the three macroscopic dimensions of space that we know, plus seven more tiny, curled-up spatial dimensions. If I were trying to explain to a Linelander what a soda straw was like, I would say it is like a line, except that every point on that line is not a point but actually a tiny circle. If we had two extra dimensions of space, that would be a tiny two-dimensional surface: not a circle, but perhaps the surface of a tiny donut. In M-theory, the seven curled-up dimensions are in some tiny pretzel shape, one that should explain the strong, weak, and electromagnetic forces. Everywhere you think there is a point in space, there is actually a tiny seven-dimensional, curled-up pretzel shape. Many shapes are possible. The goal is to find the right shape, the one that will explain the particle physics that we observe.

It's rather like the conundrum Watson and Crick faced when they were trying to find the structure of the DNA molecule. Many structures seemed possible, but what was the right one? When they finally solved the problem, the resulting structure could explain how chromosomes could divide and produce separate but identical copies. The answer was the double-helix geometry of DNA that could unzip and attract complementary base pairs to form two identical helixes. Likewise in physics, we are hoping to find the microscopic geometry of extra space dimensions that will explain the physics we see. Many people are working on this today, traveling down a path laid down by Kaluza and Klein. Lisa Randall and her colleague Raman Sundrum have explored how a highly curved extra dimension might explain why gravity is so weak relative to other forces. If someone finds a version of M-theory with testable predictions that agree with observations, then that person will have achieved Einstein's dream of finding the unified theory of particle physics, and he or she will move up into the company of Newton and Einstein. It is an exciting prospect.

Having examined the microscopic universe, we are now ready to look at the macroscopic universe. We would like to make a single map that would encompass the entire universe, that would show us interesting things from the Hubble Space Telescope in low Earth orbit, to the Sun and planets, to stars and galaxies, to distant quasars and the cosmic microwave background (CMB) radiation, the most distant thing we can see. The problem is that our galaxy is tiny compared with the visible universe, and the solar system is a microscopic dot relative to our galaxy. It is a challenge, therefore, to map the universe on one comprehensive map and display everything of interest to us.

Figure 22.2 is a cross-sectional map of the entire visible universe, looking out from Earth's equator. Earth is at the center of the map. We are at the center of the visible universe not because we are at any special position, but because we are, not surprisingly, at the center of the region we can see. In the same way, if you go to the top of the Empire State Building, you will see a circular region bounded by the horizon that is centered on the Empire State Building. From the top observation deck of the Eiffel Tower, you will see a circular region centered on it. In this map of the visible universe, the most

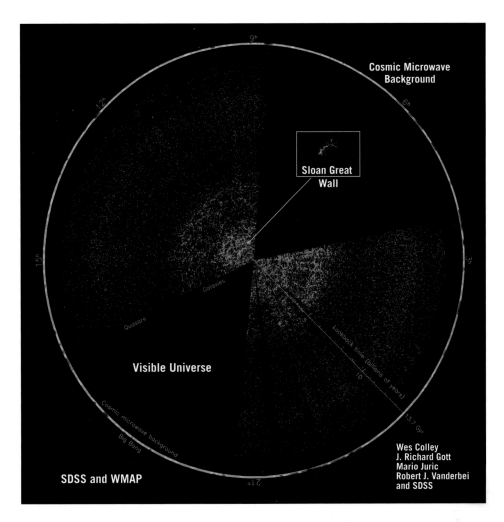

FIGURE 22.2. Equatorial cross-section through the visible universe. We are at the center of the region we can see. Each dot represents a galaxy (green) or quasar (orange) with a redshift measured by the Sloan Digital Sky Survey. (The central portion of this diagram was previously illustrated in figure 15.4.) The cosmic microwave background forms the perimeter. *Photo credit:* J. Richard Gott, Robert J. Vanderbei (*Sizing Up the Universe*, National Geographic, 2011)

distant thing we can see, shown around the circumference, is the CMB (as observed by the WMAP satellite). Inside this circle we see, plotted as dots, 126,594 galaxies and quasars from the Sloan Digital Sky Survey. The two fan-shaped areas filled with dots show a cross-section of the regions the survey covered. The blank fans are regions not covered by the survey. You can see the Sloan Great Wall (discussed in chapter 15) in the picture. Quasars are seen out to greater distances than galaxies. As you know, when we look out in space, we look back in time. The look-back time in billions of years is shown in the figure. Our Milky Way is just a dot at the center of this picture—and the locations of nearby stars and the planets in our solar system are all invisible, being microscopic.

The map we really want is one like that famous *New Yorker* cover by Saul Steinberg, called "The View of the World from 9th Avenue." It shows a New Yorker's view of the world. The buildings of Manhattan loom large in the foreground. The Hudson River is smaller, with "Jersey" a mere strip on the opposite side. The Midwest is compressed to about the width of the Hudson River, and the Pacific ocean is an equally narrow strip, bordered by Asia beyond. The things important to a New Yorker are shown at large scale, whereas more distant lands are shown as tiny. It is just the sort of view we want our map of the entire visible universe to have. We want objects in the solar system that are important to us to be shown large and more distant objects to be shown at a reduced scale.

When I was a graduate student, in the 1970s, I developed a map projection to do just this. I have made different versions of it over the years. I made a pocket version of it in the 1990s.

This Map of the Universe is a conformal map of the universe. *Conformal* means that it preserves shapes locally, as the Mercator map of Earth does. Iceland has as good a shape on the Mercator map as does Cuba. Local regions are shown in their true shapes, being neither squashed nor stretched in one direction. That's why the Mercator map is used on Google Maps. If you enlarge a little region for closer examination, it will have the proper shape. But sizes are wrong; Greenland appears to be about the same size as South America on a Mercator map, but on the globe, its true area is about 1/8 as large. My map is similar, since objects farther away from Earth are depicted at smaller scale, but with the correct shapes.

Mario Jurić and I made a large professional version of this map in 2003, which ended up being picked up by *New Scientist* and *The New York Times* and reprinted 1.5 million times. It was published in the *Astrophysical Journal* in 2005. The *Los Angeles Times* compared it with Mercator's map and Babylonian maps and called it "arguably the most mind-bending map to date." Bob Vanderbei and I have made a full-color, large-scale version of the map (which is displayed, rotated by 90°, on the next three double-page spreads—figure 22.3). Turn the book 90° counterclockwise, and flip the pages to see the bottom, middle, and top third of the map.

From left to right it is a 360° panorama looking out from Earth's equator. The horizontal coordinate is celestial longitude. The vertical coordinate shows distance from Earth, and each large tick mark represents a factor of 10 farther away from the center of Earth. Objects that are 10 times farther away are shown at 1/10 scale and so forth. The farther away an object is, the smaller it is depicted. One can see Earth's surface at the equator as a straight line. You can see the Moon, Sun, and planets. Much farther up are the stars, starting with Proxima Centauri, Alpha Centauri, and Sirius. Still farther up we see the bulk of the Milky Way. Beyond that are the galaxies M31 and M81. Then the galaxy M87. The Great Wall, discovered by Margaret J. Geller and John Huchra, is a large filament, or chain, of galaxies. Beyond, as a line at the top of the map, is the CMB, the most distant thing we can see, which surrounds us, encompassing 360°.

This map is a snapshot of the visible universe at 4:48 Greenwich Mean Time, August 12, 2003, in a slice 4° wide centered on Earth's equatorial plane (although we also show some famous objects outside those limits). The satellites and planets are shown in their positions at this time, and galaxies are shown at the distances from us they would have attained by this time—that is, they are shown at their co-moving distances. We show all known Kuiper belt objects at the time. We show all known asteroids within 2° of the equatorial plane. Below Earth's surface would be its mantle and core. The atmosphere is shown as a thin blue line above Earth's surface, stretching out to the ionosphere. We show all 8,420 artificial satellites orbiting Earth. You can see the International Space Station (ISS) as well as the Hubble Space Telescope. The moon is full, located 180° away from the Sun. Mars is shown at its closest approach to Earth in its orbit. The planets Mercury, Venus, Jupiter, Saturn, Uranus, and Neptune are also depicted. Ceres, the largest asteroid (at 945 km), is shown. Quaoar, a Kuiper belt object discovered well after Pluto, is shown, along with Pluto itself. The map includes some stars having planets, such as HD 209458, which has a Jupiter-sized planet in a close orbit. It includes the seven-solar-mass black hole Cygnus X-1 and the galaxy M87, which harbors a 3-billion-solar-mass black hole in its nucleus. The Hulse–Taylor binary pulsar, which we mentioned in chapter 11, is a system of two neutron stars locked in a tight orbit; they are slowly spiraling inward, because the system is emitting gravitational waves, just as Einstein predicted. Hulse and Taylor won the 1993 Nobel Prize in Physics for this discovery. Near the top of the map are the 126,594 galaxies and quasars from the Sloan Digital Sky Survey. They appear in two vertical bands with blank regions in between, representing regions that

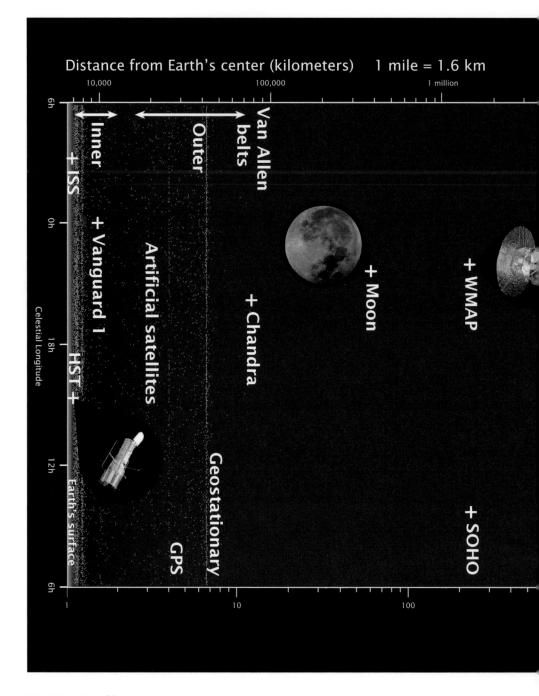

FIGURE 22.3. Map of the universe.

Photo credit: Adapted from J. Richard Gott and Robert J. Vanderbei (*Sizing up the Universe*, National Geographic, 2011)

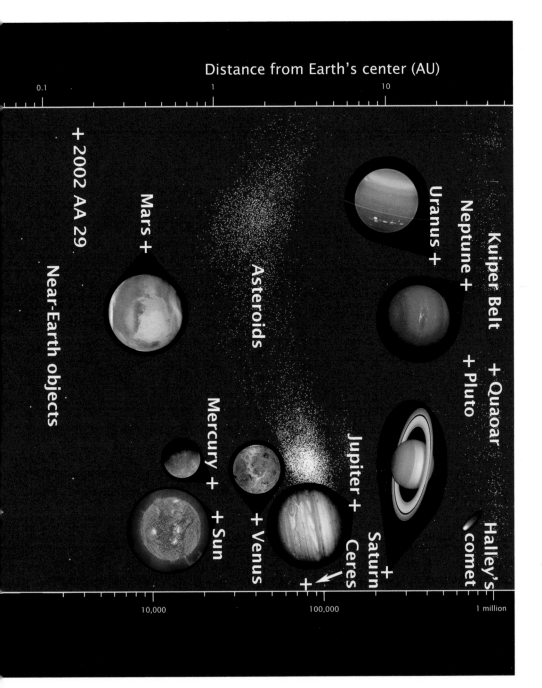

Distance from Earth's center (AU)

0.1   1   10

+ 2002 AA 29

Mars +

Asteroids

Uranus +

Neptune +

Kuiper Belt   + Quaoar

+ Pluto

Near-Earth objects

Mercury +

Jupiter +

Saturn

+ Ceres

Halley's comet

+ Sun

+ Venus

10,000   100,000   1 million

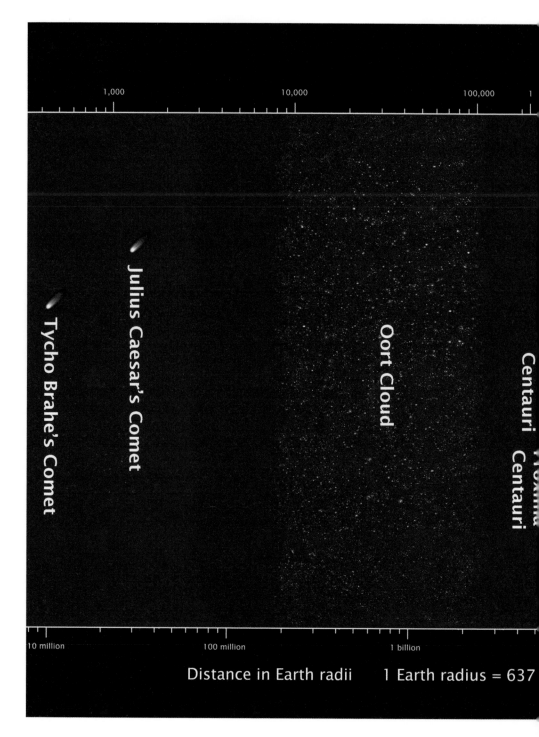

Tycho Brahe's Comet

Julius Caesar's Comet

Oort Cloud

Centauri
Centauri

Distance in Earth radii   1 Earth radius = 637

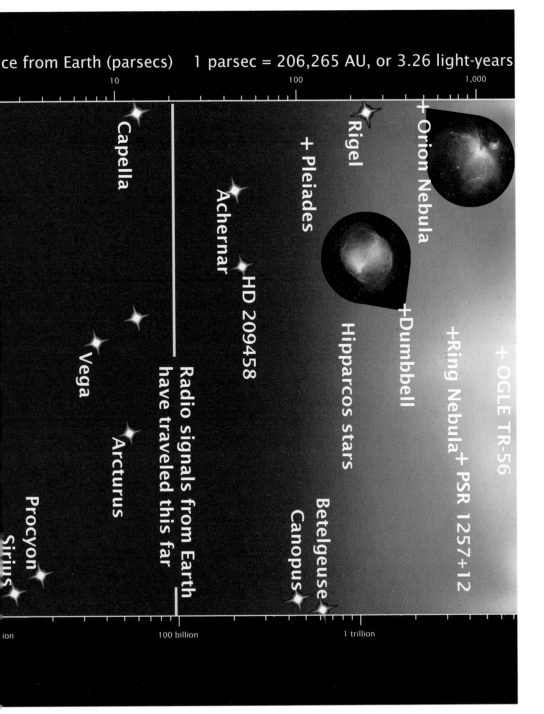

ce from Earth (parsecs)   1 parsec = 206,265 AU, or 3.26 light-years

10   100   1,000

Capella

Rigel

+ Pleiades

+ Orion Nebula

Achernar

HD 209458

+Dumbbell

Hipparcos stars

+ OGLE TR-56

+Ring Nebula+ PSR 1257+12

Vega

Radio signals from Earth
have traveled this far

Arcturus

Betelgeuse
Canopus

Procyon

Sirius

ion   100 billion   1 trillion

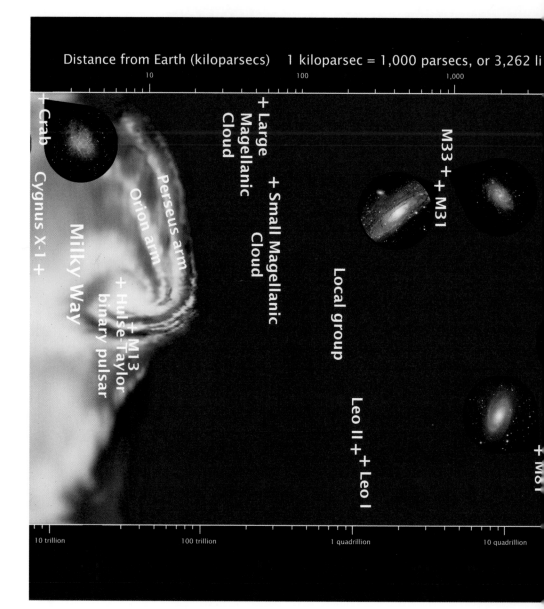

Distance from Earth (kiloparsecs)　　1 kiloparsec = 1,000 parsecs, or 3,262 li

10　　　100　　　1,000

+ Crab

Cygnus X-1 +

+ Large
Magellanic
Cloud

+ Small Magellanic
Cloud

Perseus arm

Orion arm

Milky Way

+ M13

+ Hulse-Taylor
binary pulsar

M33 + + M31

Local group

Leo II + + Leo I

+ M8T

10 trillion　　　100 trillion　　　1 quadrillion　　　10 quadrillion

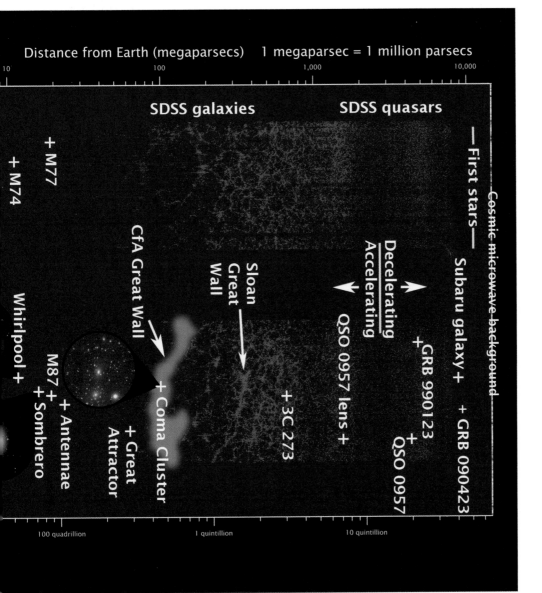

Distance from Earth (megaparsecs)   1 megaparsec = 1 million parsecs

10   100   1,000   10,000

SDSS galaxies   SDSS quasars

— First stars —

Cosmic microwave background

+ M77

+ M74

Subaru galaxy +   + GRB 090423

CfA Great Wall

Decelerating
Accelerating

Sloan
Great
Wall

QSO 0957 lens +

+ GRB 990123

+
QSO 0957

Whirlpool +

M87 +
+ Antennae
+ Sombrero

+ Coma Cluster

+ Great
Attractor

+ 3C 273

100 quadrillion   1 quintillion   10 quintillion

the survey did not cover. These are the same fan-shaped regions shown in figure 22.2, just replotted according to this new map.

This map includes the Sloan Great Wall of Galaxies, which Mario Jurić and I measured in 2003 to be 1.37 billion light-years long and found to be the largest structure in the universe known at the time. It is about twice as long as the Great Wall of Geller and Huchra. But because it is three times farther away, it is shown at one-third scale on the map. Thus, on the map it looks about two-thirds the size of the Great Wall of Geller and Huchra, although it is actually twice as large. The Sloan Great Wall was listed in the 2006 *Guinness Book of Records* as the largest structure in the universe. I never expected to be find myself in the *Guinness Book of Records*, and I didn't even have to eat 68 hotdogs in 10 minutes or collect the largest ball of twine! It held the record until 2015, when surpassed by a longer wall from a deeper survey.

The map shows 3C 273, the first quasar whose distance was measured, as discussed in chapter 16. We show the Subaru galaxy, the most distant galaxy known at the time, and GRB 090423, a gamma-ray burster, the most distant object detected at the time (most likely a supernova). At the very top of the map is the CMB, the most distant thing we can see. I got interested in astronomy when I was 8 years old. At that time there were no Kuiper belt objects known (except Pluto), no exoplanets, no pulsars, no black holes, no quasars, no gamma-ray bursters, and no observation of the CMB. This map shows how much we have progressed in just one astronomical generation.

Now let's talk about the geometry of the universe on large scales. When Einstein completed his equations of general relativity, he wanted to apply them to cosmology. His equations told how energy density and pressure caused space-time to curve. One of the solutions to his equations was flat empty spacetime, but he wanted to find a cosmological solution (i.e., one that would apply to the universe as a whole). The trouble was that his equations would not produce a static solution. Newton conceived of a static universe, with stars arrayed in infinite space, at a more or less constant number density. Each star felt a gravitational force from each of the other stars, but since these forces were pulling it equally in all directions, they canceled out, and the stars would each stay put where they were. This led to a static model, which people believed was a correct description of the universe. They didn't know about galaxies in Newton's day. Such an idea, of forces operating in different directions but canceling each other out, might work if you had a notion of absolute space as Newton did. But in Einstein's theory, if you tried to produce a model that was initially static, the attraction of all the galaxies for one another caused the universe to start to collapse. Yet Einstein too thought the universe was static (remember, this was soon after his development of general relativity in 1915; Hubble's work on the nature of galaxies and the expansion of the universe was a decade or more away). Einstein knew only of stars (in the Milky Way), and they had velocities with respect to our Sun that were small relative to the speed of light—basically static, he thought. To address this problem, Einstein did something very unusual: he added an extra term to his equations! It is called the *cosmological constant,* and it acts to counter the tendency of the universe to contract under gravity.

Today, physicists would say that this is equivalent to Einstein proposing that the vacuum of empty space actually had a small positive energy density. (Georges Lemaître first made this point in 1934.) What do I mean by that? If you were to take all the stuff out of your room—any people, chairs, or atoms in the air filling the room—and you got rid of all the photons and other particles too, you would be left with empty space, a vacuum. We would expect its energy density to be zero. But suppose the vacuum of empty space had a positive energy density. Then, for astronauts traveling in rocket ships at different speeds to each measure the same energy density—for there to be no preferred frame of rest—it must be true that the vacuum must also have a negative pressure, operating equally in each of the three directions of space. This vacuum pressure has to have a negative sign (opposite to that of the energy density). Recall that in the equation $ds^2 = -dt^2 + dx^2 + dy^2 + dz^2$, the term corresponding to the time direction ($-dt^2$) has a sign opposite to that of the terms corresponding to the three dimensions of space. This equation for ds^2 takes the same form for a moving astronaut. It has no preferred standard of rest. In the same way, neither does a vacuum with positive energy density (associated in Einstein's theory with the time dimension) and a negative pressure of the same magnitude operating in the x, y, and z directions. Now if you could put some of this vacuum inside a box, the negative pressure would pull the sides of the box together, tending to make it collapse. But if it were spread uniformly, you wouldn't notice it. In weather, it is pressure differences that cause forces; they enable wind to knock things down. But if the pressure is uniform, you don't notice it. In your room, the air pressure is about 15 pounds per square inch, but you don't notice it. Because it's uniform, it's not pushing you around. Likewise, since the pressure of the vacuum is uniform throughout space, it creates no hydrodynamic forces. However, it does have a gravitational influence.

Energy density is attractive. It pulls things together. In Einstein's equations, pressure gravitates as well as energy density. This is something Newton wouldn't have thought of, but in Einstein's equations, it is the stress–energy tensor $T_{\mu\nu}$ that causes spacetime to curve, and this has pressure terms as well as an energy-density term. In Einstein's theory, therefore, pressure gravitates. Positive pressure attracts, whereas negative pressure is gravitationally repulsive. Since the pressure in the vacuum operates in three directions, the gravitational repulsive effects of the negative pressure outweigh the gravitational attraction of the positive energy density of the vacuum by a factor of 3 to 1, and the overall gravitational effect of the vacuum is repulsive. Today we call this nonzero vacuum energy density (with its accompanying negative pressure) *dark energy*. It is called *dark* because you can't see it, and *energy* because the vacuum has a positive energy. As Neil has emphasized, astronomers like simple names.

To create his 1917 cosmological model, Einstein supposed stars were spread uniformly in space; because they were gravitationally attractive, he balanced this with the gravitational repulsion of the cosmological constant. This would produce a static model—one with a particular geometry. The spacetime diagram of the Einstein static universe looks like the surface of a cylinder (figure 22.4).

In this diagram, we are only showing the dimension of time plus one dimension of space. We are for the

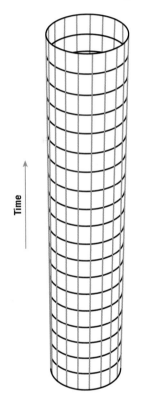

FIGURE 22.4. Einstein static universe. This is a space-time diagram. Time is the vertical dimension, with the future toward the top. We are showing only one dimension of space (around the circumference of the cylinder) and one dimension of time (the vertical direction). Worldlines of stars (or galaxies) in this model are the straight green lines (geodesics) going straight up the cylinder. The circumference of the cylinder is not changing with time—the model is static. The only thing real in this figure is the cylinder itself—the inside and outside have no significance.

Credit: Adapted from J. Richard Gott (*Time Travel in Einstein's Universe*, Houghton Mifflin, 2001)

moment leaving out the other two dimensions of space for purposes of visualization. Time is the vertical coordinate, and the cylinder is vertical. It has a circular cross-section at any given time. The circle represents one spatial dimension. It is Circleland. A Linelander might not live on an infinite line, Lineland, but might instead live on the circumference of a circle: Circleland. How would the Linelander know she lived in Circleland? Well, after she had traveled a distance $2\pi r$ in one direction, she would find herself back where she started. This is a closed cosmological model where the universe closes on itself to form a circle. The worldlines of the stars (or galaxies) are straight green lines going vertically up the cylinder. These are geodesic paths, as straight as possible. You could drive a little truck straight up the cylinder without turning its steering wheel. The galaxy worldlines are parallel. The galaxies are not getting closer together or farther apart with time. The circumference of the universe is not changing in time. This is Circleland, where the radius of the circle is not changing with time. All these properties confirm that it's a static model. The gravitational attraction of the galaxies is exactly balanced by the overall gravitationally repulsive effects of the cosmological constant (which we would now call "dark energy").

Now let us discuss the two extra spatial dimensions we have left out of the diagram. Actually the geometry of this universe is not a circle or a sphere but what we call a *3-sphere*. What is a 3-sphere? A circle is the set of points at a distance r from a central point on a Euclidean plane. A sphere is the set of points at a distance r from a central point in a three-dimensional Euclidean space. The sphere itself is a two-dimensional surface. A Flatlander might live on the surface of a sphere. He would discover that if he went straight in any direction, he would find himself back where he started after traveling a distance of $2\pi r$. He also could discover he was an inhabitant of Sphereland by drawing a triangle with three right angles in it, connecting the north pole of the sphere with two points 90° apart on its equator (as shown in figure 19.1) . This is not Euclidean plane geometry. Any cross-section of a sphere is a circle. (Interestingly, Mark Alpert and I proved that if Einstein had lived in Flatland, where point masses do not gravitationally attract one another, he could have devised a Sphereland static universe without having to introduce a cosmological constant. But Einstein did not live in Flatland—he had to use a sphere of one higher dimension!) The circle and the sphere, as we typically know them, could be called a 1-sphere and a 2-sphere, respectively. The 3-sphere is simply a one-dimension-higher version of a sphere: it is the set of points at a distance r from a central point in a four-dimensional Euclidean space. Distances between points in this four-dimensional Euclidean space are measured by $ds^2 = dx^2 + dy^2 + dz^2 + dw^2$. (There is no dimension of time here.) We have added a term for w, which is an extra spacelike dimension. The 3-sphere is the set of points where $r^2 = x^2 + y^2 + z^2 + w^2$.

Just as a circle is a curved, one-dimensional closed line, and a sphere is a curved two-dimensional surface, the 3-sphere is a curved three-dimensional volume. A circle has a finite circumferential length ($2\pi r$), the sphere has a finite surface area ($4\pi r^2$), and the 3-sphere has a finite surface volume ($2\pi^2 r^3$). If you live in a 3-sphere universe, and you set off going north, always flying straight ahead, you will eventually come back to where you started after traveling a distance of $2\pi r$. You will arrive back home from the south after having circled the universe. If you set off going east, always flying straight ahead, you will come back to your home planet from the west after traveling a distance of $2\pi r$ and circumnavigating the universe. But also if you leave home traveling straight up, you will return to your home planet from below after traveling a distance of $2\pi r$. This is a three-dimensional universe that, like ours, has three pairs of directions—north-south, east-west, and up-down—but no matter which way you start off, you will come back to where you started. An intrepid traveler in Einstein's 3-sphere universe could explore distant galaxies and be sure to return to her home galaxy, as long as she kept going straight on a geodesic route in any direction. She would always, boomerang-like, return home. The space is bounded but has no edges, or boundaries, to stop her travels.

A 3-sphere universe is closed with a finite volume and a finite number of galaxies. For example, if galaxies had a mean separation of 24 million light-years, the mean volume per galaxy would be (24 million light-years)^3. If the radius of curvature of the static 3-sphere universe were 2,400 million light-years, then the volume of the 3-sphere universe would be $2\pi^2$ (2,400 million light-years)^3. Now (2,400 million)^3 /(24 million)^3 is 100^3 or a million. That would mean that this universe would have $2\pi^2$ million galaxies; that is, about 20 million galaxies. If you lived in an Einstein static universe, you would find that galaxies were not moving apart and that there were a finite number of them. Astronomers living in such a universe could identify and count them all.

In a 3-sphere universe, there are no special observers; each galaxy location is similar to every other one, just as there are no special points on the surface of a sphere. On Earth, all observers can think of themselves as at the center (i.e., sitting on top of the sphere). To us on Earth, it seems as if we are standing on top of the world right now. We are standing straight up, so everyone else must be hanging off the sides! People in Australia must be hanging upside down! But anyone can think he or she is at the center. In Beijing, there is a circular platform that was supposed to represent the center of the world. In England, they put the 0° longitude line—the "prime meridian"—right through Greenwich, a suburb of London, where they had an observatory. All of us can think we are at the center, because all points are equivalent. Importantly, if you lived in a 3-sphere universe and you counted galaxies, you would find an equal number in all directions. The counts would be *isotropic*, that is, independent of direction—just as Hubble found.

Einstein published his static cosmology in 1917. The cosmological constant term he added to his equations provided an extra curvature to empty space, but it was very small and therefore did not interfere with any of his solar system tests of general relativity. Furthermore, adding this term did not alter the fact that the equations would preserve local energy conservation! Einstein was probably the only person at that time clever enough to figure out such a fix to produce a static universe.

Meanwhile in Russia, in 1922, Alexander Friedmann found a cosmological solution to Einstein's original field equations (without the cosmological constant). Friedmann's solution just had ordinary stars (or galaxies in it). It was a dynamical solution (not static), making it harder to solve. In his model, the geometry of the universe was a 3-sphere just as Einstein had proposed, but now the radius was allowed to change with time. He found a solution (see figure 22.5) whose spacetime diagram looked like a vertical football (teed up, ready for kickoff).

Time goes vertically in this diagram, with the future toward the top. We are showing one dimension of time and one dimension of space in this diagram. The dimension of space is shown as a circular cross-section (Circleland) whose radius changes with time. The 3-sphere universe starts with zero radius at the Big Bang (at the bottom). Then it expands to larger circumference with time until it reaches a maximum circumference in the middle of the football, and then begins to shrink, finally collapsing to zero radius at a "Big Crunch" at the end. Galaxy worldlines are green geodesic lines going along the seams in the football starting at the Big Bang and ending at the Big Crunch. These worldlines are as straight as possible. You could drive a little truck along them and not have to turn your steering wheel. This shows Einstein's equations working at their best. The mass of the galaxies is causing the spacetime to be curved, and the curvature of the spacetime causes the worldlines of the galaxies—the seams—to bend. The seams spread apart from the bottom, but the curvature of the surface of the football draws them back together at the Big Crunch. At

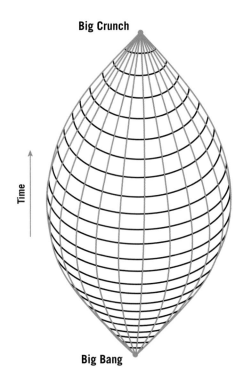

Big Crunch

Time

Big Bang

the Big Bang, the galaxies are all flying apart from each other at the beginning. But gravitational attraction (curvature) slows their expansion to a momentary halt in the middle, at the football's equator, and eventually, in the upper half of the football, it causes the galaxies to start moving toward one another. The distances between galaxies start decreasing as the circumference of the universe begins to shrink. They all crash together at the Big Crunch. You don't want to be around then! As the volume of the universe shrinks to zero, you will be crushed. You will hit a Big Crunch singularity where the curvature becomes infinite—like the singularity in a black hole.

FIGURE 22.5. Friedmann Big Bang universe. This spacetime diagram also shows only one dimension of space (the circumference of the football shape) and one dimension of time (vertical). Worldlines of galaxies are the vertical green seams in the football. They are geodesics—the straightest lines you can draw on the surface. The mass of the galaxies causes the curved shape, and the worldlines follow geodesics in the curved surface. The universe is dynamic, with a Big Bang at the beginning. The galaxies move apart at first as the circumference of the universe gets larger with time. This is an expanding universe. But eventually the gravitational attraction of the galaxies causes the universe to start contracting, and it ends with a Big Crunch at the end. The only thing real in this picture is the "pigskin" itself—the inside and outside of the football have no significance.

Credit: Adapted from J. Richard Gott (*Time Travel in Einstein's Universe*, Houghton Mifflin, 2001)

I should emphasize that the only thing that is real here is the "pigskin" itself. The inside of the football is not real, and the outside of the football is not real. We are only plotting the football in a higher dimensional space so that we can visualize it.

Time begins at the Big Bang—a singularity of infinite curvature is there. We started discussing the Big Bang in chapter 14. What happened before the

Big Bang? This question makes no sense in the context of general relativity, because time and space were created at the Big Bang. It's like asking what is south of the South Pole. If you go farther and farther south, you will eventually get to the South Pole. But you can't go farther south than the South Pole. Likewise, if you go farther and farther back in time, you will eventually get to the Big Bang. That's when time and space were created, so that's the earliest you can go. Aristotle liked a universe that was infinitely old, because you would not have to ask how it got started; if it had a beginning, a first cause, then you would have to explain what caused the first cause, he worried. Einstein and Newton liked infinitely old universes also. But Friedmann's universe started with a Big Bang at a finite time in the past, when both space and time were created.

Although Friedmann published these solutions in 1922, almost nobody paid any attention to them. Einstein thought they were an interesting mathematical solution to his field equations, but he thought his static model was what actually applied to the universe. Then, as we saw in chapter 14, Hubble discovered the expansion of the universe in 1929. Friedmann's model had predicted that the universe should be either expanding or contracting. Now Hubble had found that the galaxy worldlines were indeed moving apart. Where would that put us in Friedmann's model? We'd be in the lower half of the vertical football, during the expansion phase when the galaxy worldlines were diverging. With further data in 1931, Hubble and Humason found distant galaxies receding from us at up to 20,000 km/sec, cementing the result that the universe was expanding.

After hearing of Hubble's results in 1931, Einstein told George Gamow that the cosmological constant was the biggest blunder of his life. Why? No one had paid any attention to Friedmann's paper. But suppose Einstein had not thought of the cosmological constant; he would have had to abandon a static model and might have discovered Friedmann's model himself. If Einstein had published the same model Friedmann published, the whole world would have listened. Einstein could have been the one to predict in advance that the whole universe must not be static, but rather it must be either expanding or contracting. Then, when Hubble discovered the expansion of the universe, it would have been a great further confirmation of Einstein's theory of general relativity. It would have been Einstein's greatest triumph. Remember, no one had talked about anything like an expanding universe before. People would have asked: Expanding into *what*? But in Einstein's theory, curved space itself could be expanding. It's not expanding into anything (no inside of the football, no outside of the football—just the football itself); it's just stretching. It's the space connecting all the galaxies that is just getting bigger. Amazing. Appreciating all this, Einstein declared the cosmological constant his biggest blunder. Later, in chapter 23, we will point out that if Einstein were around today, he might have reason to revise that assessment.

Friedmann's model was not the only model using only normal matter that you could imagine (i.e., having none of that negative pressure, dark energy stuff). What is the most general model of this type you could construct? To us the universe looks isotropic (the same in all directions). Hubble observed equal counts of galaxies in all directions, and he observed galaxies fleeing from us equally in all directions. Now, as Michael discussed in chapter 14, you might think this meant that we were at the very center of a great explosion. If you were off to one side, you might expect to find more galaxies in the direction of the center than in the opposite direction. But if you were in the very center, you would expect to see equal numbers of galaxies in different directions. But after Copernicus, we were not going to believe that. No, we could not be the one special galaxy at the center when all the others were off center. If you apply the Copernican Principle that our position in the universe is not likely to be special, then the universe must look isotropic to an observer on any galaxy (otherwise we would be special). From a galaxy way over there, the universe must look isotropic as well. If all observers see a universe that is the same in all directions, then the universe must be homogeneous.

If the density of galaxies in one region were greater than in another region, an observer next to this region would see more galaxies in the direction toward the density enhancement than in the opposite direction, and his results would not be isotropic. Of course, on small scales, we do see galaxy clusters, but on large scales, we see equal counts of galaxies in different directions. Therefore, it is on large scales that the universe must be isotropic and homogeneous. The only homogeneous isotropic models in general relativity are ones having *uniform curvature*. If the curvature were greater in one region than in another, at a given epoch, it would not look the same in all directions to every observer. In an isotropic model, there are no special directions for this curvature, so the curvature must be the same in all directions and of a constant value. The 3-sphere Friedmann universe is one such solution; it has uniform positive curvature. It has positive curvature like a sphere (a 2-sphere), and the 3-sphere universe likewise has no special points or special directions.

Carl Friedrich Gauss defined the curvature of a two-dimensional surface as $1/r_1r_2$, where r_1 and r_2 are the principal radii of curvature. A sphere has a Gaussian curvature of $1/r_0^2$, where r_0 is the radius of the sphere. Both radii of curvature have the same sign, because if you are sitting on the top of the sphere, for example, geodesics going left to right as well as front to back both curve downward. Two negatives (downward bending) multiplied together give a positive, so their product r_1r_2 is positive, and $1/r_1r_2$ is positive. Thus the curvature of a spherical surface is always positive.

But there are two additional possibilities (zero curvature or negative curvature). First, the universe could have a geometry at a given epoch that was of zero curvature, or flat, like an infinite flat plane. (When we refer to such a universe as "flat," we mean "not curved" rather than 2-dimensional like Flatland. This is an infinite three-dimensional universe that obeys the laws of Euclidean solid geometry.) This universe is infinite in extent and has an infinite number of galaxies (and no center, as discussed in chapter 14).

In the third case, the curvature is *negative*. The geometry at a given epoch is negatively curved like an infinitely large Western saddle. A Western saddle curves downward, left to right, to accommodate your legs, but curves upward, front to back, to accommodate the neck and back of the horse. Thus the curvature in the two directions is opposite, and since a positive times a negative is a negative, the curvature $1/r_1r_2$ is negative. If you draw a circle on a Western saddle, its circumference is larger than $2\pi r$, as opposed to a sphere where, as we have discussed, the circumference would be less than $2\pi r$. On a Western saddle, if you go out a distance r from your location, you will go way up and way down as you trace the circumference, so the circumference is larger than the $2\pi r$ value you expect from a plane.

A negatively curved surface makes for an infinite universe that also has an infinite number of galaxies. The negatively curved case is a *hyperbolic universe* depicted in figure 22.6 as a bowl-shaped surface living in the ordinary flat spacetime of special relativity. In the figure, time is vertical with the future toward the top. We also show two dimensions of space, indicated by the horizontal red arrows.

If you were to start at the center of the bottom of the bowl and take a tape measure out to the circumference of the top of the bowl, you would find that the length of the radius drawn along the surface is unexpectedly short relative to the circumference. That's because in addition to moving out in space, your

FIGURE 22.6. Hyperbolic negatively curved space (blue) in ordinary spacetime. Time is vertical, with the future toward the top. We also show two spacelike dimensions—horizontal axes. *Credit:* Adapted from Lars H. Rohwedder

measuring tape is also moving up in time as it hugs the surface of the bowl. The distance measured is shortened because of that $-dt^2$ term, which subtracts from the value of ds^2 you will get as you measure with the measuring tape. If the radius of a circle constructed on the bowl is short relative to its circumference, that means the circumference is large relative to the radius—a hallmark of negative curvature. (The Western saddle is an analogy that captures the large circumference-to-radius ratio, but has special directions, front-back, left-right, which the hyperbolic universe does not possess: it is the same in all directions.) This hyperbolic surface stretches to infinity, has an infinite volume, and contains an infinite number of galaxies. Friedmann investigated this type of model in 1924; he found that it started with a Big Bang and expanded forever. Later Howard Robertson investigated the flat, or zero curvature, case and found that it also started with a Big Bang and expanded forever.

Let's summarize these results (table 22.1). In a positively curved universe, the sum of angles in a triangle drawn at a particular epoch is more than 180°, just like on a sphere. In a flat, or zero-curvature, universe, the sum of angles in a triangle at a given epoch is equal to 180°. In a negatively curved universe, the sum of angles in a triangle at a given epoch is less than 180°. The positively curved Friedmann universe is finite in space and finite in time. It curves back on itself in space to make a closed surface and closes off in time also—with a Big Crunch at the end. The flat and negatively curved Friedmann universes

TABLE 22.1. CHARACTERISTICS OF FRIEDMANN-TYPE BIG BANG MODELS

MODEL	3-SPHERE	FLAT	HYPERBOLIC
Curvature	Positive	Zero	Negative
Circumference of a circle	$< 2\pi r$	$= 2\pi r$	$> 2\pi r$
Sum of angles in a triangle	$> 180°$	$= 180°$	$< 180°$
Number of galaxies	Finite	Infinite	Infinite
Starts with	Big Bang	Big Bang	Big Bang
Future	Finite	Infinite	Infinite
Expansion history	Expands, then collapses, ending in Big Crunch	Expands forever	Expands forever

are infinite in space, containing an infinite number of galaxies, and are infinite in time as well, expanding forever into the future.

After Penzias and Wilson's discovery of the microwave background radiation in 1965, the search was on to find out which of these models best describe our universe. Current data from the WMAP and Planck satellites favor a zero-curvature model to an accuracy of better than 1%. But we have found that the dynamics of the universe are more complicated than Friedmann envisioned. After Hubble's observations confirmed the expanding universe, which Friedmann's models had predicted, a few mysteries remained. Was there really nothing before the Big Bang? What started the Big Bang? And how did that microwave background radiation get to be as uniform as we observe? Answering these questions would cause us to reexamine the very early history of the universe.

23

INFLATION AND RECENT DEVELOPMENTS IN COSMOLOGY

J. RICHARD GOTT

This chapter explores the very early universe—going back as far as the Big Bang and even before. As we have discussed, in 1948 George Gamow wondered what the universe would be like at its very earliest moments. Gamow reasoned that the universe would be compressed near the Big Bang and would be very hot and filled with hot thermal radiation. This radiation cools off as the universe expands.

We can explain this by thinking about the 3-sphere Friedmann universe. At each epoch, it has a finite circumference, and as this 3-sphere universe expands, its circumference increases. Imagine photons circling this circumference, like race cars going around a circular race track. The circumference of the track is getting bigger with time as the cars are continually chasing each other around the track. Suppose 12 photons are equally spaced around the circular track like the 12 numerals on a clock face. As the track expands, the cars are all racing at the same speed, the speed of light. If they start equally spaced around the track, with 1/12 of the track separating each car from the one in front of it, they will remain equally spaced around the track as it expands. Each car is equally good, so a car will not be catching up with the car in front or falling behind and running into the car behind. If the cars remain equally spaced around the track, as the circumference of the track gets larger, the distance separating the cars will increase. If the track doubles in size, the distances between the cars will double. Now imagine an electromagnetic wave circling the circumference going clockwise. Each of the 12 photons could be placed at one of the crests of the waves. The photons and the crests of the waves both travel at the speed of light, so the photons stay at the wave crests as the wave moves. Thus, as the track circumference expands, the distance between the wave crests expands by the same factor. When the circumference of the universe doubles, the wavelength (distance between the crests) doubles as well.

This explains why light will be redshifted as the universe expands: because of the stretching of space. This redshifting means that the hot thermal radiation in the early universe will cool off (become of longer wavelength) as the universe expands. Calculating the nuclear reactions occurring in the first 3 minutes and matching with the deuterium abundance we find today allowed Gamow's students Robert Herman and Ralph Alpher to calculate the temperature that the radiation would have today by estimating how much the universe would have expanded since those early times. They got a current temperature of 5 K. In the 1960s, as we saw in chapter 15, Robert Dicke at Princeton thought of the same argument, came to similar conclusions, and decided to look for the radiation. Penzias and Wilson beat Dicke's team to it.

When the Cosmic Background Explorer (COBE) satellite was launched in 1989 to measure the cosmic microwave background (CMB) in detail, it found a nearly perfect blackbody shape for its spectrum (just as Gamow would have predicted) with a temperature of 2.725 K. George Smoot and John Mather received the 2006 Nobel Prize in Physics for their work on COBE.

Gamow and Alpher's prediction of the existence of the CMB and Alpher and Herman's estimate of its temperature as 5 K together constitute one of the most remarkable predictions in the history of science to be subsequently verified. It was rather like predicting that a flying saucer 50 feet across would land on the White House Lawn and later having one 27 feet across actually show up! This is also an important vindication of the Copernican Principle, the idea that our location must not be special; with Hubble's observations of isotropy, the Copernican Principle leads us directly to the homogeneous, isotropic, Friedmann Big Bang solutions of Einstein's field equations, which Gamow and his colleagues used to predict the microwave background.

The resulting Friedmann Big Bang model has been incredibly successful, but some important questions remained. This universe has a beginning, a Big Bang, but what happened before the Big Bang? The standard answer (which we gave in chapter 22) has been that time, as well as space, was created in the Big Bang, so there was no time before the Big Bang. Still people wondered, why was the Big Bang so uniform? When we look out in different directions, the temperature of the CMB is uniform to one part in 100,000. How do these different regions "know" to be at the same temperature? When we look in one direction, we see out 13.8 billion light-years. But we are looking back in time to an epoch when the universe was only 380,000 years old. In the standard Big Bang model, that region should be influenced only by stuff no more than 380,000 light-years away from it. But if we look out 13.8 billion light-years in the opposite direction, 180° across the sky, we see another region that is at essentially the same temperature. In the standard Big Bang model, these two regions on opposite sides of the sky at an epoch 380,000 years after the Big Bang (when we are seeing them) are separated by a distance of 86 million light-years and have not had time to communicate with each other in the scant 380,000 years since their birth. Usually, if we see two regions at the same temperature, it is because they have had time to communicate with each other and reach thermal equilibrium. But in the standard Big Bang model, widely separated parts of the CMB observable in the sky have not had time to be in causal contact with each other. In the Friedmann model, different regions of the universe must have miraculously started out with a homogeneous expansion at the same temperature everywhere. How could it be so uniform?

But COBE also detected small fluctuations of one part in 100,000 from one region of the sky to another. If the universe had been perfectly uniform, no density enhancements would have been present to grow into galaxies and clusters of galaxies later. Our existence depends on the universe having small fluctuations initially, which can eventually grow by the action of gravity into the galaxies we observe today. The universe had to be almost perfectly uniform but not quite. It was a mystery. I am reminded of the old Depression Era saying: "If we had some bacon, we could have bacon and eggs for breakfast, if we had some eggs!" We needed to explain first the overall uniformity and then the small fluctuations.

In 1981, Alan Guth proposed a solution to this problem. His model proposed that the universe started with a short period of accelerated expansion he called *inflation*. In a spacetime diagram, this looks like a small trumpet bell pointing upward like a golf tee, to hold up the Friedmann football spacetime. It starts with a finite circumference near the mouthpiece of the trumpet but becomes dramatically larger as we move upward in time to the bell-like opening of the horn. The bottom tip of the Friedmann football is replaced by a little trumpet mouth with finite circumference at the bottom—perhaps as small as 3×10^{-27} centimeters (figure 23.1). The trumpet epoch lasts a little longer than the Big Bang tip of the football would alone, and this extra time allows the different regions we see today enough time to get into causal contact. In the beginning, the circumference is so small that the different regions, benefiting from that little bit of extra time, come into casual contact, and then the accel-

FIGURE 23.1. Inflationary beginning (trumpet) to start a Friedmann Big Bang universe (football).
Photo credit: J. Richard Gott

erated expansion during the trumpet epoch pulls them far apart; it only looks as though they have had insufficient time to be in communication.

What was Guth's basis for this model? He thought that in the early universe there may have been a vacuum state having a high energy density—and therefore a high negative pressure—mimicking the curvature of empty space implied by Einstein's famous cosmological constant. But Guth wanted a very high value for this cosmological constant. We are used to thinking that empty space should have a zero density. It has, after all, been cleared of all particles and radiation. But the vacuum of empty space may have an energy density due to fields like the Higgs field filling the universe. The amount of vacuum energy present depends on the laws of physics. Guth argued that in the early universe, the weak and strong and electromagnetic forces would have been united in a single superforce, and the vacuum energy at that time (when the laws of physics were different) might have been much higher than the tiny value seen today. Thus the cosmological constant was not really a constant (as Einstein had supposed) but could change with time. In the very early universe, the vacuum energy density could have been quite high. Accompanying this high energy density was a large negative pressure, ensuring, by the laws of special relativity, that the vacuum energy as seen by different observers traveling at different velocities through space would all be the same. As we have discussed, the vacuum energy density produces an attraction, but the negative pressure operating in three directions produces a gravitational repulsion three times larger. This, according to Einstein's equations, would have started the universe off on the accelerated expansion Guth wanted. It was this gravitational repulsion that produced the initial expansion we call the "Big Bang."

In fact this trumpet-like solution to Einstein's field equations had already been found by Willem de Sitter in 1917. He solved Einstein's equations for the case of empty space with a cosmological constant and nothing else. With no ordinary matter to balance the repulsive effects of the cosmological constant, this solution produced a universe whose expansion was accelerated. The whole solution is called *de Sitter space*. This spacetime is a 3-sphere universe that starts off with infinite radius in the infinite past. It is contracting at nearly the speed of light. But the repulsive effects of the cosmological constant begin to slow the contraction until it stops at a minimum radius—a waist of minimum circumference—and then begins to expand. It expands faster and faster as the repulsive effects of the cosmological constant continue. It ends up expanding at a rate closer and closer to the speed of light, and in the infinite future this universe expands to infinite size. The spacetime diagram of de Sitter spacetime looks like a corset with a narrow waist (figure 23.2). The diagram shows one dimension of space around the horizontal circumference and the dimension of time vertically. The future is toward the top. The skirt at the bottom shows the contracting phase, and the waist at the middle shows the minimum radius

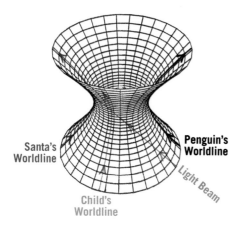

Santa's Worldline

Penguin's Worldline

Light Beam

Child's Worldline

FIGURE 23.2. A spacetime diagram of de Sitter space. As in figures 22.4 and 22.5, this figure shows one dimension of space and one of time. *Credit:* J. Richard Gott

of the universe. Then it fans out at the top to make a trumpet horn.

As with the Friedmann spacetime model, the only thing to pay attention to here is the corset-shaped surface itself. Forget the inside and outside. The surface alone is real. This corset-shaped spacetime has horizontal circular cross-sections at individual time slices. These show the circumference of the 3-sphere universe at particular instants of cosmic time. These circles are large at the bottom, reach a minimum at the waist, and then get larger again at the top, showing the size of the 3-sphere universe as it contracts and then expands. The vertical "corset stays" represent possible worldlines of particles. These are straight geodesic lines that little trucks could follow by traveling straight ahead on the corset's surface. The corset stays are coming together in the bottom half, reach a minimum separation at the waist, before spreading apart in the upper half. In the upper half, the curvature of the spacetime is causing these particles to accelerate away from each other. As the particles fly apart, their clocks begin to slow down exponentially as they approach the speed of light. Their clocks are s_l__o___w___i___n____g____d_____ o_____w_____n. During later ticks on these clocks, the circumference expands enormously. Although the diagram shows the space to be expanding approximately linearly at nearly the speed of light at late times (a cone opening out at nearly 45°), as measured by the exponentially slowing clocks carried by the particles themselves, the circumference seems to be doubling in each successive time interval, increasing as 1, 2, 4, 8, 16, 32, 64, 128, 256, 512, 1,024, . . . , resulting in an exponentially accelerated expansion. It's like currency inflation, which is why Guth called the model *inflation*.

Look at the waist. It is a circle representing the 3-sphere universe at its point of maximum contraction. Keep in mind that it is really a 3-sphere. We can designate the point on the far left of this circle as a "north pole" of this universe. Santa would live there. Consider the red corset stay at the left: it is the worldline of Santa sitting at the north pole of the 3-sphere universe. The corset stay at the right, 180° away, is the black worldline of a penguin at the south pole. Santa, whose worldline is at the north pole, will never see the penguin living at the south pole. A light beam starting from the penguin in the infinite past will travel straight at 45° upward and to the left. It will pass diagonally upward across the front of the corset like a diagonal sash, but it will never quite reach Santa's worldline at the left. There are event horizons in this universe. Santa never sees anything that happens to the penguin—he never sees anything upward and to the right of that sash. Consider a child living near Santa, whose green worldline is shown in figure 23.2. Light beams from that child can reach Santa. Santa will see the child at late times accelerating away from him. Light from the child will become more and more redshifted. If the child sends Santa a message that says "THINGS ARE GOING FINE," Santa will receive "THINGS A_R__E." But Santa will never receive the signals "GOING FINE." The signal "GOING" travels along that 45° sash and never arrives. It looks to Santa just as though the child were falling down a black hole. When the child's worldline crosses that 45° sash, which is Santa's event horizon, the child's signals no longer arrive. The space between Santa and the child is simply stretching so fast that the signal "FINE," emitted on the other side of the sash, cannot traverse the ever-widening distance between Santa and the child. This does not violate special relativity. The latter just says someone else's spaceship cannot pass you at a speed faster than the speed of light. But general relativity still allows the space between two particles to stretch so fast that light cannot cover the ever-widening gap between them. De Sitter spacetime explains how particles can get into communication and thermal equilibrium near the waist and then be spread apart to great distances.

Guth was ultimately proposing to start the de Sitter universe at the waist, with a small circumference that today we might estimate to be perhaps as small as approximately 3×10^{-27} centimeters. He was eliminating the infinite contraction phase (the lower half of the full spacetime). He just needed a little bit of high-density vacuum state at the beginning. The repulsive effects of the large negative pressure would cause the spacetime to start expanding, and then to expand faster and faster, with the size of the universe doubling every 10^{-38} seconds. The universe would become very large. As the universe expanded, the energy density of the vacuum state would stay the same. The cosmological constant would stay constant. A small region of high energy density would expand to become a large region having the same high energy density.

Curiously, this would not violate local energy conservation. If I had a box of high-density, negative-pressure fluid, as I expanded the walls of the box, I would have to do work to pull the walls apart against the negative pressure that was resisting expansion. The work I was doing pulling the walls apart against this negative pressure (or suction) would add energy to the fluid—just enough to keep its energy density at the same high level as the volume of the box expanded. Thus energy would be conserved locally. But in the universe, what is pulling on the walls of my box? It is just the negative pressure from the other little similar boxes of spacetime next to it. As long as the pressure is uniform throughout the universe, the expansion itself is doing the work.

In general relativistic cosmology, there is no global energy conservation, because there is nowhere flat (approximating the spacetime of special relativity) on which to stand and establish an energy standard. Thus the total energy content of the universe can go up with time if there is negative pressure. This enabled Guth to start off his inflation model with a little piece of high-density vacuum and then let it grow naturally into a large universe with a vacuum state of the same density. In this way the vacuum state was "self-reproducing," growing exponentially large from a tiny beginning. Because of this, Guth said the universe "is the ultimate free lunch." Eventually, the vacuum state would decay, as the strong and weak and electromagnetic forces decoupled. As the energy density in the vacuum of empty space dropped to a low value, its vacuum energy would be dumped into the form of elementary particles. The universe would fill with a thermal distribution of elementary particles.

This is where the inflationary trumpet at the beginning of the universe joins to the bottom of the football-shaped Friedmann Big Bang model. The expansion of the universe then starts to decelerate, as in the football model. The pressure is now just the ordinary thermal pressure of particles, which is positive. Worldlines that have said "goodbye" to each other during the accelerating inflationary trumpet phase (like Santa and the child), will say "hello again" after the decelerating Friedmann phase begins. Inflation showed how the initial conditions of the Friedman Big Bang model could be naturally produced. The repulsive gravitational effects of the initial vacuum state (through its negative pressure) started the Big Bang! The Big Bang did not have to start with a singularity, but instead could start with a small, high-density vacuum region. Inflation could explain why the universe was so large, and why it was so uniform. Any wrinkles would be flattened out as the universe stretched to enormous size. It could also explain the small fluctuations of 1 part in 100,000 that we observe. These are small random quantum fluctuations due to Heisenberg's uncertainty principle. The universe was doubling in size every 10^{-38} seconds in the beginning; on these short timescales, the uncertainty principle ensures random fluctuations in the energy of any field. In fact, the spongelike pattern of galaxy clustering that we see in the universe today—the *cosmic web*, as well as the pattern of hot and cold spots in the CMB, indicate that the initial conditions appear to have been random in precisely the way expected from the random quantum fluctuations predicted by inflation (see my book, *The Cosmic Web* [2016]).

Inflation had one problem, however, which Guth recognized. The high-density vacuum state at the beginning would not be expected to decay into elementary particles all at once. This high-density inflating sea would decay into bubbles of low-density vacuum, a phenomenon investigated by Sidney

Coleman. It is like boiling water in a pot. The water does not turn to steam all at once. Bubbles of steam form in the water. But this was not a uniform distribution—not the uniform universe we were hoping for. So Guth mentioned that as a problem. In 1982, I proposed that inflation would produce bubble universes—each bubble would expand to make a separate universe like ours (figure 23.3).

In my theoretical model, we live inside one of the low-density bubbles. I noticed that if, after the bubble formed, it took a while for the vacuum energy to decay, it would decay on a hyperbolic surface, making a uniform, negatively curved, hyperbolic Friedmann cosmology (recall figure 22.6). From inside the bubble, we are looking out in space and back in time, so we just see our own bubble universe and the uniform inflating sea before it was created. Everything looks uniform to us—solving Guth's nonuniformity problem. The bubble expands at nearly the speed of light. But the inflating sea expands so fast that the bubbles never percolate to fill the entire space. New bubble universes are continually forming, and the inflating sea is expanding between them to provide space for even more new bubble universes to form. I envisioned an infinite number of bubble universes forming in an ever-expanding inflating sea—what we now call a *multiverse*.[1] These bubble universes would have negative curvature inside and would expand forever—they would be hyperbolic Friedmann universes. Surfaces of constant epoch would be hyperbolas nestled inside the expanding bubble. A *surface of constant epoch* is one for which alarm clocks on individual particles all go off showing a constant time since the bubble formation event. Its shape is hyperbolic, because particles going faster have clocks that tick more slowly and therefore, the point at which their alarm clocks go off is delayed (compare with figure 22.6). This produces a hyperbolic shape that is infinite in extent as it bends upward inside the expanding bubble wall (see figure 23.3). Eventually, as the bubble expands to infinite volume in the infinite future, an infinite number of galaxies would be produced. Thus, an infinite number of infinite bubble universes could be produced from an initially very small, high-density de Sitter space.

This seems odd. How can one get an infinite number of universes each ultimately infinite in size, from just a finite beginning? De Sitter spacetime looks like a trumpet with its mouth opening upward. A horizontal slice through the waist of de Sitter space at the mouth of the trumpet is a circle. This is a small 3-sphere universe of finite circumference and finite volume, like the one

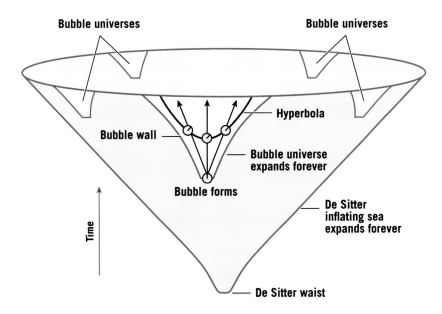

FIGURE 23.3. Bubble universes forming in an inflating sea—A multiverse.
Credit: Adapted from J. Richard Gott (*Time Travel in Einstein's Universe*, Houghton Mifflin, 2001)

Einstein considered. But the top of the trumpet resembles a cone, and one can slice a cone in a circle, a parabola, or a hyperbola, depending on how you slice it. If you slice de Sitter spacetime horizontally, you get a circle—that's a 3-sphere universe. If you slice it on a 45° slant, you get a parabola and an infinite flat universe. If you slice it with a vertical plane, you will get a hyperbola—that makes an infinite negatively curved universe. It is like the old fable about the blind men and the elephant. One man feels the trunk and says the elephant is shaped like a snake. Another feels the leg and says that the elephant is shaped like a tree trunk. Still another feels the side of the elephant and says it is like a wall. In the same way, the shape of de Sitter space depends on how you slice it. Make one hyperbolic slice inside a bubble universe that extends to infinity, and you have a slice of space that is infinite, marking the epoch where the inflating de Sitter vacuum ends and dumps its energy into particles, and the Friedmann model begins. Just like a loaf of bread, which can be cut into different slices to make American slices or French slices, the real thing is the loaf itself. If we look at the spacetime geometry of de Sitter space for the inflationary model, we can see that it starts as a finite 3-sphere universe at the waist and expands forever, becoming infinitely large. This remarkable spacetime geometry, in which inflation continues forever and the space becomes infinitely big, allows creation of an infinite number of infinite bubble universes in an ever-inflating sea.

Different bubble universes could have different laws of physics in them, if different bubbles corresponded to tunneling and rolling down into different valleys in the landscape, where the values of various fields could be different. The laws of physics we see in our universe could be only local bylaws, as emphasized by Andrei Linde and Martin Rees.

It is important for the de Sitter inflationary universe to begin at the waist. We do not want the infinite contraction phase that precedes it. Borde and Vilenkin showed why: bubbles would form in the contracting phase too, and there the bubbles would be expanding in a space that was contracting; the low-density bubbles would collide with one another and fill up the space, ending the inflating sea and preventing it from ever reaching the waist and the expansion phase. You would just get a Big Crunch singularity; the bubbles would not have negative pressure inside to cause a turnaround at the waist. So, Borde and Vilenkin concluded that the inflationary multiverse starts off as a finite piece of inflating sea at the beginning. This could be small, as small as 3×10^{-27} centimeters. It isn't nothing, but it is perhaps as close to nothing as you could get.

The vacuum energy density can be viewed as the altitude in a landscape. The altitude represents the vacuum energy density, the energy density of empty space. Different places in the countryside correspond to different values of the fields (such as the Higgs field) that are creating the vacuum energy. Different locations (different values of the fields) correspond to different altitudes (different values of vacuum energy density). Today we have a very low vacuum-energy density—we are near sea level. But in the early universe the vacuum energy density would be high, like being trapped in a high mountain valley (figure 23.4).

A ball trapped in a high mountain valley is ultimately unstable: it has a lower energy state it can go to—sea level. But it can be trapped if it is surrounded by mountains on all sides. In Newton's universe, it would have no way to roll down, but the quantum mechanical process called *quantum tunneling* allows it to tunnel through a surrounding mountain and then roll down to sea level.[2]

Quantum tunneling is a process discovered by George Gamow. It explained the radioactive decay of uranium. Uranium nuclei decay by emitting an alpha particle (a helium nucleus with two protons and two neutrons). The alpha particle is trapped inside the nucleus by the strong nuclear force attracting it to other protons and neutrons. This strong force acts like the mountain range surrounding the mountain valley, trapping the alpha particle inside the nucleus.

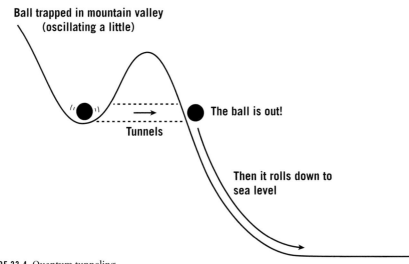

FIGURE 23.4. Quantum tunneling.

Credit: Adapted from J. Richard Gott (*Time Travel in Einstein's Universe*, Houghton Mifflin, 2001)

But the strong nuclear force is a short-range force; if the alpha particle could somehow get out of the nucleus, beyond the influence of the strong-force attraction, it could escape. The alpha particle is positively charged and would then be repelled by the positively charged main nucleus. It would roll down the hill, away from the nucleus, and the kinetic energy it picked up would be due to electrostatic repulsion. From the energy an emitted alpha particle is measured to have when uranium decays, scientists could calculate how high up the hill it was when it began. It turned out that it was emitted *outside* the uranium nucleus! How did it get out there? Quantum mechanics tells us that just as light has both a wave and a particle nature, so too do objects we usually refer to as "particles," such as alpha particles. The wave nature of an alpha particle means that it is not well localized, in a sense captured by Heisenberg's uncertainty principle. Gamow found there was a small probability that the alpha particle could "tunnel" through the mountain that was holding it inside the uranium nucleus and find itself suddenly far outside the nucleus, where it would roll down the hill away from the nucleus due to electrostatic repulsion. It reminds me of the Zen koan: *How does the duck get out of the bottle* (whose neck is too narrow to allow the duck to escape)? Answer: *The duck is out!* So the alpha particle quantum tunnels through the mountain and "the alpha particle is out." Here is another instance where Gamow might have gotten a Nobel prize.

In the bubble universe case, the mountain valley represents the initial inflationary universe (at the waist of de Sitter space) with its high vacuum energy density. It would be happy to stay in that high-density ever-expanding state forever, but after a long time, there is a chance it will tunnel through the mountain, where it will roll down to sea level, releasing the energy of the vacuum into kinetic energy and into the creation of ordinary elementary particles. This tunneling represents the instantaneous formation of a small bubble with a vacuum energy density slightly less than the vacuum energy density outside. The negative pressure outside the bubble is stronger than the negative pressure inside the bubble, and the difference pulls the bubble wall outward. It expands faster and faster, eventually approaching the speed of light. Meanwhile, inside the bubble, the vacuum energy density is slowly rolling down the hill toward sea level. Inflation continues for a while inside the bubble as it rolls down the hill. When it has rolled to sea level and deposited its vacuum energy in the form of particles, inflation stops, and the Friedmann phase begins. It was this type of scenario that Andrei Linde and also Andreas Albrecht and Paul Steinhardt independently published shortly after my paper came out. Outside the bubble, the vacuum state remains up in the mountain valley and an endless inflating sea continues its rapid accelerated expansion. I had discussed the geometry and general relativity involved in the formation of bubble universes to make what today we call a "multiverse," while Linde and Albrecht and Steinhardt independently proposed detailed particle physics scenarios that would actually allow bubble universes to be formed. I required that inflation continue in the bubble universe for a while to create the universe we live in. In the Linde and Albrecht and Steinhardt models, this did naturally occur as the vacuum energy density in the bubble took some time to roll slowly down the hill toward sea level. Later in 1982, Stephen Hawking published a paper adopting the bubble universe idea and showing that initial quantum fluctuations would be expanded by inflation to appear on cosmological scales, having just the form needed to successfully seed the formation of galaxies and clusters of galaxies in the universe.[3] The structure that was subsequently observed in both the CMB and the galaxy distribution, which we described in chapter 15, is in beautiful accord with predictions from inflation.

Although it is possible for a neighboring bubble universe to collide with ours in the far future (perhaps 10^{1800} years from now, creating a sudden hot spot in the sky whose radiation would probably kill any life around at the time), most of these other universes in the multiverse are forever hidden from our view by an event horizon. They are so far away that the light from them can never cross the ever-inflating region between us and them. It seems clear today that once it gets started, inflation is hard to stop. It will continue expanding forever, creating a multiverse with an infinite number of universes like ours. In 1983, Linde proposed *chaotic inflation,* which would also produce a multiverse of low-density pocket universes in an ever-expanding inflating sea. Linde's chaotic inflation model relied on quantum fluctuations to move you randomly on the landscape. There was a chance that a quantum fluctuation would move you up into the hills or mountains, where the vacuum energy density was high. The higher the altitude, the higher the energy density, and the shorter the doubling time for expansion. In the high-altitude regions, more space of high vacuum energy density is being created at a rapid rate by the high rate of inflation. Regions at high altitude thus reproduce more rapidly. It's as if people had more children the higher altitude at which they lived. After a few generations, almost everyone would be living in the mountains. The whole multiverse would be inflating at a high rate. Then individual regions could roll down into the valley, creating individual pocket universes like ours. Most of the volume of space would be in the rapidly expanding mountain regions, but there would be patches (pocket universes) always forming by rolling down to sea level. So, we don't really need to start in mountain valleys. In a general landscape, we expect always to be forming low-density universes like ours in an ever-inflating multiverse.

Even though we can't see these other universes in the multiverse, we have reason to think they exist, because they seem to be an inevitable prediction of the theory of inflation, which explains a wealth of observational data.

Inflation got a great boost when the WMAP and the Planck satellite produced their results. The strength of the temperature fluctuations seen at different angular scales in the CMB matches exactly the pattern expected from inflation (recall figure 15.3). The WMAP and Planck satellite observations also showed that the universe has approximately zero curvature. In a positive curvature universe, we would see fewer spots in the microwave background map, because the circumference of a large circle is smaller than the $2\pi r$ that we expect from Euclidean geometry. If it were negatively curved, the circumference would be larger than $2\pi r$, there would be more spots, and the spots would be smaller in angular size than expected from Euclidean geometry. The observations show temperature fluctuations that peak in strength at about 1° in angular scale. This agrees with the prediction of a zero curvature universe.

This means that we don't really know sign of the curvature. The curvature of the universe is just so low that we cannot measure it. Our current data show that the visible universe is flat to an accuracy of somewhat better than 1%. In the same way, a basketball court looks flat even though we know it follows the curvature of Earth. It's just that the radius of Earth is very much larger than the basketball court, ensuring that the curvature in the basketball court is not noticeable. We know that early people thought Earth was flat, because the tiny part of Earth they could see was approximately flat. All we really know is that the radius of curvature of the universe is much larger than the 13.8-billion-light-year radius out to which we can see—out to the CMB. Guth emphasized that no matter what shape the universe was initially (whether positively or negatively curved), inflation in the simplest models would usually yield enough expansion to make the universe much larger than the part we can inspect. Guth predicted that we would find an approximately flat universe, and he was right. If our universe is a bubble universe, it simply means that it continued to inflate for a long time inside the bubble, as the vacuum state was rolling down the hill after tunneling. A "long" period of inflation, say 1,000 doublings in size as seen from inside the bubble, could be accomplished in just 10^{-35} seconds, if the doubling time was 10^{-38} seconds. That would make the current radius of curvature of the universe 10^{274} times larger than the part we can see, so it would look flat.

Cosmological models today are defined by two parameters: Ω_m and Ω_Λ. The values of these parameters determine the expansion history of the universe, and whether it is finite (like a 3-sphere) or infinite in extent. The first parameter describes matter density and is given by $\Omega_m = 8\pi G\rho_m/3H_0^2$, where G is Newton's gravitational constant, ρ_m is the average density of matter in the universe today (including both ordinary matter and dark matter), and H_0 is the Hubble constant today, quantifying how fast the universe is expanding. The numerator ($8\pi G\rho_m$) describes the density in the universe (the amount of gravitational attraction), while the denominator of the fraction ($3H_0^2$) describes the kinetic energy in the expansion. In simple Friedmann models where matter only is involved, Ω_m tells us whether the universe will expand forever or not: if $\Omega_m > 1$, the gravitational attraction overcomes the kinetic energy of expansion and the universe ultimately collapses: this is the 3-sphere Friedmann football-shaped spacetime illustrated in figure 22.5. If $\Omega_m < 1$, the kinetic energy of expansion overcomes the gravitational attraction, and we get the negatively curved

Friedmann universe that expands forever. If $\Omega_m = 1$, the kinetic energy balances the gravitational attraction and the model is flat; it expands slower and slower forever as the density goes down and the kinetic energy of expansion lessens with time. These Friedmann models all have $\Omega_\Lambda = 0$, where there is no vacuum energy density in empty space—they lie along the bottom edge of figure 23.5.

If there is vacuum energy present today, we must also consider the value of the second parameter, which characterizes the vacuum energy density and is given by $\Omega_\Lambda = 8\pi G \rho_{vac}/3H_0^2$, where ρ_{vac} is the vacuum energy density (the energy density of dark energy) in the universe today. We use the subscript Λ to remind us that dark energy behaves like Einstein's cosmological constant term Λ. We can display all possible cosmological models on a plane. The horizontal coordinate indicates the value of Ω_m (matter density), while the vertical coordinate represents Ω_Λ (vacuum energy–dark energy). A particular cosmological model is represented by a point in the plane in figure 23.5 with horizontal and vertical coordinates $(\Omega_m, \Omega_\Lambda)$, representing a particular combination of values of matter density and dark energy density today.

If Ω_Λ is not zero we get models that fill the diagram. The red diagonal line shows the set of models with $\Omega_0 = \Omega_m + \Omega_\Lambda = 1$, indicating they are flat, as inflation predicts. Models to the left of that red line are saddle shaped and infinite in extent, models to the right of the red line are 3-sphere universes. The black dotted necktie-shaped region shows the models consistent with data on the CMB from the Boomerang high-altitude-balloon telescope project in Antarctica, a key early experiment. It runs directly along the red line, showing that the CMB data favor the flat model. We can get another constraint on the cosmological model by directly measuring the expansion history of the universe from observations of the relationship between redshift and distance to distant objects. Scientists use so-called Type Ia supernovae, which are good standard candles; the area in the $(\Omega_m, \Omega_\Lambda)$ plane allowed by the supernovae Ia observations is shown in light green. These data show the expansion of the universe is accelerated, and for this discovery, Saul Perlmutter, Brian Schmidt, and Adam Riess shared the 2011 Nobel Prize in Physics. Models with $\Omega_\Lambda > \Omega_m/2$ have an expansion that is accelerating today, the gravitational repulsion of the dark energy being greater than the gravitational attraction of the matter. The green area from the supernova data satisfies this inequality and indicates acceleration today. (Models with $\Omega_\Lambda < \Omega_m/2$ would have been decelerating today.) The black necktie region intersects the green region in a small overlap

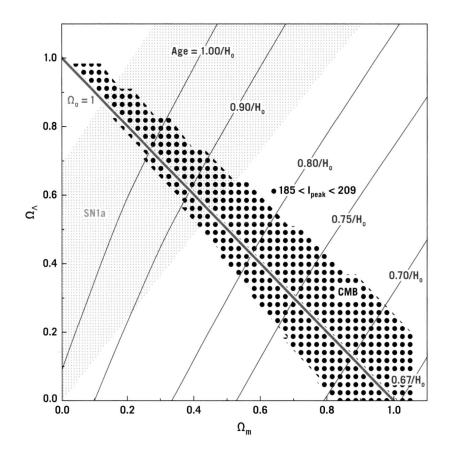

FIGURE 23.5. Cosmological models (Ω_m, Ω_Λ). Each point in this diagram represents a particular cosmological model with a particular value of matter density (corresponding to its horizontal coordinate Ω_m), and dark energy density (corresponding to its vertical coordinate Ω_Λ). The green dotted area covers models allowed by Supernova Ia observations (SN1a) showing the expansion of the universe is accelerating. The black dotted area covers models allowed by the cosmic microwave background (CMB) from the Boomerang Balloon Project in the year 2000, one of the first papers showing CMB plus supernovae observations imply a flat universe ($\Omega_0 = \Omega_m + \Omega_\Lambda$ = 1), with $\Omega_m \approx 0.3$ and $\Omega_\Lambda \approx 0.7$. Dark energy is 70% of the stuff of the universe. Subsequent observations from the WMAP and Planck satellites have greatly strengthened this conclusion.

Credit: Reprinted by permission from MacMillan Publishers Ltd: *Nature*, 404, P. de Bernardis, et al. April 27, 2000

region around $\Omega_m \approx 0.30$ and $\Omega_\Lambda \approx 0.70$. These values are consistent with both the CMB data and the supernova data.

Interestingly, this overlap region agrees with the value of $\Omega_m \approx 0.30$ from dynamical arguments based on the masses of clusters of galaxies, individual motions of galaxies, and the growth of structure in the universe. This includes both ordinary matter (baryons—protons and neutrons) and dark matter. Knowing the Hubble constant is approximately 67 (km/sec)/Mpc, it turns out that we can determine Ω_m and Ω_{baryon} directly by measuring the relative heights of the even and odd peaks in figure 15.3. The answer is $\Omega_{baryon} \approx 0.05$ and $\Omega_m \approx 0.30$.

This result from the CMB agrees with the answer $\Omega_{baryon} \approx 0.05$ from Gamow-type nucleosynthesis arguments discussed in chapter 15 and tells us that most of the matter in the universe is in the form of dark matter ($\Omega_{darkmatter} \approx 0.25$), which cannot be made of ordinary matter (baryons). The search is on to discover the detailed nature of the dark matter, as Michael has described.

The blue lines in figure 23.5 show the age of the universe in terms of $1/H_0$. The favored cosmological model is near the line marked "age = $1/H_0$."

Since the Boomerang results in 2000, the WMAP satellite has measured the CMB with high precision and refined these estimates to produce a standard cosmological model that explains all the observational constraints. The Planck satellite has further refined these estimates: $H_0 = 67$ (km/sec)/Mpc; age of the universe = 13.8 billion years; and a value of $\Omega_m + \Omega_\Lambda = 1$ within the observational errors, to an accuracy of better than 1%, consistent therefore with a flat model.

WMAP's results, when combined with supernova and other data, were even able to track the expansion history of the universe, and through application of Einstein's equations establish the *ratio of the pressure to the energy density* in the dark energy, a key measure that is simply called w. The value WMAP found for $w = -1.073 \pm 0.09$, which is equal to the value of -1 predicted by Einstein's cosmological constant model to within the observational errors. The Planck satellite produced a similar estimate. Recently, the Sloan Digital Sky Survey has measured the current value of w to be $w_0 = -0.95 \pm 0.07$, using data on galaxy clustering and a fitting formula developed by Zack Slepian and myself. Using the same data and formula, but adding data from the magnitude of gravitational lensing of background by foreground galaxies, the Planck satellite team has found $w_0 = -1.008 \pm 0.068$. All these estimates are consistent, within the observational errors, with the value of $w = -1$ expected from vacuum energy (dark energy). We know the energy density of dark energy is positive, because positive energy density, above and beyond that of ordinary matter and dark matter, is required to make the universe flat (which we observe). We know that the pressure of dark energy is negative, because, given that the energy density of dark energy must be positive, only a negative pressure for the dark energy could produce the gravitational repulsion required to cause the accelerating expansion of the universe that we observe. We can even accurately measure the amount of this negative pressure and find that it is equal to -1 times the energy density of the dark energy to within the observational errors. Einstein would be happy! His cosmological constant term was not a blunder after all!

Sometimes people say that dark energy is a mysterious force that is causing the current acceleration of the expansion of the universe, or that we know nothing about dark energy. That's not really true. The force that is causing the accelerated expansion of the universe is just gravity. And it's repulsive because of the negative pressure associated with dark energy. We strongly suspect that dark energy appears on the right side of Einstein's equations with the stuff of the universe, rather than appearing on the left side of the equations as part of the law of gravity, because we suspect that a different (higher) amount of dark energy was present in the early universe, producing inflation. We suspect that dark energy is a form of vacuum energy produced by a field or fields, but we don't know which one or ones. We know that the amount of dark energy is approximately constant with time, but we don't know whether it is slowly falling (rolling down the hill) or rising (rolling up the hill). This is the focus of current research.

The Sloan Digital Sky Survey was able to make an accurate estimate of the Hubble constant, using a characteristic scale found in galaxy clustering corresponding to the oscillations seen in the fluctuations in the CMB in figure 15.3. In this way, they could replace the Cepheid variable ruler for establishing the overall scale, while using supernova data to chart the detailed changes in the Hubble constant with time. They found a value of $H_0 = 67.3 \pm 1.1$ (km/sec)/ Mpc. This means that the density of dark energy is about 6.9×10^{-30} grams per cubic centimeter. If we were to draw a sphere centered on us with a radius equal to that of the Moon's orbit, the mass equivalent of the amount of dark energy contained within this sphere would be 1.6 kilograms—inconsequential relative to the mass of Earth–which is so small that we don't notice its slight gravitational effect or the slight gravitationally repulsive effect of its negative pressure on the orbit of the Moon. But its effects on cosmological scales, where the average density of matter is only 3×10^{-30} grams per cubic centimeter, is profound.

Establishing this cosmological model with small errors is quite an accomplishment. The WMAP and Planck satellites have produced detailed measurements of the power of fluctuations as a function of angular scale in the CMB, which agree in extraordinary detail with the results predicted from inflation (as shown in figure 15.3). This is a dramatic experimental vindication of inflation. And the dark energy we see today is of exactly the form required for inflation in the early universe, but just of very low density.

A new independent test for inflation has recently been proposed. If inflation causes the universe to double in size approximately once every 10^{-38} seconds, early on one could only see out to a distance of 10^{-38} light-seconds or 3×10^{-28} centimeters. This distance is tiny, and due to Heisenberg's uncertainty principle of quantum mechanics, this causes fluctuations in the geometry of spacetime (ripples), which according to Einstein's equations propagate at the speed of light—namely, gravitational waves. These would leave a characteristic swirling pattern in the polarization of the microwave background radiation which can be measured in principle. So far, its detection has proved elusive. The current best upper limits from the Planck satellite, plus ground-based experiments called Keck and BICEP2, lie somewhat below those of the simplest Linde chaotic inflation model. The amplitude of the gravitational waves produced depends on the detailed shape of the hill you are rolling down (see figure 23.4). The inflationary model the Planck team thinks fits the data best is one by Alexei Starobinsky; its doubling time is 3×10^{-38} seconds at the end of the inflationary epoch, compared with the 5×10^{-39} seconds doubling time in the simplest Linde model. This six-times-less-violent expansion would produce gravitational waves of six times lower amplitude, safely below the current upper limits. A number of observational efforts, including high-altitude balloons and ground experiments in Antarctica, are underway to lower the observational errors and further test inflationary models. Astronomers are anxiously waiting to see whether these observations can open a new window on the early universe.

With respect to the current universe, among the early astronomers working on cosmology in the twentieth century, the one who came closest to the truth was Georges Lemaître. In 1931, he proposed a model in which the universe started with a Big Bang and expanded like a Friedmann model, until it entered a coasting phase, during which the cosmological constant almost exactly balanced the matter density, approximating an Einstein static model for a while, after which time it expanded further, and the cosmological constant began to dominate as the matter thinned out. The spacetime diagram of this model looks like the lower half of a football at the bottom (Friedmann phase), then a cylinder (the Einstein static phase), and finally, the flared opening of a trumpet (de Sitter-space phase). Except for the coasting phase in the middle, Lemaître got it right. Lemaître was the first to calculate an expansion rate for the universe by combining Hubble's distances to galaxies with Slipher's redshifts. He was also the first to suggest that Einstein's cosmological constant could be viewed as a vacuum state having positive energy density and negative pressure. Pretty good for one career!

Inflation has been very successful at explaining the structure of the universe that we see. We don't really know how inflation got started, because inflation "forgets" its initial conditions as the universe exponentially expands, thinning out any initial components. But there are some speculations as to how inflation may have started.

Inflation can start with just a tiny de Sitter 3-sphere "waist" universe with a circumference of perhaps only 3×10^{-27} cm, which will then start expanding. But where did *that* come from? Alex Vilenkin thought it might originate via quantum tunneling, a process analogous to that occurring with the formation of bubble universes. This time, the ball at rest in the mountain valley would correspond to a 3-sphere universe of zero size. It would then tunnel through the mountain to find itself suddenly outside on the slope. This would correspond to a finite-sized 3-sphere universe—the de Sitter waist. Then as it rolled down the hill, that phase would correspond to the de Sitter funnel. What would the spacetime diagram of this universe look like?

Vilenkin showed it would look rather like a badminton shuttlecock (figure 23.6). The point at the bottom is the pointlike zero-sized universe at the beginning. The feathered, funnel-shaped top of the shuttlecock is the de Sitter expansion at the end. Connecting the point at the bottom to the flaring funnel at the top is a black hemispherical ball shape. This represents the geometry during the tunneling through the mountain. Being "underground" in the tunnel causes the minus sign in front of the time dimension to flip sign: time becomes just another spacelike dimension. The hemisphere is half of a 4-sphere with four dimensions of space and no dimension of time. No clocks tick in this region: the tunneling occurs in a single instant. The ball is in the mountain valley and then suddenly it's out. James Hartle and Stephen Hawking considered a model like this and added the idea that in this hemispherical bottom, the point at the beginning—the south pole—was no different from other points on the surface. It was exactly like the South Pole on Earth, which is likewise no different from other points on Earth's surface. This universe has no boundary at the bottom—what Hawking calls the *no boundary condition*. Hawking has spoken of this early region as having *imaginary* time. The imaginary number i is the square root of –1. Normally, $ds^2 = -dt^2 + dx^2 + dy^2 + dz^2$, so if we had

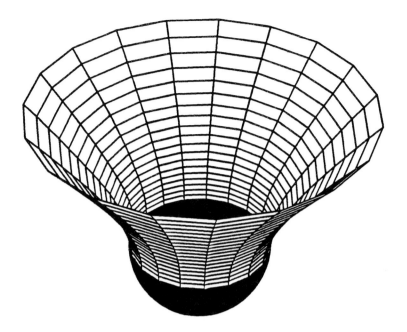

FIGURE 23.6. The spacetime diagram of a universe that has tunneled from nothing.
Credit: J. Richard Gott (*Time Travel in Einstein's Universe*, Houghton Mifflin, 2001).

imaginary time, *it*, because $i^2 = -1$, the quantity $-d(it)^2$ would become $+dt^2$, and we would have $ds^2 = dt^2 + dx^2 + dy^2 + dz^2$. Imaginary time sounds spooky, but it just makes time into another ordinary dimension of space. We would have four dimensions of space in this region instead of three dimensions of space and one dimension of time.

Quantum tunneling is certainly weird. We are looking for something weird to happen at the beginning of the universe, because what happened then was truly remarkable. Maybe it could have been quantum tunneling. But you don't really start with nothing. You start with a quantum state corresponding to a universe of size zero that knows all about the laws of physics and quantum mechanics. How does *nothing* know about the laws of physics? The laws of physics are simply the rules by which stuff behaves; if you have no stuff, what do the laws of physics mean? This is one of the problems with trying to make a universe out of nothing.

Meanwhile, Andrei Linde had noted that an inflating universe can give birth to another inflating universe via a quantum fluctuation. A de Sitter inflating trumpet horn could give birth to another inflating trumpet horn, which would sprout and grow off it, like a branch grows off a tree. In fact, this branch will inflate and grow to be as large as the trunk and sprout branches of its own. Branches will continue forming branches, making an infinite fractal tree of universes, all from one original trunk. Each individual branch is a funnel that could form bubble universes (as in figure 23.3). We would be living in a bubble universe in one of the branches, but still, you might ask yourself: where did the trunk come from?

Li-Xin Li and I tried to answer this question. We proposed that one of the branches curled back in time and grew up to be the trunk. Our model is illustrated in figure 23.7. Along the top, we see four funnel-shaped de Sitter inflating universes labeled 1, 2, 3, and 4, from left to right. Universe 2 gives birth to Universe 1. Universe 2 gives birth to Universe 3. Universe 3 gives birth to Universe 4. Universe 4 is the granddaughter universe of Universe 2. These branches will continue to expand and give birth to additional branches ad infinitum. These funnels do not hit one another—imagine them missing each other in some higher dimensional space. In this spacetime diagram, as in previous diagrams, only the surface itself is real.

Now we come to the most surprising feature of the model: Universe 2 also gives birth to another branch that curls back in time and grows up to become the trunk. It creates a little time loop in the beginning that looks like the loop in the number "6." Universe 2 is its own mother! As we have discussed, general relativity allows for loops in spacetime. There are no curvature singularities in this model. We were able to find a quantum vacuum state for this universe that was self-consistent and stable. The time loop has a Cauchy horizon marking the boundary where the time travel ends. It cuts at 45° across the trunk just above where the branch leaves the tree. You can continue to circle the loop in the "6" at the bottom as many times as you want, but when you move out beyond the branch into the top of the "6," there is no going back. If you are before the Cauchy horizon, you can go back out the branch and back in time for another loop to visit yourself in the past, but once you cross the Cauchy horizon, you are beyond the branching point, and you just keep going on upward into one of the funnels at the top. This universe has a little time machine at the beginning that shuts down. Curiously, exiting such a time machine is stable, which actually makes building one at the beginning of the universe easier.

That's interesting, because locating the time machine at the beginning of the universe puts it just where you want it to explain the first-cause problem. Every event in this universe has events that precede it. If you are anywhere

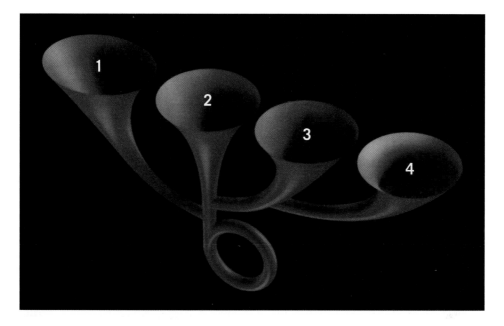

FIGURE 23.7. Gott–Li self-creating multiverse. The loop at the bottom represents a time machine; the universe gives birth to itself. *Photo credit:* J. Richard Gott, Robert J. Vanderbei (*Sizing Up the Universe*, National Geographic, 2011).

in the time loop, there are always events counterclockwise from you that are before you and give rise to you in the usual causal way. This multiverse is finite to the past but has no earliest event. This can occur in a curved spacetime of general relativity.

This theoretical model would also seem to fit in well with superstring theory. Superstring theory or M-theory posits an eleven-dimensional spacetime consisting of one macroscopic dimension of time, three macroscopic dimensions of space, and seven additional spatial dimensions, which are curled up and microscopic, as Kaluza and Klein would have liked. The complex microscopic shape determines the laws of physics. Interestingly, inflation suggests that the three macroscopic spatial dimensions we see today were originally roughly as small as the microscopic Kaluza–Klein dimensions: a de Sitter waist of perhaps 3×10^{-27} cm. This microscopic de Sitter circumference has inflated greatly as the universe has expanded. Originally, there were ten curled-up, microscopic spatial dimensions; seven have remained curled-up and tiny, while three have simply ballooned up in size since the beginning. Our (Gott–Li) model proposes that originally time was also curled up in a microscopic time loop. The time loop may have had a circumference in time (clockwise around the loop) as short as anywhere from 5×10^{-44} to 10^{-37} seconds if it had the self-consistent quantum vacuum state we have proposed. In the time loop, all ten dimensions of space as well as the dimension of time are curled up and tiny.

One of the wonderful things about inflation is that a small piece of inflating vacuum state expands to create a large volume, each little piece of which looks exactly like the piece you started with. If one of those little pieces is the piece you started with, then you have a time loop. Therefore, in our theory, the universe isn't created from nothing but rather it is created from something, from a little piece of itself. Then the universe can be its own mother. Time travel is something unusual that appears to be allowed by general relativity; perhaps it is just what we need to explain how the universe got started.

Today, I would say the theory of inflation is in very good shape. It explains the fluctuations we see in the CMB in detail (recall figure 15.3). If you doubt that inflation occurred, remember that we see a low-grade inflation going on today. The universe's expansion is accelerating, caused most likely by a low-density vacuum state (dark energy) with a density of 6.9×10^{-30} grams per cubic centimeter. Inflation just relies on a large amount of dark energy in the early universe. Inflation seems inevitably to produce a multiverse of universes. Just how sure are scientists of this? Once Sir Martin Rees (the Astronomer Royal) was asked at a conference how sure he was that we lived in a multiverse. He said he wouldn't be willing to bet his life on it, but he would go so far as to bet the life of his dog. Linde rose to say that since he had spent decades of his life working on the multiverse idea, he had proven that he would bet his life on it. Nobel Laureate Steven Weinberg said he would be willing to bet Linde's life on it, and Martin Rees's dog's!

How did inflation begin? We don't know. Did it emerge by quantum tunneling from nothing (perhaps the most popular model), or even stranger, was there a little time loop at the beginning? Pedro González-Díaz has speculated that when we have a true theory of quantum gravity, those two models might even turn out to be the same. Another speculation by Paul Steinhardt and Neil Turok is that the Big Bang occurred when two universes floating in an eleven-dimensional spacetime collided, heating them up suddenly. Repeated bangs could occur. (This would be rather like two pieces of paper—representing flatland universes—slapping together repeatedly in three-dimensional space. Such things could happen in principle in M-theory.) Lee Smolin thinks our universe could have been born inside a black hole in a previous universe. As a star collapsed to form a black hole, its interior density grew and grew until a high-density vacuum state was created, whose gravitationally repulsive nature caused it to bounce at a de Sitter waist and produce an expanding inflationary state that could spawn a multiverse, as pointed out by Claude Barrabès and Valeri Frolov. All this would occur inside the black hole that formed, the smile singularity in the Kruskal diagram being replaced by the start of a de Sitter expanding phase.

These are some of the speculative ideas that physicists are exploring to answer the ultimate question: how did the universe begin? Of these alternatives, probably the tunneling from nothing model is the most popular at the moment, but we simply don't know which is correct. We may learn the answer when we find a "theory of everything," which will unite general relativity and quantum mechanics and the strong, weak, and electromagnetic forces to explain all the laws of physics. When we have the equations of the "theory of everything," we will see what cosmological solutions they produce. This is why we are studying fundamental physics. We are looking for clues about how the universe works, and maybe even how it began.

24

OUR FUTURE IN THE UNIVERSE

J. RICHARD GOTT

This chapter is about the future of the universe. I am going to put salient events in the universe's history, both past and future, on a timeline. This will involve some enormous times in the far future, and also some very short times in the early universe. What is the earliest time we can speak of in the early universe?

To answer that, we need to answer two related questions: What is the shortest time we can measure? What is the fastest possible clock I can imagine? Every clock, even a quartz watch, has to have something moving back and forth, like the pendulum in a grandfather clock. If I want the fastest possible clock, I just need the fastest thing to move back and forth. What should I use? Light! It's the fastest thing I can send back and forth. In fact, all I need is that light clock in figure 17.1 with two mirrors and a light beam bouncing back and forth between them. If I want to have it tick faster, what do I do? Bring the two mirrors closer together. The closer together the two mirrors are, the faster the clock will tick. I am going to have one photon in my clock bouncing up and down.

What happens when I make my clock very small? I have a problem. At least one wavelength λ of my photon must fit inside my clock. If the distance between the mirrors in my clock is L, the smallest my clock can be is $L = \lambda$. The wavelength and the frequency of the photon are related by $\lambda = c/v$. The smaller the wavelength is, the higher the frequency will be. As I decrease the size of my clock L, I must reduce the wavelength of the photon so it can fit inside, and so I must increase my photon's frequency. Increasing its frequency means increasing its energy, because the photon has an energy $E = hv$. And we must not forget Einstein's equation $E = mc^2$. The energy of the photon corresponds to a certain amount of mass. So, as I make my clock smaller, the energy of the photon goes up, and the mass of the clock increases. Eventually the mass of the clock becomes so large and is compressed into such a small size L, that it falls within its own Schwarzschild radius and forms a black hole! If I try to make a clock that ticks too fast, it will collapse in this way and form a black hole when the length of the clock is about $L = 1.6 \times 10^{-33}$ centimeters and when it ticks once every 5.4×10^{-44} seconds. This time is called the *Planck time*. It is the shortest time one can measure. The length $L = 1.6 \times 10^{-33}$ cm is one you have heard of before. I have said that the size of the singularity at the center of the Schwarzschild black hole is not exactly zero—it is actually about 1.6×10^{-33} centimeters across, being blurred by quantum effects. This length is called the *Planck length*, and it is the shortest length one can measure. When I explained that the circumferences of those extra spatial dimensions predicted by string theory might be of order 10^{-33} cm, this is also the Planck length.

You can't measure a time shorter than the Planck time. The length of that little time loop that Li-Xin Li and I were talking about in the beginning of the universe might be as short as this (see chapter 23). In fact, if you look at ordinary spacetime on scales of 1.6×10^{-33} cm and times of order 5×10^{-44} seconds, the geometry of spacetime should become uncertain, according to the uncertainty principle. Spacetime should become spongelike and multiply connected when seen at this scale. We can calculate the value of the Planck length, $L_{\text{Planck}} = (Gh/2\pi c^3)^{1/2} = 1.6 \times 10^{-33}$ centimeters, using the fundamental constants. Here we see all our old friends: Newton's gravitation constant G, used to calculate the Schwarzschild radius of a black hole; Planck's constant h, used in calculating the energy of a photon $E = hv$; and c, the speed of light, used to calculate the mass equivalent of the energy of the photon ($E = mc^2$). The Planck time $T_{\text{Planck}} = L_{\text{Planck}}/c$ is equal to the amount of time it takes a light beam to cross the Planck length. Ignoring factors of order 2 and π, this is the minimum size of the fastest clock before it collapses into a black hole. The mass of this little, fastest clock is the *Planck mass*, or 2.2×10^{-5} grams, and the density of this little clock is the *Planck density*, or 5×10^{93} grams per cubic centimeter. This is the sort of density you might get up to in the singularity in a black hole, before quantum mechanics begins to smear things out. The Planck scales are where quantum mechanics comes into play in general relativity, and as mentioned before, we do not yet have a unified model for quantum gravity. Thus the Planck scale (in length or time) represents a limit beyond which we cannot go with our current understanding.

The Planck time, 5×10^{-44} seconds, is the shortest time one can measure, and it is the earliest time we can speak of in the universe. As I have discussed, our universe may be just one bubble (or patch) in one inflating funnel, making up one branch in an infinite fractal tree of universes, making up a multiverse, which may be arbitrarily old. But I am counting time after our little bubble universe has formed. Table 24.1 shows what is happening at each epoch.

As inflation ends, at about 10^{-35} seconds, the vacuum state that filled the early universe with high-density dark energy decays into thermal radiation. This thermal radiation is very hot and includes not only photons (carriers of the electromagnetic force), but quarks, antiquarks, electrons, positrons, muons, antimuons, taus (a heavier counterpart to the muon), antitaus, neutrinos, anti-neutrinos, gluons (carriers of the strong nuclear force), X-bosons (hypothetical particles, predicted in some theories, whose asymmetric decays produce the excess of matter over antimatter in the universe seen today), W and Z particles (carriers of the weak force), Higgs bosons (the particle associated with the Higgs field that gives particles their mass), and gravitons (carrier of the gravitational field, just as the photon is the carrier of the electromagnetic field). And if the theory of supersymmetry is correct, there would be supersymmetric partners for each of these particles listed above.

A word about gravitons: Einstein found that gravitational waves, ripples in the geometry of spacetime that travel through empty space at the speed of light, were a solution of his field equations of general relativity. In a similar fashion, Maxwell had previously found that electromagnetic waves traveling at the speed of light through empty space were a solution of his field equations of electromagnetism. We have indirect evidence for gravitational waves (which would consist of gravitons) from Taylor and Hulse's binary pulsar, which is inspiraling toward a tighter and tighter orbit exactly as Einstein predicted would occur from the emission of gravitational waves as the neutron stars orbited. On September 14, 2015, the LIGO experiment made the first direct detection of gravitational waves. A laser interferometer measured with extreme accuracy (1/1,000 of the diameter of a proton) the distance between pairs of mirrors and noted the tiny oscillations in the distances between the mirrors

TABLE 24.1. EPOCHS IN THE UNIVERSE

TIME SINCE THE BEGINNING	WHAT'S HAPPENING
5×10^{-44} seconds	Planck time
10^{-35} seconds	Inflation ends; random quantum fluctuations seeding galaxy formation already established; matter is made; quark soup
10^{-6} seconds	Quarks condense into protons and neutrons
3 minutes	Helium synthesis; light elements are made
380,000 years	Recombination; electrons combine with protons to form hydrogen atoms; cosmic microwave background
1 billion years	Galaxy formation
10 billion years	Life forms on Earth
13.8 billion years	We are here
22 billion years	Sun finishes main-sequence lifetime and becomes white dwarf
850 billion years	Universe cools to Gibbons and Hawking temperature
10^{14} years	Stars fade; last red dwarfs die
10^{17} years	Planets detach; stellar encounters strip planets away from their home stars, destroying white dwarf or neutron star solar systems
10^{21} years	Galactic-mass black holes form; most stars and planets ejected
10^{64} years	Protons should have decayed by now; black holes, electrons and positrons, photons, neutrinos, and gravitons are left
10^{100} years	Galactic-mass black holes evaporate

as the gravitational wave rippled by. How fitting that gravitational waves, predicted by Einstein, would eventually be detected using lasers, since Einstein discovered the principle of the laser as well. The source of these gravitational waves was a 29-solar-mass black hole and a 36-solar-mass black hole in a tight binary orbit that inspiraled toward each other and merged to form a single black hole of 62 solar masses. So, gravitational waves exist, and the results are consistent with gravitons traveling at the speed of light. Because gravity is such a weak force, we have not detected any individual gravitons yet, but we expect they must exist because we have detected gravitational waves, and we expect a wave-particle duality for these, just as we do for electromagnetic waves and photons.

We call this epoch, when all these elementary particles are buzzing around, *quark soup*. Quarks are traveling freely and are not confined in tight triples. Because of the uncertainty principle, in some regions the quantum vacuum state decays slightly later and in other regions it decays slightly earlier, causing random density fluctuations in the thermal radiation that is created when the quantum vacuum state decays.

These density fluctuations are present at 10^{-35} seconds, when inflation ends. They form the seeds, which will, by the action of gravity over the course of 13.8 billion years, ultimately lead to the formation of the galaxies and great clusters of galaxies we see today. The spongelike pattern of galaxies we see (figure 15.4), in which great clusters of galaxies are connected by filaments (or chains) of galaxies, is called the *cosmic web*, and represents the (greatly expanded) fossilized remnants of these early quantum fluctuations made when the universe was just 10^{-35} seconds old.[1]

As the universe expands, this hot soup cools and massive particles decay into lighter ones. Initially the universe contains equal amounts of matter and antimatter, but it is thought that asymmetric decays of heavy X-bosons, favoring matter over antimatter, will lead to slightly more matter than antimatter in the decay products. As matter and antimatter particles collide and annihilate in equal numbers to create photons, the remainder becomes dominated by matter. The galaxies we see today are made out of matter. Antimatter particles in the universe today are rare, and always in danger of meeting up with one of the many matter particles and annihilating. Antimatter particles are greatly outnumbered by matter particles today.

At 10^{-6} seconds, the radiation has become so cool that the quarks bind with other quarks to form protons and neutrons. Quarks come in six different flavors: up, down, strange, charm, top, and bottom. The lightest quarks are the up and down quarks. A proton is formed by two up quarks and one down quark; they are held together by interchanging three gluons between them. A neutron is formed by two down quarks and one up quark, also held together by three gluons. (A way to remember this is that the proton has more up quarks, and up has a "p" in it for proton, while the neutron has more down quarks, and down has an "n" in it for neutron.) The up quark has an electric charge of $+2/3$, while the down quark has an electric charge of $-1/3$. Thus the proton ends up with an electric charge of $+1$ while the neutron is neutral with a charge of 0.

At 3 minutes, helium synthesis occurs, as discussed in chapter 15. The universe has cooled to the point where protons and neutrons can fuse to make light elements. The most common element is hydrogen (a proton), but in addition, an appreciable amount of helium is made, as well as small amounts of deuterium, and lithium. This is the epoch Gamow and his students were using to predict the existence of the CMB.

At 380,000 years, the universe has cooled to about 3,000 K. At this point, electrons can bind to protons to produce hydrogen atoms. This process, as we have discussed, is called *recombination*. The universe goes from being an electrically charged plasma of mostly electrically charged protons (+) and electrons (−) to being an electrically neutral gas of mostly hydrogen, where each proton has captured an electron to produce an electrically neutral hydrogen atom. Before this epoch, photons were constantly being deflected by either an electrically charged proton or electron—making them execute a random or "drunken" walk. Photons didn't get very far, being deflected all the time. After the epoch of recombination, photons can travel unimpeded in straight lines over long distances. Because of this change to freely moving photons, we can see directly back to this epoch, when we observe the CMB radiation.

At 1 billion years, galaxies typically begin to form. The high-redshift quasars discussed in chapter 16 come from early-forming galaxies that are seen at a time a little before this.

The universe today is 13.8 billion years old.

By 22 billion years, the Sun will have finished its main-sequence lifetime and will have become a white dwarf. The Andromeda galaxy will have crashed into the Milky Way.

At about 850 billion years, the universe will cool to a constant temperature, due to a process described by Gibbons and Hawking. As discussed in chapter 23, observations indicate that the universe is filled with dark energy characterized by a pressure equal in magnitude to its energy density but *negative* (dynamically equivalent to Einstein's cosmological constant). As the matter of the universe thins out due to the expansion, while the dark energy remains at the same density, the universe becomes ever more dominated by dark energy in the far future. Thus the geometry of the universe in the future should resemble that of de Sitter space, a spacetime funnel. It should be ever expanding. Two galaxies that can communicate today will flee from each other faster and faster. Eventually the space between the two galaxies will expand so fast that light cannot cross the ever-increasing distance between them. Event horizons form. A distant galaxy will look to us just like it is falling into a black hole. It will get redder and redder. If extraterrestrials in the distant galaxy were sending a signal saying "THINGS ARE GOING OK," it would seem to us that they are saying "THINGS A. . . . . R . . . . E." We would never receive the end of the signal, "GOING OK." Events occurring at late times in the distant galaxy would be beyond our event horizon, and we would never see them (recall figure 23.2).

Hawking showed that event horizons create Hawking radiation. Gibbons and Hawking calculated that in de Sitter space at late times, any observers present would see the resulting thermal radiation, appropriately termed *Gibbons and Hawking radiation*. This thermal radiation, seen in the future of our universe, would have a characteristic wavelength (λ_{max}) of about 22 billion light-years. The CMB radiation continues to increase in wavelength as the universe expands exponentially, doubling its size every 12.2 billion years. After 850 billion years, the CMB thermal radiation will have a characteristic wavelength longer than 22 billion light-years and will become unimportant compared with the Gibbons and Hawking radiation produced by the event horizons. At that point we should see the temperature of the universe stop falling and become constant at a Gibbons and Hawking temperature of about 7×10^{-31} K. That's very cold, but still above absolute 0 K.

These ideas are actually testable. Gibbons and Hawking radiation is also produced in the early inflationary phase of the universe. This includes both electromagnetic radiation and gravitational radiation. If such gravitational radiation from the early universe is eventually detected via the imprint left on the polarization of the CMB, as discussed in chapter 23, to my mind it would constitute an important experimental verification of the Hawking radiation mechanism. These gravitational waves are not made by moving bodies, as were the gravitational waves detected by LIGO, these gravitational waves would be produced by something different, the Hawking mechanism—a quantum process. So this would be something new and exciting.

The Gibbons and Hawking radiation we expect to see in the far future is ultimately bad for intelligent life. Freeman Dyson once showed that intelligent life could last indefinitely on a finite reserve of energy if it could dump its waste heat in an ever-colder temperature bath. If I showed a movie in a theater at a temperature of 300 K, using visible light photons, it would take a certain amount of energy to show the movie. But suppose we slow everything down in the theater. Suppose the movie was shown using infrared photons having twice the wavelength of visible photons; I could show the same movie using half the energy (each photon would take half the energy) but the movie would last twice as long (because the photons have twice the wavelength). The wavelengths of the photons in the thermal radiation in the theater would also be twice as long, so the temperature in the theater would be 150 K instead of the usual 300 K. Intelligent life could conserve energy by thinking and communicating ever more S . . . L . . . O . . . W . . . L . . . Y. One could even have an infinite number of thoughts using a finite amount of energy by continuing to slow down one's thinking. This is allowed if one can dump one's waste heat (which all biological processes, including thought processes, generate) into the ever-cooling microwave background, hibernating from time to time and operating at ever lower temperatures as time goes on. As long as the CMB continues to cool off toward absolute 0 K, that works. But at 850 billion years, the universe will reach an equilibrium temperature equal to the Gibbons and Hawking temperature, and its temperature will remain constant after that. Then one cannot operate at temperatures lower than that to save energy. One would need refrigeration, which uses up the remaining energy very quickly. Furthermore, the other galaxies will have fled beyond the event horizon, leaving only a finite energy reserve at one's disposal; intelligent life begins to be in energy trouble and will ultimately die out.

Here's another trouble. At 10^{14} years, the stars fade as the last low-mass stars run out of hydrogen fuel and die. The universe becomes dark. Only stellar remnants are left—white dwarfs, neutron stars, and black holes. Some planets may still circle them. But by 10^{17} years, enough close encounters of stars will have occurred to rip the planets from their orbits and fling them into interstellar space.

At 10^{21} years, galactic-mass black holes form. Two-body gravitational interactions slingshot some stars out of galaxies, while the rest fall into the central black hole. Gravitational radiation causes stars close to the black hole to spiral in.

By 10^{64} years (if it hasn't happened already), according to Hawking, protons should decay through a rare process of temporarily falling inside a Planck-sized black hole (via the uncertainty principle), and then having the black hole decay quickly by Hawking radiation. The black hole does not conserve baryons (protons or neutrons)—it does not remember whether it was made of a proton or a positron—but it does remember its electric charge. For that reason, a positron (which is lighter than the proton) can be emitted as one of the decay products of the black hole into which the proton has disappeared. As protons decay, we are left with electrons and positrons as the most massive particles. Protons may decay even earlier than this; perhaps on a timescale of 10^{34} years, but it's likely they would have decayed in any case by 10^{64} years.

At 10^{100} years, galactic-mass black holes evaporate via Hawking radiation.

What happens after that? The standard picture physicists have is that dark energy, which today is causing an exponential expansion of the universe, represents a vacuum state with a constant positive energy density (and a negative pressure). Steven Weinberg likens our current situation to living in a valley slightly above sea level—our altitude indicating the amount of dark energy present in the vacuum. We have rolled to the bottom of this valley and are just sitting there. The amount of energy in the vacuum—the dark energy—is not changing with time. This will keep the universe doubling in size every 12.2 billion years for a very long time.

Given enough time, it would be likely that our vacuum state, which causes dark energy, could quantum tunnel (through the valley walls) into a lower energy state (of lower-altitude terrain beyond our valley). This would cause a bubble of lower-density vacuum state to form somewhere in our visible universe. The negative pressure outside the bubble would be more negative than the pressure inside, which would pull the bubble wall outward. After a short time, the bubble wall would be traveling outward at nearly the speed of light. It would expand forever. The laws of physics would be different inside the bubble, and you would be killed when the bubble wall hit you.

One can calculate the probability per unit time to quantum tunnel out of the valley to a lower-altitude region outside. We could possibly see bubbles of lower-density vacuum forming in as "little" as 10^{138} years due to a known instability in the Higgs vacuum. But many physicists think the Higgs vacuum will be stabilized by higher-energy effects. In that case, according to speculative calculations by Andrei Linde, bubbles of lower density vacuum should start forming only after $10^{\wedge}(10^{\wedge}34)$ years! These bubbles would form, and just like the bubble universes in figure 23.3, they would never percolate to fill the entire space. The ever-expanding vacuum state would continue to double in size every 12.2 billion years and have a volume that would increase endlessly—an ever-inflating sea punctuated by forming bubbles. At late times, our universe would be like eternally fizzing champagne.

Even more rarely, as Linde and Vilenkin have suggested, a quantum fluctuation could cause the entire visible universe to jump up to a high-vacuum energy density and create a new, rapidly inflating high-density inflationary universe. This would be like the high-energy inflation we saw at the beginning of our universe and would initiate a new multiverse. It might be 10^(10^120) years before this happens!

Alternatively, we don't live in a valley at all, but on a slope, and we will slowly roll down to sea level. This is called *slow-roll dark energy*. As Bharat Ratra, Jim Peebles, Zack Slepian, and I, and many others, have explored, this would cause the amount of dark energy to slowly dissipate over billions of years, rolling down ultimately to a vacuum state of zero energy density. Just such a rolling down occurred once before with inflation, where a very high-density dark energy state rolled down to the low-energy vacuum that we see today. That could occur again, allowing us to ultimately roll down to sea level—a vacuum energy of zero. These scenarios can be investigated by measuring the expansion history of the universe up to now in detail. This allows us, using Einstein's equations, to measure the ratio of the pressure to energy in the dark energy, a ratio we call w. If w turns out to be exactly –1, dynamically equivalent to Einstein's cosmological constant, that favors the "trapped in a valley" scenario, and the dark energy will remain at its present value, and the universe will keep doubling in size every 12.2 billion years forever. If w is less negative than –1, however, we will roll slowly down to sea level, and the accelerated expansion should eventually give way to an approximately linear expansion rate. The universe will continue to expand forever but at a linear rate. In this case the universe's expansion goes like 1, 2, 3, 4, 5, 6, . . . with time.

A radical proposal, by Robert Caldwell, Mark Kamionkowski, and Nevin Weinberg, is that w could be more negative than –1. This is called *phantom energy*. It would produce a vacuum energy that increased with time as the universe expanded, leading to a runaway expansion and creating a singularity in the future (a Big Rip) that would tear galaxies, stars, and planets apart in perhaps as little as a trillion years. This "phantom" energy would require a negative kinetic energy in the rolling motion of the field that controls dark energy, which seems to me to be unlikely on physical grounds. That scenario would make the dark energy we see today nothing like the dark energy that was present earlier in inflation. So, although that remains a possibility, it seems less likely to me than the other two scenarios. But many physicists take "phantom energy" quite seriously.[2]

As discussed in chapter 23, the best estimate of the current value of w (from the Planck satellite team using all available data, including that from the Sloan Digital Sky Survey) is $w_0 = -1.008 \pm 0.068$. Remarkably, within the errors, this is consistent with the simple value of –1 (approximating Einstein's cosmological constant), which corresponds to the model where we are sitting at the bottom of a valley. This result strongly supports the general idea that dark energy represents a vacuum state with positive energy and negative pressure, but these observations are not yet able to truly distinguish between models in which we are sitting still at the bottom of a valley from those in which we are slowly rolling down (or up) a hill. In the latter cases, w_0 would be close to, but not exactly equal to –1, slightly above or below it. If future precision measurements of w show that it is unambiguously different from –1, we could learn whether the slow-roll dark energy or phantom energy picture was favored. But, if, as measurements continue to improve and the errors continue to go down, we continue to be consistent with $w_0 = -1$ within the errors, we may well pronounce the "sitting at the bottom of the valley" model triumphant. There are a number of experimental programs either underway or proposed for the future that can potentially lower the errors in w_0 by more than an order of magnitude; it is hoped these programs can illuminate the ultimate fate of the universe.

Now you have our best predictions of what the universe is likely to be doing in the future. But what about our future in the universe? What's likely to happen to us? How is our species *Homo sapiens* likely to fare in the far future? This is a question we would very much like to answer.

First, I would point out that we are living in a very habitable epoch. The universe has cooled off enough to be habitable, carbon and other elements essential to life have had enough time to form, and the stars are shining nicely today, providing warmth and energy. This is an epoch at which we might expect to find intelligent observers. After the stars have faded, it will be much more difficult for intelligent life. If we look at table 24.1, we find ourselves in a habitable epoch. The *Weak Anthropic Principle*, an idea proposed by Robert Dicke and later given its name and precise formulation by Brandon Carter, says that intelligent observers should, of course, expect to find themselves at habitable locations—in a habitable epoch in the universe. (Logically, they couldn't be alive to be asking the question in an *uninhabitable* epoch!) In fact, we do find ourselves in the middle of what looks like the most habitable epoch in the universe.

But as the only intelligent observers we have encountered in the universe so far, we would like to know how long our future longevity as a species is likely to be. How might we think about this question?

In 1969, I visited the Berlin Wall, which separated the two sectors of the city belonging to East and West Germany. People at that time wondered how long the Berlin Wall would last. Some people thought it was a temporary aberration and would be gone quickly. But others thought the wall would remain a permanent feature of modern Europe.

Figure 24.1 shows a picture of me at the Wall in 1969. To estimate the Wall's future longevity, I decided to apply the Copernican Principle. I remember thinking: I'm not special. My visit is not special. I am just coming to Europe after college—it was "Europe on 5 dollars a day" back then. I'm coming by to see the Berlin Wall just because I'm in Berlin and the Wall happens to be there. I could have seen it at any point in its history. But if my visit is not special, my visit should be located at some random point between the Wall's beginning and its end. (The end comes either when the Wall ends or when there is no one left alive to see it, whichever comes first.) There should be a 50% chance, then, that I am located somewhere in the middle half of its existence—in the middle two quarters. If I were visiting at the beginning of that middle 50%, then I would have been 1/4 of the way through its history, with 3/4 still in the future. In this case, the Wall's future longevity would be 3 times its past longevity. In contrast,

FIGURE 24.1. Rich Gott at the Berlin Wall in 1969. My right foot is in East Berlin, my left foot is in West Berlin, and the Berlin Wall is vertical behind me. *Photo credit:* Collection of J. Richard Gott

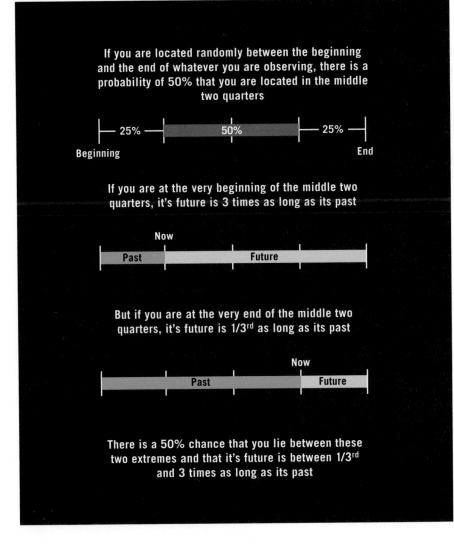

If you are located randomly between the beginning and the end of whatever you are observing, there is a probability of 50% that you are located in the middle two quarters

25% — 50% — 25%

Beginning   End

If you are at the very beginning of the middle two quarters, it's future is 3 times as long as its past

Now

Past   Future

But if you are at the very end of the middle two quarters, it's future is 1/3rd as long as its past

Now

Past   Future

There is a 50% chance that you lie between these two extremes and that it's future is between 1/3rd and 3 times as long as its past

FIGURE 24.2. The Copernican formula (50% confidence level).
Credit: J. Richard Gott

if I were at the end of that middle 50%, 3/4 of its history would be past and 1/4 would remain in the future, making its future 1/3 as long as its past.

I therefore reasoned that there was a 50% chance that I would be between these two limits and that the future longevity of the Wall would be between 1/3 and 3 times as long as its past (figure 24.2). At the time of my visit, the Wall was 8 years old. While standing at the Wall, I predicted to a friend, Chuck Allen, that the future longevity of the Wall would be between 2.66 years and 24 years.

Twenty years later I'm watching television and I call up my friend and say: "Chuck, you remember that prediction I made about the Berlin Wall? Well turn on your television, because NBC news anchor Tom Brokaw is at the Wall now and it's coming down today!" Chuck did remember the prediction. The Berlin Wall had come down 20 years later, inside the range of 2.66 years to 24 years that I had predicted. My visit was in the middle of the Cold War, so an atomic bomb could have taken it (and me) out in the next millisecond. In contrast, some famous walls, such as the Great Wall of China, have lasted for thousands of years. My predicted range was rather narrow but it still gave me the right answer.

Scientists generally prefer to make predictions that are more than 50% likely to be right. They like to make predictions that have a 95% chance of being correct. That is the usual 95% confidence level used in scientific papers. How does this change the argument? When applying the Copernican Principle, keep in mind that if your location in time is not special, there is a 95% chance that you are somewhere in the middle 95% of the period of observability of whatever you are observing—that is, neither in the first 2.5% nor in the last 2.5% (figure 24.3).

Expressed as a fraction, 2.5% is 1/40. If your observation falls at the beginning of that middle 95%—only 2.5% from the start—then 1/40 of the history of what you are observing is past, and 39/40 of it remains in the future. In this case, the future is 39 times as long as the past. If you are only 2.5% from the end, then 39/40 of it is past and 1/39 remains. The future is 1/39 as long as the past. If you are in the middle 95%, between these two extremes (there is a 95% chance of that), that makes its future between 1/39 and 39 times as long as its past. Thus:

the future longevity of whatever you are observing will be between 1/39 of its past longevity and 39 times its past longevity (with 95% certainty).

I decided I'd like to apply this to something important, to the future of the human species, *Homo sapiens*. Our species is about 200,000 years old. That goes back to Mitochondrial Eve, in Africa, from whom we are all descended. The formula would say with 95% confidence that, if our location in the timeline of the history of our species is not special, the future longevity of our species *Homo sapiens* should be at least 5,100 more years (that's 200,000/39) but less than 7.8 million more years (that's 200,000 × 39).[3] We do not have actuarial data on

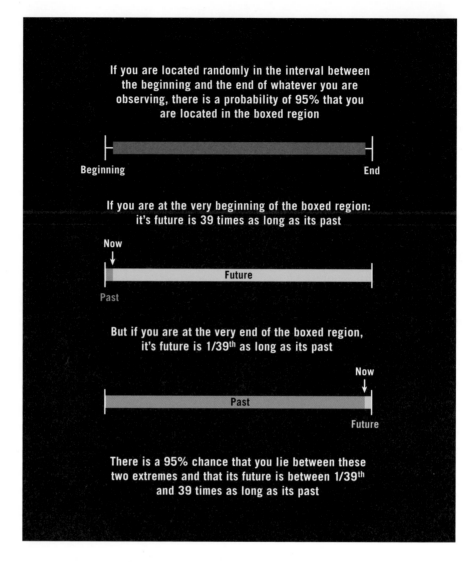

FIGURE 24.3. The Copernican formula (95% confidence level).
Credit: J. Richard Gott

other intelligent species (those able to ask such questions), so arguably this is the best we can do. The range of predicted future longevity is as large as this, because one wants to be 95% certain of being correct. Yet many experts who offer their own estimates make predictions outside this range. Some apocalyptic predictions say we are likely to be extinct in less than 100 years. But if that were true, we would be very unlucky to be located at the very end of human history. Some optimists think we will go on to colonize the galaxy and last for trillions of years. But if that were true, we would be very lucky to be located at the very beginning of human history. Thus, even with its broad range, the Copernican-based formula is still highly informative, limiting the possibilities to a tighter range than those considered by many others.

Certainly everything we have learned in astronomy tells us that we should take the Copernican Principle (that your location is not likely to be special) seriously. We started out thinking that we occupied a special place at the center of the universe. But then we realized that ours was just one of a number of planets orbiting the Sun. Then we found out that the Sun was just an ordinary star, not at the center of our galaxy but orbiting at a random location about halfway out. We learned that our galaxy was in an ordinary group of galaxies in a run-of-the-mill supercluster of galaxies. The more we have discovered, the less special our location has turned out to be.

The Copernican Principle is one of the most successful scientific hypotheses of all time, proving itself over and over in a variety of contexts. Christiaan Huygens used it to predict the distances to the stars. He asked why the Sun should be the brightest light in the universe. The stars, he reasoned, were just other suns like our Sun. If the other stars were as intrinsically bright as the Sun (assuming the Sun was not special), then the fact that the stars appear much dimmer in the sky than the Sun must mean they are far away. He figured that the brightest star in the sky, Sirius, was the closest. From his estimated brightness of Sirius relative to the Sun, he calculated it must be 27,664 times as far away as the Sun. He actually got the distance right to within a factor of 20, a remarkable accomplishment, given the large uncertainties involved. Huygens correctly found that the distances between the stars were vast compared to the size of our solar system.

When Hubble saw other galaxies receding from us equally in all directions, he could have concluded that we were at a special place at the center of a massive explosion. But after Copernicus, we were not going to fall for that notion. With so many galaxies, we could not be so lucky as to live in the *one* in the center. If it looked that way to us, it must look the same way to observers in all galaxies—otherwise we would be special. This led to the homogeneous, isotropic, Big Bang models of general relativity. Gamow, Herman, and Alpher used these to predict the existence of the CMB radiation 17 years before its discovery by Penzias and Wilson. It was one of the greatest predictions to be verified in the history of science. This success was achieved in large measure by taking the Copernican Principle seriously and then following wherever it would lead.

Interestingly, the total longevity of our species predicted by the Copernican formula agrees remarkably well with the actual longevities of other species on Earth. My 95% confidence prediction for the total longevity of *Homo sapiens* was between 205,100 years and 8 million years (that's just the 200,000 years we have already had plus the additional 5,100 to 7.8 million years we are likely to have in the future). *Homo erectus*, our parent species, lasted 1.6 million years, and the Neanderthals lasted only about 300,000 years. Mammal species have an average longevity of 2 million years, and other groups of species on Earth have average longevities of between 1 and 10 million years. Even the fearsome *Tyrannosaurus rex* went extinct after existing for only 2.5 million years. It was knocked out by an asteroid strike about 65 million years ago.

Keep in mind that my Copernican prediction is only based on our past longevity as an intelligent species—that is, one that is self-conscious and able to ask questions like this—a species able to do algebra, as Neil would say. If we were really going to last another trillion years, we would be very lucky to find ourselves at such an early epoch in the history of our species, a mere 200,000 years from the beginning and, furthermore, at just such an epoch that our past longevity would predict a total longevity in line with other species. If we were observing at some random location in that trillion year history, say, 400 billion years from now, we would already know that our species had lasted far longer than other species and would rightly project a long future longevity for ourselves as well. I would be far more optimistic about our future if the human race were already 400 million years old rather than 200,000 years old, our actual observed longevity so far.

Homo sapiens could, in principle, have a far greater longevity than other species simply because we are an intelligent species. But we are still mammals, and we have a Copernican-predicted longevity quite in line with the longevities of other mammals. Even though mammals are much smarter than the average species, their longevity is not appreciably longer, and hominids (like *Homo erectus* and the Neanderthals) lasted no longer than typical mammal species. Intelligence and longevity do not seem to be correlated. This should give us pause.

Indeed, if we simply used actuarial data on other mammal species to forecast our future longevity, we would find a future longevity of between 50,600 years and 7.4 million years (with 95% confidence). These limits are within the limits implied by the Copernican Principle based only on our past longevity as an intelligent species. As long as we stay on Earth, we are subject to the same dangers that have caused other species to go extinct, and the fact that we have been around only 200,000 years should make us worry that our intelligence will not necessarily improve our fate relative to other species. Einstein was very smart, but he did not last longer than the rest of us. Intelligence may not be all that helpful for species longevity.

Now you might think that's fine. Sure, *Homo sapiens*, will go extinct, but that's okay, because we will give birth to an even more intelligent species in the future to replace us. Darwin noted, however, that most species leave no descendant species at all. A few species leave many descendant species; they propagate well. But most species die off without progeny. In this regard, note that all other species in our hominid family (including the Neanderthals, *Homo heidelbergensis*, *Homo erectus*, *Homo habilis*, and *Australopithecus*) have gone extinct. We are the only hominid species left. The rodent family, by comparison, has 1,600 species alive today. They are doing well and have many chances for survival. In a wonderful book called *After Man*, Dougal Dixon speculated on what might happen after another 50 million years of evolution, and it was not to our liking. We humans were gone in a million years. Fifty million years from now rabbits were still prevalent but they had grown up to be as big as deer and were hunted by packs of ratlike creatures, descended from present-day rodents. The scary thing about this book and the future world it imagines is that it seems so reasonable, and yet it's clearly not what we would like to hear. None of the animals left on Earth were intelligent observers, able to ask questions like, "How long will my species last?" Of course the chances are low that Dougal Dixon's specific animal predictions will come true, because there are many ways for evolution to proceed, but the book points out that quite plausibly, most of these ways do not include intelligent observers in the far future. Stephen Jay Gould made a similar point, calling us "just one bauble" on the Christmas tree of evolution.

The same Copernican argument applies to our entire intelligent lineage: our species plus any intelligent species which might descend from us in the future. We are the first intelligent species in our lineage (able to ask questions like this), so the current age of our entire intelligent lineage is only 200,000 years (only 1/65,000 of the age of the universe) and, therefore, our entire intelligent lineage is not likely to go on forever,[4] and its future lifetime should have the same limits as those we found for our species. We might well be the only intelligent species in our lineage—given that we observe that we are the first. This accords with Darwin's observation that most species leave no descendants when they go extinct.

There are some times you shouldn't use the formula. Don't use it at a wedding one minute after the vows have been said to forecast that the marriage has only 39 minutes left! You have been invited to the wedding at a special time to witness its beginning. But most of the time you can use the Copernican formula. Since I introduced it, the formula has been tested many times and successfully predicted future longevities of everything from Broadway plays and musicals, to governments, to reigns of world leaders.[5] Another exception: do not use it to estimate the future longevity of the universe. You may live in a special (habitable) location, because you are an intelligent observer. (Intelligent observers were not present in the hot early universe and may die out when main sequence stars burn out.) But among intelligent observers your location in spacetime should not be special. In general, the Copernican formula works because out of all the places for intelligent observers to be, there are by definition only a few special places, and many nonspecial places. You are simply likely to live in one of those many nonspecial places. Also, your current observation is not likely to be in a special location relative to the full array of observations made by intelligent observers.

We do not have actuarial data on the longevities of intelligent species in the universe. We have no data on how long they may last. But we do know our own past longevity as an intelligent species, and we should not ignore that salient piece of data. The Copernican formula tells us how to use that information to make a rough estimate of our future longevity with 95% confidence.

If you are not special, you should expect yourself to be born somewhere randomly on the chronological list of human beings. Approximately 70 billion people have been born in the past 200,000 years. The Copernican formula gives a 95% chance that the number born in the future should be somewhere between 1.8 billion and 2.7 trillion. I found out from the referee of my 1993 Nature paper, Brandon Carter of Anthropic Principle fame, that he and John Leslie and Holgar Nielson had also pointed out that it was unlikely that you were in the first tiny fraction of all human beings ever to be born. Carter (and later Leslie elaborating on Carter's work) used Bayesian statistics to come to this conclusion, whereas Nielson had independently arrived at the same conclusion using the idea that you should occupy a random position on the chronological list of human beings—like my own line of reasoning. I had found like-minded colleagues.

You are likely to come from a country with a population higher than the median. Half of the 190 countries in the world have populations of less than 7 million. But since more people live in the more populous countries, about 97% of all the people in the world live in countries having populations above the median. Were you born in a country with a population of more than 7 million? Just as you are likely to live in a country with a population above the median, you are likely to live in a high-population century. Indeed, you live in a century whose population is the highest it has ever been. You expect to live after some event (like the discovery of agriculture) that caused the population to soar, but before some event that causes the population to fall. You expect to live in a population spike, where the population is larger than that in a median century. This spike could occur at any random point in human history. If you want to know how many people will live after you, ask how many have lived before. If you want to know how long humanity will live in the future, ask how long it has lived in the past.

You are likely to live in an intelligent civilization that is above the median population for intelligent species in the universe—for the same reason that you are likely to live in a high-population country. Most intelligent observers live in civilizations that are above the median, and you are likely to be one of those many observers, rather than among the few who come from civilizations that are below the median in population. That means that our current population on Earth is likely to be above the median population for intelligent species in the universe. That's not the usual situation in science fiction stories, where a big galactic civilization of extraterrestrials arrives to attack our puny Earth. Although this makes good drama—we are David facing an extraterrestrial Goliath—it does not fit the probabilities. We ourselves are likely to be one of the more successful civilizations in terms of population! High-technology civilizations are likely to have large populations, so we may expect to be one of those.

In 2015, Fergus Simpson of the University of Barcelona pointed out an interesting corollary: since we are likely to come from a planet with a larger-than-median population, most planets inhabited by intelligent observers are likely to be smaller than Earth. Thus, searches for intelligent life, or any kind of life, should focus more attention on planets smaller than Earth—where most of the examples may lie.

We can also make a Copernican, 95% confidence level upper limit for the mean longevity of radio-transmitting civilizations in the galaxy, a value to plug into the Drake equation discussed in chapter 10. This is based on the idea that you are not likely to be special among the intelligent observers living in radio-transmitting civilizations. You are likely to live in one of the longer-lived radio-transmitting civilizations, because they contain more intelligent observers over time. Also, you are unlikely to live at the start of our radio-transmitting epoch. Still, a few radio-transmitting civilizations can always be longer lived than we are, and they contribute to the average. Imagine adding the civilizations to make one long timeline whose length is equal to the total longevities of all civilizations added together. Rank the radio-transmitting civilizations in order of longevity, with the longest-lived radio-transmitting civilization at the end of the long timeline. If you are not special, you should be located randomly in that long timeline, and randomly within the *Homo sapiens* time segment (giving the total longevity of our radio-transmitting civilization). I was able to use this idea, and some fancy algebra, to set a 95%-confidence upper limit on the mean longevity of radio-transmitting civilizations: 12,000 years. If the mean longevity were longer than that, my 1993 paper would appear either unusually early in our radio-transmitting civilization, or unusually early in the timeline of all radio-transmitting civilizations. This yields a Copernican estimate you can plug into the Drake equation: $L_C < 12,000$ years (with 95% confidence). Neil has used this estimate in chapter 10.

If you think intelligent species typically evolve into intelligent machine species or genetically engineered species, then you must ask yourself—why am I not an intelligent machine? Why am I not genetically engineered?

If you think intelligent species typically colonize their galaxy, then ask yourself—why am I not a space colonist? In 1950, Enrico Fermi asked a famous question about extraterrestrials: Where are they? Why haven't they colonized Earth already, long ago? The Copernican Principle offers an answer to Fermi's question: A significant fraction of all intelligent observers must still be sitting on their home planets (otherwise you would be special). Colonization must not occur that often. Importantly, this means that we are allowed to apply the Drake equation in the first place: it estimates the number of intelligent civilizations that arise independently on their home planets. If colonization is not common, that is approximately equal to the total number of extraterrestrial civilizations we will find.

Suppose you thought a priori that each of the following two hypotheses was equally probable:

H1. Humans would stay on Earth until they went extinct.
H2. Humans would colonize 1.8 billion habitable planets in the galaxy in the future.

Bayesian statistics says that you must multiply your prior probabilities for hypotheses H1 and H2 by the likelihood of observing what you are observing, given either H1 or H2. Under the H1 hypothesis that we stay on Earth, then there is a 100% likelihood that you as a human being would observe that you were on Earth. But if humans colonize 1.8 billion planets (i.e., if H2 is true), then as a human being, there is only 1 chance in 1.8 billion that you would find yourself on the first planet out of 1.8 billion that humans lived on. Therefore, even if you viewed the odds as 1:1 initially that we would colonize the galaxy rather than stay on Earth, after considering that you are living on Earth, Bayesian statistics would require you to reevaluate the odds as 1.8 billion:1 against colonization of the galaxy. My Copernican argument would simply say that if you are not special, there is only one chance in 1.8 billion that you would find yourself in the first 1/(1.8 billion) of all planets inhabited by humans and therefore only one chance in 1.8 billion that we would go on to colonize 1.8 billion planets, given that you are on the first one. Nevertheless, our colonizing a few more planets in the future, starting with Mars, would not be that improbable and could give us more chances to survive. We should be doing that quickly, while we still have a space program.

The goal of the human spaceflight program should be to improve our survival prospects by colonizing space. This could be achieved at reasonable cost. For example, you could begin by sending eight astronauts, both men and women, to Mars, and they could multiply there using indigenous materials. You just have to find a handful of astronauts who would be willing to take a one-way trip to Mars and stay there to have their children and grandchildren—people who would rather be founders of a Martian civilization than return to be celebrities back on Earth. It is easy to find such daring people. Story Musgrave, the astronaut I know best, once told me that he would happily volunteer for a one-way trip to Mars. The Mars One group has found a hundred serious candidates who would like to be Mars colonists. Frozen egg and sperm cells could be taken along for genetic diversity. (In this way, even though only a handful of astronauts would actually be sent, many people born on Earth could ultimately have descendants on Mars.) Mars has reasonable gravity (1/3 that of Earth), an atmosphere, water, and all the chemicals necessary for life—unlike the Moon. The atmosphere is CO_2, from which oxygen for breathing can be obtained, and water is plentiful in permafrost and in Mar's polar caps. Radiation levels could be tolerable if the colony were placed 10 meters below ground and the colonists made only brief forays on the surface. Our ancestors lived in caves—so could our Martian colonists. Our orbiters have even found some nice cave mouths on Mars worth checking out.

I have shown that placing such a colony on Mars would require us to put into orbit only as much mass in the future as we have already done in the past, not too much to ask for. According to Robert Zubrin, taking eight astronauts to Mars and supplying them with emergency return vehicles (hopefully not to be used) would require launching 500 tons into low-Earth orbit. From there they would be launched on a trajectory to Mars and aerobrake into its atmosphere prior to landing. According to Gerard O'Neill, a space colony requires as little as 50 tons per person to create a "life in a closed system" biosphere. Delivering these 400 tons to the Martian surface requires launching about 2,000 tons into low-Earth orbit. Thus a self-supporting Mars colony of eight colonists would require launching 2,500 tons into low-Earth orbit. By comparison, the *Saturn V* rockets of the Apollo program and the U.S. Space Shuttles have launched over 10,000 tons into low-Earth orbit, with the Russian and Chinese human space-flight programs adding even more. NASA is currently considering building a heavy-lift vehicle capable of putting a payload of 130 tons into low-Earth orbit (a *Saturn V*-class vehicle). Twenty launches would be sufficient to build the colony (compared with the 18 *Saturn V* rockets that were built for the Apollo program). Four of these rockets could be constructed at a time in the vertical assembly building at the Kennedy Space Flight Center. If it takes a decade to develop such a rocket, and four are launched in every 26-month launch cycle, the Mars colony could be completed in an additional 9 years. Starting now, the colony would take only 19 years to finish. The human spaceflight program is 55 years old as I write this; the Copernican Principle suggests that funding for the human spaceflight program has at a 50% chance of lasting for at least another 55 years—long enough to establish a Mars colony. Asking for such a Mars colony is not unreasonable. Elon Musk, head of Space X, is interested in privately funded efforts to colonize Mars. I once shared a podium with him at a Mars conference organized by Robert Zubrin. I told of my reasons the human race should want to colonize Mars in the near future, and Elon told how he would go about doing it! Neil has made the case for going to Mars in his book *Space Chronicles*. Planting a colony on Mars would change the course of world history, in fact you couldn't even call it "world" history anymore! Stephen Hawking has recently added his voice, saying, in an interview with bigthink. com: "I believe that the long-term future of the human race must be in space. It will be difficult enough to avoid disaster on planet Earth in the next hundred years, let alone the next thousand, or million. The human race shouldn't have all its eggs in one basket, or on one planet. Let's hope we can avoid dropping the basket until we have spread the load."

If Martian couples had four children on average, the population could double every 30 years and reach 8 million after 600 years. (Small populations can grow—the entire original aboriginal population of Australia is thought to have descended from as few as 30 individuals who landed there by raft from Indonesia 50,000 years ago. This population had grown to between 300,000 and 1 million by the time of European settlement.) If you are worried about the funds for the space program being canceled, planting a self-supporting colony is just what you want. Don't send astronauts to Mars and then bring them all back to Earth. Instead, leave them there, where they can help our survival prospects. Colonizing Mars would give our species two chances instead of one, and might as much as double our long-term survival prospects. It would be a life insurance policy against any catastrophe that might overcome us on Earth, from climatological disasters, to asteroid strikes, to surprise epidemics. It might also double our chances of ever getting to Alpha Centauri. Colonies can found other colonies. The first words spoken on the Moon were in English not because England sent astronauts to the Moon but because they founded a colony in North America that did.

If we look around, we can see the universe showing us what we should be doing. We live on a tiny speck in a vast universe. The universe tells us: spread out and increase your habitat to improve your survival prospects. We live on a planet littered with the bones of extinct species, and the age of our species is tiny relative to that of the universe as a whole. We should spread out before we die out. We have a space program only about half a century old that is capable of sending us to other planets. We should make the wisest possible use of it before it is gone. Will we venture out, or turn our backs on the universe? The

FIGURE 24.4. The *Apollo 11* liftoff.

Photo credit: J. Richard Gott

fact that we are having this conversation on Earth is a warning that there is a significant chance that we will end up trapped on Earth.

In the summer of 1969, I did more than just visit the Berlin Wall. I visited Stonehenge. At that time, Stonehenge was about 3,870 years old. It's still there! I also went to Florida specifically to see the *Saturn V* rocket take off, sending Neil Armstrong, Buzz Aldrin, and Michael Collins to the Moon on *Apollo 11*. At that time, *Saturn V* rockets had been taking off for the Moon for 7 months. In another 3.5 years, such *Saturn V* launches to the Moon would be over. The sight of the *Saturn V* rocket launching was spectacular (see my photo of it in figure 24.4). As it rose higher and higher, it looked like some magic sword, trailing a plume of fire much longer than itself. I had never seen anything like it. A crowd of about a million people had come to see it. They watched the launch in perfect silence, but after the rocket disappeared into a high layer of cirrus clouds, the crowd let loose with tremendous cheering. Colonizing space is what we should be doing.

Our intelligence gives us great potential, the potential to colonize the galaxy and become a supercivilization, but most intelligent species must not achieve this—or you would be special to find yourself still a member of a one-planet species. We control energy sources that are far less powerful than that of our own Sun. We are not very powerful, and we have not been around for very long. But we are intelligent creatures and we have learned a lot about the universe and the laws that govern it—how long ago it started, how its galaxies and stars and planets formed. It is a stunning accomplishment whose story we have told here.

ACKNOWLEDGMENTS

This book, and the course that initially inspired it, came to fruition due to the hard work of many people. To start with, we thank our fellow faculty at Princeton University, from whom we have learned so much over the years, and who have given us such a productive and congenial atmosphere in which to work. We especially thank Professor Neta Bahcall, who had the initial vision that brought the three of us together.

We thank our students, including Cullen Blake, Wes Colley, Julie Comerford, Daniel Grin, Yeong-Shang Loh, Justin Schafer, Joshua Schroeder, Zack Slepian, Iskra Strateva, and Michael Vogeley. We thank Ramin Ashraf, Sorat Tungkasiri, Paula Brett, Sofia Kirhakos Strauss (Michael's wife), and Kathy Gryzeski for help along the way, as well as Lucy Pollard-Gott (Rich's wife), who copyedited the entire book. We thank Robert J. Vanderbei for sharing some of his astrophotographs, and Li-Xin Li for help with graphics. We also thank Adam Burrows, Chris Chyba, Matias Zaldarriaga, Robert J. Vanderbei, and Don Page for helpful conversations.

At Princeton University Press, we thank our production editor, Mark Bellis; our copyeditor, Cyd Westmoreland; and, for her extraordinary faith and vision, our editor, Ingrid Gnerlich.

MICHAEL A. STRAUSS

NEIL DEGRASSE TYSON

J. RICHARD GOTT

APPENDIX 1

DERIVATION OF $E = MC^2$

Suppose you had a laboratory with a particle moving slowly from left to right inside it with velocity v much, much less than c (i.e., $v \ll c$). Newton's laws will apply, and if the particle has a mass m, it will have, according to Newton, a momentum $P = mv$ pointed toward the right. The particle gives off two photons, each of energy $E = h\nu_0$ in opposite directions: one to the right and one to the left. The particle loses an amount of energy $\Delta E = 2h\nu_0$, equal to the energy the particle sees carried off by the two photons. Einstein showed that the momentum of a photon is equal to its energy divided by the speed of light c. The particle sees the two photons carry away equal amounts of momentum but in opposite directions, making the total momentum carried off by the two photons zero as seen by the particle. The particle "thinks" it is at rest (by Einstein's first postulate), and it gives off two equal photons in opposite directions. By symmetry, a particle at rest that gives off two equal-frequency photons in opposite directions stays at rest. The recoils from the two photons cancel out. The particle's worldline remains straight: it does not change in velocity. (Refer to figure 18.4.)

The photon going to the right will eventually slam into the right wall of the lab. It hits the wall, and the wall is pushed a tiny bit toward the right. This is the effect of radiation pressure: the wall absorbs the momentum of the photon, and this pushes the wall to the right. An observer sitting on the right wall will see the photon headed to the right hitting the right wall with a frequency that is higher than the emitted frequency (it will be shifted toward the blue end of the spectrum), because the particle is approaching the right wall. This is an instance of the Doppler effect. In contrast, an observer sitting on the left wall of the lab will see a redshifted photon traveling to the left hit the left wall with a lower frequency than emitted, because the particle is going away from him. A higher-frequency (bluer) photon carries a larger momentum than does a lower-frequency (redder) photon. So, the right wall receives a harder kick (to the right) than the left wall receives (to the left). The two kicks do not cancel out, and the lab receives an overall kick to the right. Let's calculate how big that overall kick is.

The time between wave crests in the emitted photons (seen as light waves) as measured by the particle is Δt_0. The time between the emission of the two wave crests, Δt_0, is equal to 1 over the frequency of the light ν_0 as seen by the particle. If the light has a frequency of 100 cycles per second; for example, the time between wave crests is 1/100 of a second. Thus, $\Delta t_0 = 1/\nu_0$. Let v be the velocity of the particle relative to the lab. The particle's clock will tick (as measured in the rest frame of the lab) at a rate of $\sqrt{[1 - (v^2/c^2)]}$ that of the lab clock, as we have discussed. But in this calculation, we are supposing that $v \ll c$, and so we are going to ignore all terms that are of order (v^2/c^2), and only keep terms that are of order (v/c). (For example, if $v/c = 10^{-4}$, corresponding to the 30 km/sec orbital speed of Earth around the Sun, then $v^2/c^2 = 10^{-8}$; this second term is so small that it can be neglected relative to the first.) Since we are working in the limit where $v \ll c$, the rate at which the particle's clock ticks is essentially the same as the rate at which the lab clock ticks. That means that the time interval between ticks as seen by the particle (Δt_0) and the lab ($\Delta t'$) are essentially the same, because the particle is moving so slowly.

An observer at rest with respect to the lab, therefore, also sees a time $\Delta t' = \Delta t_0 = 1/\nu_0$ pass between the emission of the first wave crest and the next wave crest emitted by the particle. (Refer to figure 18.4, where the time interval $\Delta t'$ is shown as a vertical dashed line.) At the moment the next wave crest is emitted toward the right by the particle, it lags behind the first wave crest by a distance $d = (c - v)\Delta t'$. That is equal to the distance the light beam has traveled in the time $\Delta t'$ (which is $c\Delta t'$) minus the distance the particle has traveled (which is $v\Delta t'$). The two wave crests are both traveling to the right at speed c (by Einstein's second postulate); thus they travel in parallel, and the distance between them remains fixed at $d = (c - v)\Delta t'$. The wavelength λ_R of the light seen by an observer sitting on the right wall of the lab is equal to this distance between wave crests, so $\lambda_R = (c - v)\Delta t'$. The spacetime diagram in figure 18.4 illustrates the thought experiment. This distance λ_R between wave crests is measured at an instant of lab time (along a horizontal line in the spacetime diagram.)

The time interval between the arrival of the two wave crests at the right wall is therefore $\Delta t_R = \lambda_R/c = (c - v)\Delta t'/c$, and the frequency of the photon going to the right is observed to be $v_R = 1/\Delta t_R = c/[(c - v)\Delta t'] = v_0 c/(c - v)$. Now for $v \ll c$, the quantity $c/(c - v)$ is approximately $[1 + (v/c)]$ keeping only terms of first order in v/c. (For example, if $v/c = 0.00001$, $c/(c - v) = 1/0.99999 = 1.00001$ to high accuracy—try it on your calculator). Thus, an observer sitting on the right wall of the lab sees the photon headed to the right hit the wall with frequency $v_R = v_0 [1 + (v/c)]$. He sees a higher frequency than the emitted frequency v_0 by a factor of $[1 + (v/c)]$ due to the Doppler effect, where v is the velocity of the particle. This is the standard Doppler-shift formula for blueshifted light hitting the right wall of the lab from a particle moving at low velocity v toward the wall.

When the photon going to the right hits the right wall, it imparts a rightward momentum of $h v_R/c = h v_0[1 + (v/c)]/c$ to the wall.

The particle also emits a photon traveling to the left. It will eventually hit the left wall. An observer sitting on the left wall of the lab sees this photon going to the left hit this wall with a frequency of $v_L = v_0[1 - (v/c)]$. The sign of the velocity in the formula is reversed, because the observer on the left wall sees the particle moving away from him with a velocity v. He sees a frequency that is lower than the emitted frequency because of the Doppler effect. The total rightward momentum imparted to the lab by the two photons is therefore equal to the amount of momentum imparted by the rightward photon, $h v_0[1 + (v/c)]/c$, minus the momentum imparted by the leftward particle, $h v_0[1 - (v/c)]/c$, which is going in the opposite direction. This gives $2 h v_0(v/c^2)$ for the total rightward momentum imparted by the two photons to the lab. This total rightward momentum is imparted to the lab because the high-frequency (bluer) photon traveling toward the right imparts a bigger punch, which is not offset by the smaller punch imparted by the low-frequency (redder) photon traveling toward the left. Now $2 h v_0 = \Delta E$, is just the energy given off by the particle in the form of the two photons. So the rightward momentum acquired by the lab is $\Delta E\, v/c^2$. The factor v/c^2 comes from a factor v/c due to the Doppler shifts and a factor of $1/c$ due to the ratio of momentum to energy carried by photons.

Conservation of momentum requires that the amount of rightward momentum acquired by the lab be equal to the rightward momentum lost by the particle. The rightward momentum of the particle is mv (since $v \ll c$, Newton's formula for momentum is accurate). Because the particle's velocity is unchanged, the only way for it to lose rightward momentum mv is for it to lose mass. Its loss of rightward momentum must be $v\Delta m$ where Δm is the mass lost by the particle.

Setting $\Delta E v/c^2 = v\Delta m$, we find $\Delta E/c^2 = \Delta m$. The small velocity v of the particle cancels out! As long as $v \ll c$, the answer does not depend on v. Multiplying both sides of the equation by c^2 gives $\Delta E = \Delta m c^2$. The particle loses mass. The amount of mass lost, Δm, multiplied by c^2 is equal to the amount of energy in the two photons given off, ΔE. Get rid of the delta (Δ) signs on both sides of the equation, and you have $E = mc^2$. The energy given off by the two photons is equal to the mass lost by the particle multiplied by c^2. When the particle loses mass, it gives off an amount of energy given by $E = mc^2$. Many books explain the significance of the equation and how it works, but they don't tell you how you could derive it. Now you know how to do it.

APPENDIX 2

BEKENSTEIN, ENTROPY OF BLACK HOLES, AND INFORMATION

Current hard drives 6 inches in diameter can store about 5 terabytes or 4×10^{13} bits of information. How many bits of information could you possibly pack into a hard drive 6 inches in diameter? First, since this is a thought experiment, make it spherical to pack the most volume inside that diameter—it's about the size of a grapefruit with a radius of 7.5 cm. Bekenstein showed that a black hole had a finite entropy proportional to the area of its event horizon. In the end, the entropy of a black hole horizon (S) turned out to be exactly ¼ of the area of the event horizon when the area is measured in Planck lengths squared (the exact value being obtained ultimately by Hawking). Measured in Planck units, the surface area of a black hole of radius 7.5 cm is $4\pi(7.5 \text{ cm}/1.6 \times 10^{-33} \text{ cm})^2 = 2.76 \times 10^{68}$. One quarter of that is an entropy of $S = 6.9 \times 10^{67}$. A specific amount of entropy (increase in disorder) corresponds to a specific amount of destruction of information. The number of bits of information corresponding to an entropy S is $S/\ln 2$. The natural logarithm of 2 (denoted as "ln 2" in the formula) is 0.69. The 2 comes into it, because one bit of information is the answer to one yes-or-no question, which has 2 possibilities. (A game of 20 questions, which answers 20 yes-or-no questions, gives you 20 bits of information. If you know I'm thinking of a number between 1 and 2^{20}, which is about a million, your first question would be: is it in the upper half? Keep dividing the allowed region by two. After 20 questions, you will have guessed my number.) The creation of a black hole of radius 7.5 cm is thus an increase in the disorder of the universe equal to the destruction of 10^{68} bits of information. There are 2^(10^68) different ways to make such a black hole, taking 10^{68} bits of information to describe, and the information on what the black hole was made of is lost when the black hole forms. If the 7.5-cm-radius hard drive contained more than 10^{68} bits of information, more than 10^{68} bits of information would be lost if you collapsed it (i.e., crushed it to make it smaller and smaller until it formed a black hole smaller than 7.5 cm). But that's not allowed, because if more than 10^{68} bits of information are lost when a black hole forms, the black hole that forms must have a radius of more than 7.5 cm. That's a contradiction. So what actually happens is that as you try to pack more and more information into your hard drive with a fixed radius of 7.5 cm, its mass will increase, until, when it contains 10^{68} bits of information, its mass will be 8.4 times the mass of Earth, and it will collapse to form a black hole. Thus 10^{68} bits of information (1.16×10^{58} gigabytes) is the upper limit on the amount of information a 6-inch-diameter hard drive could store.

NOTES

CHAPTER 1: THE SIZE AND SCALE OF THE UNIVERSE

1. Technically, a megabyte is $2^{20} = 1,048,576$ bytes, and a gigabyte is $2^{30} = 1,073,741,824$ bytes. But colloquially, these are rounded to an even million and a billion.

CHAPTER 3: NEWTON'S LAWS

1. D. T. Whiteside, "The Prehistory of the 'Principia' from 1664 to 1686." *Notes and Records of the Royal Society of London* 45, no. 1 (January 1991): 38.

CHAPTER 9: WHY PLUTO IS NOT A PLANET

1. Comet Hale-Bopp, for example, 35 kilometers in diameter, was discovered only 2 years before its closest approach to the Sun. If it had been headed for us, it would have hit Earth with the explosive force of 4 billion megatons of TNT, more than 60 million times the most powerful H-bomb ever exploded.

CHAPTER 10: THE SEARCH FOR LIFE IN THE GALAXY

1. Perhaps there is a get-out-of-jail-free card for the screenwriter. At the beginning, Jodie says there are 400 billion stars in our galaxy alone, but at the end she says there are millions of civilizations out there. Did that mean, out there in the galaxy, as anyone might have thought, or could it have meant out there in the universe? Let's try that out. There are 130 billion galaxies in the visible universe (Jodie is looking for extraterrestrials, and you can at most only look in the visible universe). In that case, you would have to multiply 0.0000004 civilizations by 130 billion, that gives 52,000 civilizations in the visible universe, not millions. So even that won't work.

CHAPTER 14: THE EXPANSION OF THE UNIVERSE

1. That historical limit has just now been extended by the European Space Agency's Gaia space-craft, currently making the best measurements of stellar parallaxes, which will allow the distances of stars to tens of thousands of light-years to be determined.

CHAPTER 17: EINSTEIN'S ROAD TO RELATIVITY

1. It is important that scientific hypotheses be falsifiable, according to the criteria established by philosopher Karl Popper.
2. I observe an astronaut's light clock as illustrated in Figure 17.1. In the general case, the astronaut moves past me at a velocity v. I observe the astronaut's light clock as she moves by me from left to right. While the light makes progress of 1 foot along the diagonal line, the rocket makes progress of v/c feet from left to right. During this time, the light makes vertical progress of $\sqrt{[1 - (v^2/c^2)]}$. That's because a right triangle with a diagonal hypotenuse of length 1, a horizontal side of length v/c, and a vertical side of $\sqrt{[1 - (v^2/c^2)]}$ satisfies Pythagoras's theorem for right triangles. The square of $\sqrt{[1 - (v^2/c^2)]}$ is just $[1 - (v^2/c^2)]$, and that plus (v^2/c^2) equals 1^2. Pythagoras is happy. During the time the light beam in my clock is moving 1 foot upward, I see her light beam making upward progress of only $\sqrt{[1 - (v^2/c^2)]}$ feet. If I age 10 years, she ages 10 years times $\sqrt{[1 - (v^2/c^2)]}$.

CHAPTER 18: IMPLICATIONS OF SPECIAL RELATIVITY

1. J. Richard Gott, "Will We Travel Back (or Forward) in Time?" *Time* magazine, April 10, 2000, 68–70.

CHAPTER 19: EINSTEIN'S GENERAL THEORY OF RELATIVITY

1. The Riemann curvature tensor $R^{\alpha}{}_{\beta\gamma\delta}$ in four dimensions has 256 components. Each of its indices (superscripts or subscripts) α, β, γ, and δ can take on any of four values corresponding to one of the four dimensions of spacetime (t, x, y, and z). That gives $4 \times 4 \times 4 \times 4 = 256$ components.
2. $T_{\mu\nu}$ is the *stress energy tensor*, which describes the stuff at a particular location in spacetime: mass-energy density, pressure, stress, energy flux, and momentum flux. The metric $g_{\mu\nu}$ (which we have encountered before: in flat spacetime, it is given by $ds^2 = -dt^2 + dx^2 + dy^2 + dz^2$) tells us how distances in space and time are measured. $R_{\mu\nu}$ and R can be calculated from the components of the Riemann curvature tensor. The tensors in Einstein's equations have two indices that can take on any of four values and so represent $4 \times 4 = 16$ equations. Ten of these equations are independent.
3. From a lecture at the University of Glasgow, June 20, 1933. Published in Albert Einstein, *The Origins of the Theory of Relativity*, reprinted in *Mein Weltbild* (Amsterdam: Querido Verlag, 1934), 138; and in *Ideas and Opinions* (reprint, New York: Broadway Books, 1995), 289–290.

CHAPTER 20: BLACK HOLES

1. Private communication from Don Page, Hawking's student. He has recounted this story in "Hawking Radiation and Black Hole Thermodynamics" Don N. Page, Alberta University, September 2004. Published in *New Journal of Physics* 7 (2005): 203, ALBERTA-THY-18-04, DOI: 10.1088/1367-2630/7/1/203, e-Print: hep-th/0409024 | PDF, This account is in accord with Hawking's own account of the events in his book, *A Brief History of Time*, 99–105.

CHAPTER 22: THE SHAPE OF THE UNIVERSE AND THE BIG BANG

1. Mark Alpert and I investigated how general relativity would work in Flatland. We found that the geometry around a point mass looks conical, and that distant objects would not attract one another in Flatland, because empty space is locally flat (i.e., a cone can be made out of a flat piece of paper by cutting out a slice and taping the edges together). This work on Flatland would eventually inspire my work on cosmic strings. To get an exact solution for a cosmic string, all I had to do was add a vertical coordinate to our exact Flatland solution for a point mass. In this case, our exploration of a fanciful world led to some solutions of interest in the real world. The fact that point masses in Flatland do not gravitationally attract one another would make aggregating mass to form planets in Flatland more difficult.
2. This concept was updated by A. Dewdney in his 1984 book *Planiverse*. In 2007, an animated movie version of *Flatland* starred Martin Sheen and Kristen Bell, who voiced the characters of Arthur Square and his granddaughter Hex, a hexagon. One of my mentors from my undergraduate days at Harvard, Thomas Banchoff, added illuminating mathematical commentary to the movie as an extra feature on the DVD.

CHAPTER 23: INFLATION AND RECENT DEVELOPMENTS IN COSMOLOGY

1. In my 1982 *Nature* paper, I said, "our Universe is one of the normal vacuum bubbles."
2. In that same paper, following Sidney Coleman's work on bubble formation, I identified quantum tunneling as the process that would create the bubble universes: "Thus we can see the formation of our Universe as a quantum tunneling event."
3. The title of Hawking's 1982 paper was "The Development of Irregularities in a Single Bubble Inflationary Universe" and included references to the papers by Linde, Albrecht, and Steinhardt, and me, among others. The events of that year are recounted in *Physics News in 1982*, published by the American Institute of Physics, which used a key diagram from my paper on its cover.

CHAPTER 24: OUR FUTURE IN THE UNIVERSE

1. I describe all this in great detail in my book, *The Cosmic Web* (2016).
2. I have discussed these three scenarios, $w > -1$, $w = -1$, and $w < -1$, and their implications, in even more detail in *The Cosmic Web*.
3. I published this in the scientific journal *Nature* on May 27, 1993, in a paper titled "Implications of the Copernican Principle for our Future Prospects."

4. Is our intelligent lineage (*Homo sapiens* and its intelligent descendants) likely to last forever? Our intelligent lineage has been around for 200,000 years. That is very short relative to the age of the universe, 1 part in 65,000. As our intelligent lineage grows older, the ratio of its age to the age of the universe must approach 1; if our intelligent lineage lasts forever, most of its observers must find its age to be of the same order of magnitude as the universe itself. You don't observe that, so that would make you special. We can quantify this idea. Imagine plotting a two-dimensional diagram where the vertical coordinate y represents the age of the universe when our intelligent lineage starts, and the horizontal coordinate x represents the age of the universe when you observe. Each point in the plane thus represents a possible observation by you. But there are constraints. Both ages x and y are positive (limiting your observation to the upper right quadrant of the plane). Since your observation must occur after our intelligent lineage starts, it must be true that $x > y$. That limits your observation to half of that quadrant or 1/8 of the entire plane—the east to northeast octant of the plane. You can visualize this as a 45°-wide region fanning out from the origin to infinity—since we are assuming our intelligent lineage lasts forever. Your observation point (with the values of x and y that you observe) could be a point anywhere in this 45°-wide fan. If your observation is not special, there should be only a 1/45 chance your point of observation lies within 1° of the bounding diagonal line $x = y$, for example. But in fact, you are even closer to that diagonal line. You observe $x = (1 + [1/65{,}000])\, y$. That (x, y) point, measured from the origin, is only 0.00044° from the upper edge (the line $x = y$). The probability of being that close to the edge by chance if an observation is not special is $P = 0.00044°/45° = 10^{-5}$. Thus, if our intelligent lineage were to last forever, and your observation is not special, it would be highly unlikely (a probability of only 10^{-5}) for you to find our intelligent lineage only 1/65,000 as old as the universe itself, or less. The Copernican Principle says that it is highly unlikely ($P = 10^{-5}$) for you to find yourself in a situation where your location is special to one part in 100,000 (in this case in an intelligent lineage that lasts forever.) Thus, the Copernican Principle, in accord with common sense, tells us that it is highly unlikely ($P = 10^{-5}$) that our intelligent lineage will last forever. If it has an end, the Copernican formula predicts (with 95% confidence) when that end will be.

5. The Copernican formula can be tested. For example, on the date of publication of my paper, there were 44 plays and musicals open on Broadway. Those that had been open only a short time tended to close after a short time: for instance, *Marisol,* which had been open for 7 days, closed after another 10 days. That's within a factor of 39, in agreement with my prediction. The formula worked just as well for long-running plays. The famous musical *The Fantastics* had been open for 12,077 days and closed after another 3,153 days, again within a factor of 39. Overall, of the plays and musicals on my original list that have closed, I have gotten 42 out of 42 correct, with two left to be decided. I could even be wrong about those two and still get at least 95% right.

As of the same date there were 313 world leaders in power—heads of state and heads of government of independent countries. Most are out of power now; if none remaining in power continues in office past age 100, the success rate for the formula will be more than 94% (extraordinarily close to the expected 95%). In agreement with Copernican expectations, Henry Bienen and Nicholas van de Walle concluded in their book *Of Time and Power* (after a detailed statistical analysis of 2,256 world leaders): "The length of time a leader has been in power is a very good indicator of how long that leader will stay in power. Of all the variables examined it is the predictor that gives the most confidence."

On September 30, 1993, in *Nature*, P. T. Landsberg, J. N. Dewynne, and C. P. Please used my formula to predict how long the Conservative government in Britain would continue in power. They estimated with 95% confidence that, having been in power for 14 years, it would continue for at least 4.3 more months but less than 546 more years. It went out of power 3.6 years later, in agreement with the prediction.

I used the United Nations actuarial tables to calculate that if every person in the world in 1993 had applied my formula to calculate their future longevity, then for 96% of those people, the formula would have proven correct.

Philosophers Bradley Montond and Brian Kierland defended my core thesis in a 2006 article in *The Philosophical Monthly*. They argued that my formula can be used to forecast future longevity in any timescale-free problem, or in cases where the timescale is not known empirically. Any probability problem can be given a Bayesian formulation. Bayesian reasoning explains how you should revise your prior views when new data become available. My Copernican formula is equivalent to adopting what is called a *vague (Jeffreys) Bayesian prior* (also called a *public policy prior*, because it is designed for anyone to use). You then revise your prior views after considering the past longevity you have observed. This type of prior view is agnostic, weighting each order of magnitude of total longevity equally. If you do not have actuarial data on intelligent species (i.e., ones able to ask questions like this), this is arguably the best you can do, and you get exactly my Copernican formula results. Any observer can apply it, and among such intelligent observers, you should not be special.

SUGGESTED READING

Abbott, E. A. *Flatland*. New York: Dover, 1992.

Bienen, H. S., and N. van de Walle. *Of Time and Power*. Stanford, CA: Stanford University Press, 1991.

Brown, M. *How I Killed Pluto and Why It Had It Coming*. New York: Spiegel & Grau/Random House, 2010.

Ferris, T. *The Whole Shebang*. New York: Simon and Schuster, 1997.

Feynman, R. *The Character of Physical Law*. Cambridge, MA: MIT Press, 1994.

Gamow, G. *One, Two, Three . . . Infinity*. New York: Dover, 1947.

Goldberg, D. *The Universe in the Rearview Mirror*. Boston: Dutton/Penguin, 2013.

Goldberg, D., and J. Blomquist. *A User's Guide to the Universe*. Hoboken, NJ: Wiley, 2010.

Gott, J. Richard. *Time Travel in Einstein's Universe*. Boston: Houghton Mifflin, 2001.

———. *The Cosmic Web*. Princeton, NJ: Princeton University Press, 2016.

Gott, J. Richard, and R. J. Vanderbei. *Sizing Up the Universe*. Washington, DC: National Geographic, 2010.

Gould, S. J. *Wonderful Life*. New York: W. W. Norton, 1989.

Greene, B. *The Elegant Universe*. New York: Vintage Books, 1999.

Hawking, S. W. *A Brief History of Time*. New York: Bantam Books, 1988.

Kaku, M. *Hyperspace*. New York: Doubleday, 1994.

Lemonick, M. D. *The Light at the Edge of the Universe*. New York: Villard Books/Random House, 1993.

———. *The Georgian Star*. New York: W. W. Norton, 2009.

———. *Mirror Earth*. New York: Walker & Company, 2012.

Leslie, J. *The End of the World*. London: Routledge, 1996.

Misner, C. W., Thorne, K. S., and J. A. Wheeler. *Gravitation*. San Francisco: Freeman, 1973.

Novikov, I. D. *The River of Time*. Cambridge: Cambridge University Press, 1998.

Ostriker, J. P., and S. Mitton. *Heart of Darkness*. Princeton, NJ: Princeton University Press, 2013.

Peebles, P. J. E., Page, L. A., Jr., and R. B. Partridge. *Finding the Big Bang*. Cambridge: Cambridge University Press, 2009.

Pickover, C. A. *Time: A Traveler's Guide*. New York: Oxford University Press, 1998.

Rees, M. *Our Cosmic Habitat*. Princeton, NJ: Princeton University Press, 2001.

——. (ed.). *Universe*. Revised edition. New York: DK Publishing, 2012.

Sagan, C. *Cosmos*. New York: Random House, 1980.

Shu, F. *The Physical Universe*. Sausalito, CA: University Science Books, 1982.

Taylor, E. F., and Wheeler, J. A. *Spacetime Physics*. San Francisco: W. H. Freeman, 1992.

Thorne, K. S. *Black Holes and Time Warps*. New York: Norton, 1994.

Tyson, N. deG. *Death by Black Hole*. New York: W. W. Norton, 2007.

——. *The Pluto Files*. New York: W. W. Norton, 2009.

——. *Space Chronicles*. New York: W. W. Norton, 2012.

Tyson, N. deG., and D. Goldsmith. *Origins*. New York: W. W. Norton, 2004.

Tyson, N. deG., C. T.-C. Liu, and R. Irion. *One Universe*. New York: John Henry Press, 2000.

Vilenkin, A. *Many Worlds in One*. New York: Hill and Wang/Farrar, Straus and Giroux, 2006.

Wells, H. G. *The Time Machine* (1895), reprinted in *The Complete Science Fiction Treasury of H. G. Wells*. New York: Avenel Books, 1978.

Zubrin, R. M. *The Case for Mars*. New York: Free Press, 1996.

INDEX

puter simulations of, 314; cosmic microwave background (CMB) and, 318–19; as cosmic vacuum cleaners, 248; decay and, 314, 317, 407; Doppler shift and, 249; early universe and, 250; Einstein and, 300–303, 314, 317, 319–20; electromagnetism and, 317; electrons and, 317–20; energy density and, 318; entropy and, 315–16, 323, 431; evaporation of, 318–19; event horizons and, 305, 308–20, 337–38, 340, 344, 379–80, 386, 405–7, 431; firewalls and, 319; formation of, 252–53; future of universe and, 401, 403, 405, 407–8; general relativity and, 300–302, 314, 320; geometry and, 300, 306–11, 314, 316; gravity and, 13, 23, 122, 247, 249, 302–4, 314, 403; Hartle-Hawking vacuum and, 318–19, 344; Hawking and, 314–20; Hawking radiation and, 317–19, 340, 406–8; Hubble Space Telescope and, 247–48, 306; inflation and, 380, 398; jet streams of, 247–48; Karl Schwarzschild and, 300–302, 308; Kerr, 314–17, 337–39; lives of stars and, 123; lost information and, 316–17, 319, 431–32; Martin Schwarzschild and, 301–2; mass and, 121, 176, 196, 246–53, 300–311, 314–19, 338, 355, 403, 407–8; mass of, 121, 249–50, 252, 300–301, 303, 308; Mercury and, 300; Milky Way and, 247–49, 251; orbitals and, 314; Penrose and, 317; perturbations and, 300, 314; photons and, 310, 318; Planck length and, 302, 316, 407, 431; point-masses and, 300–301, 303, 309, 311; positrons and, 317–20; quantum theory and, 249, 302, 316–19; red giants and, 302–3; redshift and, 252–53, 318; Schwarzschild radius and, 301–10, 313–15, 318, 401; shape of universe and, 355, 362, 368; singularities and, 302, 305–6, 309–13; solar mass and, 247–49, 252, 303, 305–6, 314–16, 318–19, 338, 355, 403; spacetime and, 300–304, 309–17; spaghettification and, 305–6; speed of light and, 300, 302, 304–6, 309–

12, 316; Sun and, 237–50, 300–303, 313; supermassive, 121, 249–50, 252, 303, 308; temperature and, 247, 316–19; thermal, 314, 317–19; tidal forces and, 304, 306; time travel and, 337–41, 344; uncertainty principle and, 316–17, 381, 385, 393, 401, 403, 407; virtual pairs and, 317–18; wavelength and, 318; Wheeler and, 314–15; white dwarfs and, 303; white holes and, 312, 338; worldlines and, 309–13; wormholes and, 311–12, 319–20

blink comparator, 130, 134
BL Lacertae stars, 190
blueshift, 209, 217, 231, 245, 339, 341, 429
blue stars, 90, 93, 96–98, 180, 194, 204
Blumenthal, George, 234
Bode, Johann, 139
Bohr, Neils, 82, 92, 258
Boltzmann, Ludwig, 75, 80
Boltzmann constant, 73
Boötes, 120
Borde, Arvin, 384
Bradley, James, 263
Brahe, Tycho, 35–37, 43, 258
brightness: apparent, 169; eclipses and, 141; energy radiation and, 78–80; inverse-square law and, 78–79, 184, 188; lives of stars and, 116; quasars and, 246 (see also quasars); radio, 168; spectra and, 92; variable stars and, 189–90, 198, 200, 209–11, 213–15
Brokaw, Tom, 413
Brown, Mike, 144
brown dwarfs, 68, 117
bubble universes, 381–88, 394, 396, 402, 408, 431nn1–3
Buller, A.H.R., 326
Burney, Venetia, 130

calcium, 88–89, 169, 207
calculus, 40–41, 45, 50, 62, 74–75, 257, 261
Caldwell, Robert, 409
California Institute of Technology (Caltech), 144, 196, 242–44, 320, 334
Callisto, 139, 142

Campbell, W. W., 299

carbon: abundance of, 81–82; early universe and, 224, *251*; interstellar medium and, 181; lives of stars and, 99, 102, 105, 119–21; nucleus of, 99; search for life and, 147, 162–63, 169, 410; spectra and, 88

carbon dioxide, 132, 147, 163, 169, 422

Carnegie Observatories, 215–16

Carter, Brandon, 410, 418

Cartesian coordinate system, 278

Cassidy, Michael J., 345

Cassini, Giovanni, 115, 262

Catholic Church, 37

Cauchy horizon, 335–41, 345, 396

Cavendish, Henry, 116

Celsius temperature scale, 69

centripetal acceleration, 292

Cepheid variables, 198, 200, 209–10, 213–15, 392

Ceres, 134, 138–40, 143–44, 355

CERN particle accelerator, 351

Chang, Kenneth, 135–36

chaotic inflation, 387, 393

Charon, 13, 130–31, 137–38, 141–45

Christodoulou, Demetrios, 317

Chronology Protection Conjecture, 345

Circleland, 365, 367

circumferential radius, 306

circumpolar stars, 30

Clairaut, Alexis, 53

Coalsack Nebula, 176, *178*

Coat of the Future, 321–22, *323*

Cold War, 103, 411–13

Coleman, Sidney, 381–82, 435n2

Colley, Wes, 331

Collins, Michael, 424

colonization, 414, 420–24

color temperature, 68–70

Columbus, Christopher, 31–32

Coma Cluster, 196

comets, 48, 53, 127–28, 131–32, 134, 137, 139–42, 156, 163

computers: apps and, 35; black holes and, 314; blink comparators and, 130; first, 121; human, 93–94; megabytes and, 17; patterns and, 32; power of large, 238; simulations of early universe and, 238–40; supercomputers and, 314

Comte, Auguste, 92

conservation of angular momentum, 39

conservation of charge, 261

conservation of energy, 297, 334, 367, 380

conservation of momentum, 285, 429–30

constellations, 27–33, 57, 59, 92, 120, 123, 186, 189

Contact (film), 165–67

Contact (Sagan), 343

Copernican Principle: Drake equation and, 420; expansion and, 210; future of universe and, 411, 413, 415–16, 420, 422, 435n3; inflation and, 375; search for life and, 165; testing of, 436n5; shape of universe and, 370

Copernicus, Nicolaus, 165; Catholic church and, 37; *De Revolutionibus orbium coelestium* and, 36; Earth as planet and, 138; expansion and, 210; future of universe and, 411–22; heliocentric theory and, 36–37, 42, 139, 184–85; inflation and, 375; known universe in day of, 183–85; Newton's laws and, 42–43; night sky and, 36–37; shape of universe and, 370

Cosmic Background Explorer (COBE), 230–31, 234–35, 375–76

cosmic distance ladder, 215

cosmic microwave background (CMB): black holes and, 318–19; blueshift and, 231; cosmic strings and, 331; cosmological principle and, 230; dark matter and, 233–37; Doppler shift and, 232; early universe and, 71, 74, 216, 229, 235–40, 252, 318–19, 331, 352–55, 362, 373–76, 381, 386–93, 398, *403*, 404–7, 415; expansion and, 216; future of universe and, 404–7, 415; Gibbons and Hawking radiation and, 406–7; Hubble's law and, 236–37; inflation and, 298, 375–76, 381, 386–92; Lyman series and, 252; Map of the Universe and, 354,

362; Penzias and, 71, 229–30, 373, 375, 415; redshift and, 252; shape of universe and, 352–55, 362, 373; undulations in, 232–36; uniformity of, 230–31, 252–53; Wilson and, 71, 229–30, 373, 375, 415; WMAP and, 235–40

cosmic strings: cosmic microwave background (CMB) and, 331; Cutler and, 334–35, 339; Einstein and, 327, 331; Flatland and, 340, 435n1; general relativity and, 327, 331; gravity and, 331; length of, 328–29; mass of, 330; oscillation of, 331; quasars and, 330; shape of universe and, 351; shortcuts and, 327, 330–32; spacelike separation and, 332; speed of light and, 331; tension of, 331; time travel and, 327–37, 340

cosmic web, 404

Cosmic Web, The (Gott), 381

cosmological constant: as big blunder, 369; as correct, 391; Einstein and, 363–70, 377–78, 380, 389, 391, 393–94, 409–10; future of universe and, 409–10; inflation and, 377–78, 380, 389, 391, 393–94; phantom energy and, 409–10; shape of universe and, 363–70

cosmological models: all possible, 389; closed, 365; Einstein and, 364–65; expansion and, 221; gravity and, 364–65; inflation and, 388–92; parameters of, 388–89; standard, 13, 391

cosmological principle, 225–26, 230, 237–38, 248

Cosmos: A Spacetime Odyssey (film), 33

Cosmos (TV series), 165

Cotham, Frank, 164

Coulomb's law, 261

Crab Nebula, 123–25

Crick, Francis, 352

Curie, Marie, 288

Curtis, Heber, 198, 201

Cutler, Curt, 334–35, 339

cyanobacteria, 147, 161–62

Cygnus X-1, 355

dark energy: decay and, 402; early universe and, 236; future of universe and, 402, 405, 408–10; inflation and, 236, 389–92, 398; shape of universe and, 364–65, 370; slow-roll, 409; standard cosmological model and, 13

dark matter: amount of, 13, 195; axions and, 232; composition of, 222, 232–34; cosmic microwave background (CMB) and, 233–37; early universe and, 222, 232–34, 236–37; gravitational lensing and, 253; gravitinos and, 234; inferred existence of, 243; inflation and, 388, 390–91; Milky Way and, 195–96; photinos and, 232; selectrons and, 232; shape of universe and, 351; standard cosmological model and, 13; supersymmetry and, 232; Zwicky and, 196, 243–44

d'Arrest, Heinrich Louis, 53

Davis, Marc, 230

decay: asymmetric, 402, 404; beryllium to lithium, 224; black holes, 314, 317, 407; dark energy, 402; fission and, 287; gravitinos, 234; Hawking radiation and, 407; muons, 268–69; neutrons, 223; protons, 223, *403*, 407; quantum theory and, 403; radioactivity and, 103, 114, 218, 226, 351, 384; timescale for, 407; uranium, 114, 384–85; vacuum energy, 328, 381–82, 403

density wave, 194

deoxyribonucleic acid (DNA), 22, 66, 161, 213, 352

De Revolutionibus orbium coelestium (Copernicus), 36

de Sitter, Willem, 378, 380, 382–85, 393–99, 405–6

deuterium, 99, 117, 158, 223–25, 227, 233, 240, 375, 404

deuterons, 223–24

Deutsch, David, 326

Dewdney, A., 435n2

Dewynne, J. N., 436n5

Dicke, Robert, 228–29, 375, 410

differential calculus, 45

Einstein, Albert (*continued*)
    337, 339, 341, 345–46; uniform motion and, 264, 280
Einstein-Rosen bridge, 320
electric fields, 261–62, 290, 320, 350
electromagnetism: black holes and, 317; Einstein and, 261; energy radiation and, 66, 406; fields and, 290; Hertz and, 263; inflation and, 375, 377, 381, 399; lives of stars and, 125; Maxwell and, 11, 261–65, 290, 296–97, 349–50, 402; photons and, 402–3 (*see also* photons); radio telescopes and, 241; relativity and, 290, 294, 297; shape of universe and, 348–52; time travel and, 330
electrons: binding energy and, 86, 103–4; black holes and, 317–20; early universe and, 222–23, 227–28, 233–34; energy levels and, 82–90, 119, 179, 250, 258; ground state and, 82–84, 86, 91, 250; inflation and, 402; interstellar medium and, 179; ionization and, 86–88, 107, 113, 228, 233; lives of stars and, 98–99, 103, 107, 113, 119, 122; orbitals and, 21, 82–85, 92, 162; positrons and, 99, 222–23, 227, 317–20, 402, *403*, 407; quantum theory and, 83–84; recombination and, 236–37, *403*, 405; relativity and, 266, 268, 299; selectrons and, 232; shape of universe and, 350–51; spectra and, 81–92; time travel and, 324; weak force and, 402–3
electron shell, 81
electron volt (eV), 85–86, 103–4
electrostatic forces, 24, 98–99, 261, 385
Elements in the Theory of Astronomy, The (textbook), 139–40
eleven dimensions, 351–52, 397–98
elliptical galaxies, 202, 204, 248–49, 303
elliptical orbits: Kepler and, 37–42, 53, 156, 196, 297; planetary motion and, 37–42, 53, 131, 134, 156, 196, 271, 297
emission lines, 90, 179, 242, 245–46, 250–51
emission nebulae, 90, 179–80
e (natural logarithm), 73
endothermic process, 103–4

energy density: as attractive force, 364; black holes and, 318; future of universe and, 405, 408–9; inflation and, 377–81, 384, 386–91, 394; radiation and, 76; relativity and, 434n2; shape of universe and, 362–64; slow-roll dark energy and, 409; spacetime and, 76, 318, 327, 340, 344, 362–64, 377–81, 384–91, 394, 405, 408–9, 434n2; time travel and, 327, 340, 344
energy levels, 119, 179; equilibrium and, 85, 89, 153; excited electrons and, 81–92; ground state and, 82–84, 86, 91, 250; ionization and, 86–88, 107, 113, 228, 233; Paschen series and, 86, 90–92; relativity and, 258
Englert, François, 327–28
entropy, 315–16, 323, 431
equator, Earth's, 238, 291–95, 308, 352–55, 365
equilibrium, 407; energy levels and, 85, 89, 153; gravity and, 112–13, 122, 189; internal stellar pressure and, 112–13, 122, 189; static, 144; thermal, 376, 380
Equivalence Principle, 290, 292, 298, 310
Eratosthenes, 115
Eris, 141, 144, 355
ET (film), 165–66
Euclidean geometry, 58, 278, 290–92, 330, 365–66, 371, 387
Europa, 139–40, 142–43, 160
event horizon: firewalls and, 319; future of universe and, 405–7; Hawking radiation and, 406; inflation and, 379–80, 386; Planck length and, 431; spacetime and, 305, 308–20, 337–38, 340, 344, 379–80, 386, 405–7, 431; time travel and, 337–38, 340, 344
exoplanets, 149–51, 362
exothermic process, 102–4
expansion: acceleration and, 392, 409; Big Bang and, 218–21; Big Crunch and, 367–68, 372, *373*, 384; Copernican Principle and, 210; cosmic distance ladder and, 215; cosmological constant and, 362–70, 377–78,

and, 352; Euclidean, 58, 278, 290–92, 330, 365–66, 371, 387; expansion and, 216, 221; four-dimensional universe and, 270–76, 290, 296, 324, 326, 347, 349, 366, 394–95, 434n1; great circle and, 18, 291–94; inflation and, 383, 386–87, 393–94; Kepler and, 39; relativity and, 270, 278, 290–92, 294, 296; shape of universe and, 362, 364–65, 367, 371, 405, 435n1; spacetime and, 260, 270–76, 280–81, 285–86, 289–98, 401 (*see also* spacetime); 3-sphere geometry and, 365–70, *373*, 374, 378–79, 382–83, 388–89, 394; time travel and, 327, 329–30, 332–33, 343

George III, king of England, 138–39

Gibbons, Gary, *403*, 405–7

Glashow, Sheldon, 351

Gliese 581, 100

globular clusters, 96–97, 127, 168, 184–91, 214, 219

Gödel, Kurt, 341

Google, 20, 354

googol, 20

googolplex, 20

Gott, J. Richard, 11, *12*, 13, *122*, *411*; Berlin Wall and, 411–13, 424; black holes and, 300–320; Coat of the Future of, 321–22, *323*; cosmic strings and, 327–341; future of universe and, 400–424; *Guinness Book of Records* and, 362; Hubble constant and, 216; inflation and, 374–99; Map of the Universe and, 354–62; relativity and, 257–99; shape of universe and, 347–73; time travel and, 321–46; undergraduate course of, 12; Zwicky and, 244

Gott-Li model, 396–98

Gould, Stephen Jay, 417

grandmother paradox, 324–26

gravitational instability, 232, 238, 240

gravitational lens, 244, 253, 313, 330–31, 391

gravitational mass, 289

gravitational waves: black holes and, 13, 314, 403; cosmic strings and, 331; Einstein and, 402–3; existence of, 403; future of universe and, 402–3, 406; inflation and, 393; interstellar medium and, 182; LIGO and, 13, 314–15, 331, 402–3, 406; Maxwell and, 402; pulsars and, 355

gravitinos, 232

gravity, 13; acceleration and, 23, 47, 49, 52, 116, 158, 294; black holes and, 23, 122, 247, 249, 302–4; cosmic strings and, 331; cosmological constant and, 363–70, 377–78, 380, 389, 391, 393–94, 409–10; cosmological models and, 364–65; curved spacetime and, 260 (*see also* spacetime); dark matter and, 234 (*see also* dark matter); early universe and, 232–34, 238; Earth and, 23, 46, 49, 51–52, 111, 192; equilibrium and, 112–13, 122, 189; Equivalence Principle and, 290, 292, 298, 310; expansion and, 217, 232–34, 238; general relativity and, 260, 289–90, 292, 294–99; great circle and, 291–94; inflation and, 376, 392, 398; interstellar medium and, 180; inverse-square law and, 47, 49, 79, 116, 125, 168, 188, 213, 243; Kepler's law and, 37–39; Lagrange points and, 144; lives of stars and, 104, 107, 111–19, 122, 156, 197; Mars and, 422; measuring, 3; Milky Way and, 189, 192–95, 204; Newton's laws and, 41–45, 79, 82, 97, 192–93, 195, 246, 259, 264, 302; perturbations and, 53, 131, 160, 300, 314; Planet X and, 130–31; Pluto and, 134, 144; quantum, 339, 341, 345, 398, 402; Schwarzschild radius and, 301–10, 313, 315, 318, 401; self-gravitating objects and, 111–12; shape of universe and, 349–52, 363; singularities and, 302, 305–6, 309–13, 337–40, 344, 368, 381, 396, 399, 401, 409; spaghettification and, 305–6, 339; special relativity and, 282–83; static model and, 362–65, 369, 393; Sun and, 32, 111, 114–15; supernovae and, 100, 106–7, 121–25, 181, 216, 219, 243–44, 247, 252–53, 362, 389–92; temperature and, 134; tidal forces and, 118, 151, 160, 192, 204, 304, 306, 337, 339; time travel and, 339,

black holes and, 247–49, 251; brightness and, 184, 188; Cepheid variables and, 198, 200, 209–10, 213–15, 392; dark matter and, 195–96; dark regions of, 175–76; density wave and, 194; Drake's equation and, 193; dust and, 184–90, 196, 238; early universe and, 232–39; exoplanets and, 149–51, 362; expansion and, 209–12, 217, 221; fuzzy appearance of, 197; galactic plane and, 185–86, 313; Galileo and, 183; globular clusters and, 184–91; gravity and, 189, 192–95, 204; halo of, 192; Herschel and, 183; Hubble and, 92; infrared light view of, 186–88, 196; interstellar medium and, 173–82; inverse-square law and, 184, 188; as island universe, 198, 200; Kapteyn and, 183–85, 210; lives of stars and, 106, 125; as living ecosystem, 181–82; Local Group and, 127; luminosity and, 184, 188–90, 252; main sequence stars and, 188–89, 192; Map of the Universe and, 353–55; mass and, 192–96; number of stars in, 18, 149, 193; orbit of, 232; quasars and, 245–46; red giants and, 189; Rose Center display and, 127; RR Lyraes and, 189–90; search for life and, 149, 151–52, 159, 166, 168; shape of universe and, 363; Shapley and, 184–85, 187, 189–90; size of, 185–86, 197, 212; spiral arms of, 106, 190–92, 194; star orbits in, 193–94; stars beyond, 92; structure of, 174–76, 190–91; Sun's location in, 184–86, 194–95; supernovae and, 100, 106–7, 121–25, 181, 216, 219, 243–44, 247, 252–53, 362, 389–92; 2MASS telescopes and, 176, 188; visible light and, 186
Minkowski, Hermann, 270
Miranda, 139
Misner, Charles W., 314
Mitochondrial Eve, 413
mobile phones, 11, 64–65
molecules: atmospheric, 20–21, 25, 156; complex, 161, 163; DNA, 22, 352; energy levels and, 84, 89; formation of, 161, 163, 222; water, 65, 81–82, 161

Montond, Bradley, 436n5
Moon, 131; acceleration and, 47–48; age of, 218; atmosphere of, 50, 156, 422; center of mass of, 137; distance to, 54, 283; Earth's wobble and, 32; half-illumination from Sun and, 33; landing on, 158, 326, 423–24; lunar eclipse and, 34, 92, 115, 298; Map of the Universe and, 355; Newton's laws and, 47–51, 290; night sky and, 32–36, 40, 137–39, 241, 355; phases of, 30, 33–35, 109; relative size of, *132*; tides and, 151
Mordor, 145
Morley, Edward, 265–66
Morris, Mike, 341
Mount Wilson Observatory, 198, 201
M-theory, 351–52, 397–98
multiverse, 382–84, 386–87, 397–99, 402, 409
muon, 22, 269, 402
Musk, Elon, 422–23
mythology, 130, 138–39

NASA, 137, 144, 159, 164, 230, 235, 282, 422
nebulae: emission, 90, 179–80; expansion and, 207, 209; galaxies and, 197–201; interstellar medium and, 176–81; lives of stars and, 107–8, 119–20, 123–25; reflection, 180; spectra and, 90–91; spiral, 197–98, 200
Neptune: distance to, 80; as gas planet, 131; Great Dark Spot and, 133; historical perspective on, 139; Hubble Space Telescope and, 133; Kepler satellite and, 150; Kuiper Belt and, 136; Lowell's time and, 130; Map of the Universe and, 355; mass of, 131; moons of, 140, *142*; Newton's laws and, 53, 130; official IAU listing of planets and, 144; Planet X and, 131; Pluto's orbit across, 134; relative size of, *133*; Rose Center display and, 129, 135; Standish and, 131; statistics of, *135*; variant orbit of, 131
neutrinos, 99, 122, 222–23, 227, 351, 402–3
neutrons: atomic nucleus and, 13, 82, 99–104, 122, 181, 222–24, 227, 233–34, 349, 384, 390, *403*, 404; decay and, 223; early universe and, 222–24, 227, 233–34;

Planck length, 302, 316, 401, 431

Planck mass, 401

Planck satellite, 13, 216, 218, 236, 240, 373, 387, 391–93, 409

Planck's constant, 67, 73, 84, 92, 259, 284, 316, 401

Planck spectrum, 87, 225, *231*, 407

Planck time, 401, *403*

planetary motion: Brahe and, 35–37, 43, 258; changing speed of, 37–38; Copernicus and, 36–37, 42–43, 138–39, 165, 183–85, 210, 370, 375, 411–18, 420–22; elliptical orbits and, 37–42, 53, 131, 134, 156, 196, 271, 297; gravity and, 130 (*see also* gravity); Kepler and, 36–43, 46–47, 49, 52–53, 157, 196, 258, 297; Newton's laws and, 42–43, 46–47, 52–53; parallax and, 55–60, 115, 213–14, 262, 434n1; retrograde, 36; transit and, 130, 149–51, 213

planetary nebulae, 119–20, 123, 181, 197

planetos (wanderer), 39–40

Planet X, 130–31

Planiverse (Dewdney), 435n2

plasma, 113, 121, 227, 405

Please, C. P., 436n5

Pleiades, 96, 180

plutinos, 135

Pluto: Charon and, 13, 130–31, 137–38, 141–45; controversy over, 129, 135–37, 141–45; defining planet and, 139–40; discovery of, 130, 138; downgrading of, 39, 126–45; gravity and, 134, 144; Hubble Space Telescope and, 130; Kuiper Belt and, 134–36, 140–44; Lowell and, 129–30, 209; Map of the Universe and, 355; mass and, 131, 138, 144; moons of, 130–31; naming of, 130; *New York Times* article and, 129, 135–36; official IAU listing of planets and, 144; orbit of, 40, 134–36, 144; relative size of, 131–32, 140; Rose Center display and, 126–28, 132, 135, 141, 144; Sun and, 128–34, 139–44; Tombaugh and, 130, 138, 144–45; uniqueness of, 132, 134

Pluto (Disney character), 138

Pluto Files, The: The Rise and Fall of America's Favorite Planet (Tyson), 144

Poldowski, Boris, 320

Polshek, Jim, 126

Popper, Karl, 434n1

Positive Philosophy, The (Comte), 92

positrons: black holes and, 317–20; early universe and, 222–23, 227; formation of, 99; future of universe and, 402, *403*, 407; lives of stars and, 99

precession, 33, 297–98

President's Award for Distinguished Teaching, 12

Primack, Joel, 234

prime meridian, 367

Procyon, 100

Project Daedalus, 158

proportionality constant, 49, 67, 210, 213

protons: alpha particles and, 384; atomic nuclei and, 13, 21–24, 62, 82, 85–88, 98–105, 122, 222–24, 227–28, 233–34, 266, 282, 350, 384, 390, 402–7; as baryon, 390; decay and, 223, *403*, 407; density of in Sun, 21; early universe and, 222–24, 227–28, 233–34; fusion and, 99, 102–5, 111, 114–15, 117–18, 121, 158, 181, 222–23; lives of stars and, 98–105, 122; proton-proton collisions and, 223; quantum tunneling and, 384–85; quarks and, *403*, 404–5; relativity and, 266, 282; spectra and, 85–86, 88; strong force and, 24, 351, 384

Proxima Centauri, 59, *94*, 355

Ptolemy, Claudius, 33, 37, 210

public policy prior, 436n5

pulsars, 22, 59, 125, 182, 244, 355, 382, 402

Pythagorean theorem, 268, 278, 434n2

QSO 0957+561, 331

quantum theory: atomic nuclei and, 21; Bekenstein formula and, 20; black holes and, 249, 302, 316–19; Bohr and, 258; decay and, 403; early universe and, 403–4; electrons and, 83–84; energy levels and, 82–90, 119, 179, 250, 258; entanglement and, 319,

temperature (*continued*)

inflation and, 375–77, 387; Jupiter and, *135*; Kelvin scale, 69; lives of stars and, 93–100, 105, 113–14, 117–18; of local planets, *135*; luminosity and, 214; Mars and, *135*; Mercury and, *135*; Neptune and, *135*; radiation and, 67–78; red objects and, 251–52; red stars and, 251; Saturn and, *135*; scale of universe and, 23–24; search for life and, 151–56, 169; spectra and, 84–85, 87; Stefan-Boltzmann constant and, 75, 78, 95; Sun and, 23, 67–69, 78; terrestrial planets and, 134; Uranus and, *135*; Venus and, 132, *135*; wavelength and, 74–75; of whole universe, 71; Wien's Law and, 75

theory of everything, 13, 320, 399

Theory of Everything, The (film), 12, 320

thermal emission, 68–72

thermal radiation: black holes and, 314, 317–19; early universe and, 223, 228, 231; Earth's, 153; future of universe and, 402–3, 406; Hawking radiation and, 317–19, 340, 406–8, 435n1; inflation and, 374–75; radiation and, 71, 74–76; spectra and, 87, 89

thermodynamics, 75, 315–16, 323, 431

3C 48, 243

3C 273, 238, 241–45, 362

3-sphere geometry, 365–67, 370, *373*, 374–75, 378–79, 382–83, 388–89, 394

third-quarter Moon, 34–35

Thisted, Ronald, 257

Thorne, Kip, 306, 314, 320, 325, 341–44

thought experiments, 263–65, 273, 275, 280, *285*, 316, 325, 386, 429, 431

tidal forces, 118, 151, 160, 192, 204, 304, 306, 337, 339

Time Machine, The (Wells), 283, 345–46

Time magazine, 12, 258

time travel: black holes and, 337–41, 344; Cauchy horizon and, 335–41, 345, 396; Coat of the Future and, 321–22; cosmic strings and, 327–37, 340; Earth and, 324, 327–31, 341–43; Einstein and, 323, 326–27, 331, 337, 339, 341, 345–46; electro-

magnetism and, 330; electrons and, 324; energy density and, 327, 340, 344; event horizon and, 337–38, 340, 344; Flatland and, 340; four-dimensional universe and, 324, 326; general relativity and, 283, 321, 323, 327, 331, 334, 339, 341, 345–46; geometry and, 327, 329–30, 332–33, 343; grandmother paradox and, 324–26; gravity and, 331, 339, 341, 345; Hawking and, 336, 340, 344–45; inflation and, 396–97; information and, 323; jinn particles and, 322–24; mass and, 323–24, 329–30, 334, 336–38, 341–44; Newton's laws and, 283–84, 286, 345; Padalka and, 283; photons and, 284–86, 327, 338–39, 341; quantum theory and, 322, 325–27, 337, 339–41, 344–46; shortcuts and, 327, 330–32, 342, 344–45; singularities and, 337–40, 344; spacelike separation and, 332; spacetime and, 300–304, 309–17, 322–28, 331–45; spaghettification and, 339; special relativity and, 276–84, 326; speed of light and, 276–86, 324, 326–27, 330–33, 336, 342–43; Sun and, 324, 330; tidal forces and, 337, 339; time machines and, 205, 249, 321–46, 396–97; vacuum energy and, 327–28; warp drives and, 185, 336, 342, 344–45; worldlines and, 322–27, 334–35, 364–69; wormholes and, 327, 336, 341–44

Time Travel in Einstein's Universe (Gott), 346

Tipler, Frank, 341

Tombaugh, Clyde, 130, 138, 144

Tombaugh Regio, 145

transit, 130, 149–51, 213

Trifid Nebula, *179*, 180

Triton, 140, *142*

Truman, Harry S., 287–88

Trumpler, R., 299

Turner, Ed, 331

Twin Paradox, 280–83

Two Micron All-Sky Survey (2MASS), 176, 188

Tyson, Neil de Grasse, *12*; Hayden Planetarium and, 11; Levy on, 137; lives of stars and, 93–110; Newton's laws and, 26–41; Pluto and, 126–45; *The Pluto Files: the Rise and Fall of America's Favorite Planet* and, 144; radiation and, 54–80; scale of universe and, 17–25; search for life and, 146–69; signature sayings of, 11; *Space Chronicles* and, 423; spectra and, 81–92; Sykes debate and, 136; undergraduate course of, 12

ultraviolet catastrophe, 74
ultraviolet (UV) light, 62–69, 74, 86, 91, 107, 119–20, 179, 263
uncertainty principle, 316–17, 381, 385, 393, 401, 403, 407
uniform velocity, 43–45
United Nations, 436n5
universal law of gravitation, 40–41, 48–49
University of California, Santa Cruz, 182
uranium, 102–3, 114, 139, 181, 287, 384–85
Uranus: as gas planet, 131; gravity and, 53; Herschel and, 138, 183; historical perspective on, 138–39; Lowell's time and, 130; Map of the Universe and, 355; mass of, 131; moons of, 139, *142*; Newton's laws and, 53, 130; official IAU listing of planets and, 144; Planet X and, 131; relative size of, *133*; rings of, 133; Rose Center display and, 129, 135; Standish and, 131; statistics of, *135*
Urtsever, Ulvi, 341
U.S. Declaration of Independence, 138
U.S. Naval Observatory, 131

vacuum energy: future of universe and, 408–9; inflation and, 377–92; phantom energy and, 409–10; quantum theory and, 327–28, 364, 377–78, 381–82, 408–9; shape of universe and, 364; time travel and, 327–28
vague (Jeffreys) Bayesian prior, 436n5
Vanderbei, Bob, 354
van der Meer, Simon, 351
van de Walle, Nicholas, 436n5
variable stars, 189–90, 198, 200, 209–11, 213–15
vector fields, 350
Vega, 56–58, 343

Venus, 25; atmosphere of, 132; AU measurement and, 213; clouds of, 132; Frigga and, 40; historical perspective on, 138–39; Kepler's time and, 39–40; Map of the Universe and, 355; Newton's laws and, 130; official IAU listing of planets and, 144; relative size of, *132*; Rose Center display and, 129; statistics of, *135*; temperature of, 132, *135*; terrestrial family and, 132, 141; world-line of, 271–72
Vesta, 139–41, 143
Vilenkin, Alex, 327, 384, 394, 408
Virgo Supercluster, 127
virtual pairs, 317–18
visible light: interstellar medium and, 176; lives of stars and, 114, 117, 125; Maxwell and, 263; Milky Way and, 186; night sky and, 29; radiation and, 65–71, 77–78; search for life and, 406; spectra and, 84, 86, 90; telescopes and, 241 (*see also* telescopes); wavelength and, 263
Voyager spacecraft, 131, 133

warp drives, 185, 336, 342, 344–45
water: abundance of, 81–82; boiling point of, 69, 154–155; density of, 22, 113, 141; early universe and, 222; entropy and, 315–16; Europa and, 140; freezing point of, 69, 154–155; Mars and, 129, 133, 422; microwaves and, 65; nonuniform distribution in boiling, 382; search for life and, 146, 149, 154–61
Watson, James, 352
wattage, 78
wavelength: black holes and, 318; blueshift and, 209, 217, 231, 245, 339, 341, 429; color and, 208–9; early universe and, 225, 228–29, 231, 250–53; expansion and, 250; frequency and, 65–67, 84, 208, 259, 284–85, 400–401, 427–29; Gibbons and Hawking radiation and, 406; inflation and, 375; Kaluza–Klein dimension and, 350; Maxwell and, 263; Milky Way observations and, 188; photons and, 66; Planck and, 73–74; quasars and, 241–42, 245, 253;